T0398830

Demystifying Big Data, Machine Learning, and Deep Learning for Healthcare Analytics

Demystifying Big Data, Machine Learning, and Deep Learning for Healthcare Analytics

Edited by

Pradeep N

Professor, Computer Science and Engineering, Bapuji Institute of Engineering and Technology, Davangere, Karnataka, India

Sandeep Kautish

Professor & Dean Academics, LBEF Campus, Kathmandu, Nepal (In Academic Collaboration with APUTI Malaysia)

Sheng-Lung Peng

Professor, Department of Computer Science and Information Engineering, National Dong Hwa University, Taiwan

ELSEVIER

ACADEMIC PRESS

An imprint of Elsevier

Academic Press is an imprint of Elsevier
125 London Wall, London EC2Y 5AS, United Kingdom
525 B Street, Suite 1650, San Diego, CA 92101, United States
50 Hampshire Street, 5th Floor, Cambridge, MA 02139, United States
The Boulevard, Langford Lane, Kidlington, Oxford OX5 1GB, United Kingdom

Library of Congress Cataloging-in-Publication Data
A catalog record for this book is available from the Library of Congress

British Library Cataloguing-in-Publication Data
A catalogue record for this book is available from the British Library

ISBN 978-0-12-821633-0

For information on all Academic Press publications
visit our website at https://www.elsevier.com/books-and-journals

Publisher: Mara Conner
Acquisitions Editor: Chris Katsaropoulos
Editorial Project Manager: Ruby Smith
Production Project Manager: Selvaraj Raviraj
Cover Designer: Greg Harris

Typeset by SPi Global, India

Working together
to grow libraries in
developing countries

www.elsevier.com • www.bookaid.org

Dedicated to the almighty and our well wishers.

Contents

SECTION 2 Machine learning and deep learning for healthcare

Contributors

Shridhar Allagi
Department of Computer Science and Engineering, KLE Institute of Technology, Hubballi, India

J. Anitha
Karunya Institute of Technology and Sciences, Coimbatore, Tamilnadu, India

B. Annappa
Department of Computer Science and Engineering, National Institute of Technology Karnataka, Mangalore, India

Siddhartha Kumar Arjaria
Rajkiya Engineering College, Banda, Uttar Pradesh, India

M. Bhuvaneshwari
Karunya Institute of Technology and Sciences, Coimbatore, Tamilnadu, India

Chinmay Chakraborty
Dept. of Electronics & Communication Engineering, Birla Institute of Technology, Mesra, Ranchi, Jharkhand, India

Jincy S. Cherian
The Bhopal School of Social Sciences, Bhopal, Madhya Pradesh, India

S. Thomas George
Karunya Institute of Technology and Sciences, Coimbatore, Tamilnadu, India

Yogita Gupta
Thapar Institute of Engineering and Technology, Patiala, India

Sanjeevakumar M. Hatture
Basaveshwar Engineering College (Autonomous), Bagalkot, Karnataka, India

Aboobucker Ilmudeen
Department of Management and Information Technology, Faculty of Management and Commerce, South Eastern University of Sri Lanka, Oluvil, Sri Lanka

Nagaveni Kadakol
Basaveshwar Engineering College (Autonomous), Bagalkot, Karnataka, India

E. Grace Mary Kanaga
Karunya Institute of Technology and Sciences, Coimbatore, Tamilnadu, India

M. Karthiga
Bannari Amman Institute of Technology, Erode, Tamil Nadu, India

Sandeep Kautish
LBEF Campus, Kathmandu, Nepal (In Academic Collaboration with APUTI Malaysia)

Roshni Khedgaonkar
Computer Technology, Yeshwantrao Chavan College of Engineering, Nagpur, India

B. Lalith Bharadwaj
Department of Information Technology, VNR Vignana Jyothi Institute of Engineering and Technology, Hyderabad, Telangana, India

G. Madhu
Department of Information Technology, VNR Vignana Jyothi Institute of Engineering and Technology, Hyderabad, Telangana, India

M.V. Manoj Kumar
Department of Information Science and Engineering, Nitte Meenakshi Institute of Technology, Bengaluru, India

Pradeep N
Computer Science and Engineering, Bapuji Institute of Engineering and Technology, Davangere, Karnataka, India

S.S. Nandhini
Bannari Amman Institute of Technology, Erode, Tamil Nadu, India

Rashmi Rachh
Department of Computer Science and Engineering, Visvesvaraya Technological University, Belagavi, India

Mukesh Raghuwanshi
Computer Engineering, G.H. Raisoni College of Engineering and Management, Pune, India

Kumudha Raimond
Karunya Institute of Technology and Sciences, Coimbatore, Tamilnadu, India

Sandeep Raj
Department of Electronics and Communication Engineering, Indian Institute of Information Technology Bhagalpur, Bhagalpur, India

Megha Rathi
Dept. of Computer Science & Engineering, JIIT, Noida, Uttar Pradesh, India

Abhishek Singh Rathore
Computer Science & Engineering, Shri Vaishnav Vidyapeeth Vishwavidyalaya, Indore, Madhya Pradesh, India

A. Reyana
Department of Computer Science and Engineering, Hindusthan College of Engineering and Technology, Coimbatore, Tamilnadu, India

B. Rohit
Department of Computer Science, VNR Vignana Jyothi Institute of Engineering and Technology, Hyderabad, Telangana, India

K. Sai Vardhan
Department of Information Technology, VNR Vignana Jyothi Institute of
Engineering and Technology, Hyderabad, Telangana, India

H.A. Sanjay
Department of Information Science and Engineering, Nitte Meenakshi Institute of
Technology, Bengaluru, India

S. Sankarananth
Excel College of Engineering and Technology, Namakkal, Tamil Nadu, India

B. Sathis Kumar
VIT University, Chennai, India

Shravan B.K.
Department of Computer Science and Engineering, Visvesvaraya Technological
University, Belagavi, India

Kavita Singh
Computer Technology, Yeshwantrao Chavan College of Engineering, Nagpur,
India

S. Sountharrajan
VIT Bhopal University, Bhopal, Madhya Pradesh, India

Likewin Thomas
Department of Computer Science and Engineering, PES Institute of Technology
and Management, Shivamogga, India

Editors biography

Dr. Pradeep N is a professor of computer science and engineering at the Bapuji Institute of Engineering and Technology in Davangere, Karnataka, India, affiliated with Visvesvaraya Technological University in Belagavi, Karnataka, India. He has 18 years of academic experience, which includes teaching and research experience. He has worked at various positions, including moving from a lecturer to an associate professor. He has been appointed as a senior member of the Iranian Neuroscience Society–FARS Chapter (SM-FINSS) for a duration of 2 years (March 1, 2021, to March 2, 2023). His research areas of interest include **machine learning, pattern recognition, medical image analysis, knowledge discovery techniques, and data analytics**. He is presently guiding two research scholars on **knowledge discovery and medical image analysis**. He has successfully edited a book published by **IGI Publishers, United States**. He is also editing books to be published by Elsevier, IGI, De-Gruyter, and Scrivener Publishing, which are in progress. He has published more than 20 research articles in refereed journals and also authored six book chapters. He is a reviewer of various international conferences and a few journals, including **Multimedia Tools and Applications**, Springer. His one Indian patent application has been published and one Australian patent has been granted. He is a professional member in IEEE, ACM, ISTE, and IEI. He was named the **Outstanding Teacher in Computer Science and Engineering** during the third Global Outreach Research and Education Summit and Awards in 2019, organized by the Global Outreach Research and Education Association. Also, he is a technical committee member for Davangere Smart City, Davangere.

Sandeep Kautish is a professor and dean of academics with the LBEF Campus, Kathmandu, Nepal, in academic collaboration with the Asia Pacific University of Technology and Innovation Malaysia. He is an academician by choice and is backed by more than 17 years of work experience in academics, including more than 6 years in academic administration in various institutions in India and abroad. He has earned meritorious academic records throughout his academic career. He earned his bachelors, masters, and doctorate degrees in computer science on intelligent systems in social networks. He holds a PG diploma in management also. His areas of research interest are business analytics, machine learning, data mining, and information systems. He has more than 40 publications and his research works have been published in reputed journals with high impact factors and SCI/SCIE/Scopus/WoS indexing. His research papers can be found in *Computer Standards and Interfaces* (SCI, Elsevier) and the *Journal of Ambient Intelligence and Humanized Computing* (SCIE, Springer). Also, he has authored/edited more than seven books with reputable publishers such as Springer, Elsevier, Scrivener Wiley, De Gruyter, and IGI Global. He has been invited to be the keynote speaker at VIT Vellore (QS ranking with 801–1000) in 2019 for an international virtual conference. He filed one patent in the field of solar energy equipment using artificial intelligence in 2019. He is an editorial member/reviewer of various reputable SCI/SCIE journals such as *Computer*

Communications (Elsevier), *ACM Transactions on Internet Technology*, Cluster Computing (Springer), *Neural Computing and Applications* (Springer), *Journal of Intelligent Manufacturing* (Springer), *Multimedia Tools and Applications* (Springer), *Computational Intelligence* (Wiley), *Australasian Journal of Information Systems* (AJIS), *International Journal of Decision Support System Technology* (IGI Global USA), and *International Journal of Image Mining* (Inderscience). He has supervised one Ph.D. student in computer science as a cosupervisor at Bharathiar University Coimbatore. Presently, two doctoral scholars are pursuing their Ph.D. under his supervision in different application areas of machine learning. He is a recognized academician as a session chair/Ph.D. thesis examiner at various reputable international universities such as the University of Kufa, the University of Babylon, the Polytechnic University of the Philippines (PUP), the University of Madras, Anna University Chennai, Savitribai Phule Pune University, M.S. University, Tirunelveli, and various other technical universities.

Google Scholar—https://scholar.google.co.in/citations?user=O3mUpVQAAAAJ&hl=en.
Linkedin Profile—https://www.linkedin.com/in/sandeep-k-40316b20/.
ORCID Profile—https://orcid.org/0000-0001-5120-5741.
More details about the academic profile can be found at www.sandeepkautish.com.

Sheng-Lung Peng is a professor and the director of the Department of Creative Technologies and Product Design, National Taipei University of Business, Taiwan. He received his BS degree in mathematics from National Tsing Hua University, and his MS and PhD degrees in computer science from the National Chung Cheng University and the National Tsing Hua University, Taiwan, respectively. He is an honorary professor at the Beijing Information Science and Technology University, China, and a visiting professor at the Ningxia Institute of Science and Technology, China. He is also an adjunct professor at Mandsaur University, India. He serves as the secretary general of the ACM-ICPC Contest Council for Taiwan and the regional director of the ICPC Asia Taipei-Hsinchu site. He is a director of the Institute of Information and Computing Machinery of the Information Service Association of Chinese Colleges and of the Taiwan Association of Cloud Computing. He is also a supervisor of the Chinese Information Literacy Association and the Association of Algorithms and Computation Theory. Dr. Peng has edited several special issues at journals, such as Soft Computing, the Journal of Internet Technology, the Journal of Real-Time Image Processing, the International Journal of Knowledge and System Science, MDPI Algorithms, and so on. He is also a reviewer for more than 10 journals such as IEEE Access and Transactions on Emerging Topics in Computing, IEEE/ACM Transactions on Networking, Theoretical Computer Science, the Journal of Computer and System Sciences, the Journal of Combinatorial Optimization, the Journal of Modelling in Management, Soft Computing, Information Processing

Letters, Discrete Mathematics, Discrete Applied Mathematics, Discussiones Mathematica Graph Theory, and so on. His research interests are in designing and analyzing algorithms for bioinformatics, combinatorics, data mining, and network areas, in which he has published more than 100 research papers.

Foreword

Healthcare has assumed paramount importance in this work-a-day world, thanks to the emergence of life-threatening diseases and ailments. As the name suggests, healthcare analytics is targeted toward evolving technologies and methods to facilitate the measurement, management, and analysis of healthcare data for better diagnosis and prognosis. As such, healthcare analytics-based interpretations are expected to properly enable medical practitioners to make more effective and efficient operational and clinical decisions. Added to that, the insights gained from proper healthcare data analytics can help one to better understand the factors behind operational and clinical successes, their medical outcomes, the underlying costs of delivery, optimized healthcare resources and utilities, and other aspects essential for overall improvement.

One of the obvious fallouts of business analytics in the healthcare industry is improving the quality-of-care experience by means of facilitating the most informed decision-making around patient experience by the fruitful analysis of healthcare data.

In this competitive world, healthcare organizations need to uncover consumer care preferences by means of proper analysis. Thus, they require clear knowledge of optimized healthcare delivery. Analytics in the healthcare industry assist healthcare providers in assimilating the knowledge base regarding better and meaningful expansion, types of specialty services in which to invest, and which current services to optimize.

Given the influx of machine intelligence encompassing machine-learning and deep-learning paradigms, intelligent healthcare analytics provides a fair opportunity to healthcare providers to optimize their supply chains to have a strategic and competitive advantage over competitors as far as the optimization of the cost of delivering services is concerned. These intelligent techniques help to streamline the inherent supply chains by incorporating inventory data, supply expenses, vendor reliability, and other crucial factors that improve supply chain efficiency.

Another benefit of healthcare analytics that deserves special mention is the efficiency in the billing process, which is always complex in the healthcare industry. Deep insights into the processes and outcomes spanning multiple facilities or with multiple providers help to minimize losses by preventing false claims, redundant billing, or duplicated supply orders.

The healthcare industry faces several challenges in implementing healthcare analytics, including the privacy and security of patient data, by following regulations and protocols.

As far as the management of the massive amounts of involved data is concerned, big data analysis may be a solution to the problem by resorting to proper formatting, organization, storage, and governance monitored by authorized parties. In addition, the retention or stewardship of sensitive data is important for effective healthcare data management. Healthcare analytics can prove beneficial in a healthcare organization's data life cycle management.

This proposed volume is intended to address these challenges and issues by envisaging an efficient healthcare analytics system through introducing novel intelligent mechanisms and procedures. The contributory chapters are focused on delivering the requisite insights into healthcare analytics essentials, implementation, and future perspectives. Thus, the book presents a holistic approach to the domain of intelligent healthcare analytics from the perspectives of analysis, management, and efficient customer experience.

I would like to take this opportunity to render my heartfelt thanks to the editors to visualize this timely requirement and to come up with such a novel literary initiative which would come to the benefit of the mankind and the society.

Siddhartha Bhattacharyya
Principal, Rajnagar Mahavidyalaya, Birbhum, West Bengal, India

Preface

In this advanced digital era, healthcare industries are reshaped by the advancements in big data technologies, machine learning and deep learning. Big data and machine learning are redefining the mechanism of business operations across various industries worldwide. With the advent of machine learning, traditional business operations can now process and make use of all the large amounts of data they have accumulated over the years. The processed data can be analyzed and the inferences derived can be used to make intelligent and informed business decisions. Therefore, healthcare is also actively using big data technologies and machine learning to provide accurate diagnoses and offer superior medical treatments.

By harnessing enormous big data, machine learning is now being used in healthcare for superior patient care and has contributed positively to improved and sustained business outcomes. The machine learning algorithms can quickly process huge datasets and provide useful insights that can cater to the needs of superior healthcare services. Even though healthcare industries have been slow in adopting this technology, it is now gearing up to provide successful, preventive, and prescriptive healthcare solutions.

Presently, healthcare industries are increasingly using computational techniques to analyze their voluminous datasets and identify patterns that provide useful insights from the existing patient data to make accurate diagnosis and provide better patient care.

Through the amalgamation of big data technology with machine learning and deep learning, healthcare companies can make accurate decisions, significantly improve operating efficiencies, and eliminate unwanted costs.

Nowadays, healthcare industries are able to determine which patients are at a higher risk of contracting a certain disease. Apart from this, postdischarge outcomes can also be kept under control and the number of readmissions can be reduced substantially. The main attractive aspect is that diagnoses require much less time. Therefore, patients will be able to know at the earliest what they are suffering from.

Pradeep N,
Sandeep Kautish,
Sheng-Lung Peng

Overview

The book is divided into two sections: big data in healthcare analytics and machine learning and deep learning for healthcare. The first section contains six chapters that include the foundation of healthcare informatics to the emergence of decision support systems in healthcare. The second section contains seven chapters that include a comprehensive review on deep-learning techniques for BCI-based communication systems to disease prediction using a machine-learning approach.

Chapter 1 discusses the foundations of healthcare informatics, which include goals, applications and clinical decision support systems.

Chapter 2 discusses the different concepts of smart healthcare systems with big data, which covers topics that explains how big data analytics can be used in healthcare industries, the role of sensor technology in eHealth, and applications and challenges of big data in healthcare.

Chapter 3 elaborates different big data-based frameworks for healthcare systems.

Chapter 4 encapsulates the different aspects of predictive analysis and modeling in healthcare systems, which covers topics ranging from the basic techniques of predictive analysis and modeling to research problems in predictive modeling.

Chapter 5 explains the different challenges and opportunities of big data integration in healthcare analytics using mobile networks.

Chapter 6 will investigate the need for decision support systems in healthcare.

Chapter 7 focuses on a comprehensive review of deep-learning techniques for BCI-based communication systems, which begins with the introduction of the brain-computer interface and concludes with research challenges and opportunities based on a literature review.

Chapter 8 elaborates machine learning- and deep learning-based clinical diagnostic systems, which covers the introduction, a literature review, the methodology devised, and the future scope of the work.

Chapters 9–13 focus on the efficient diagnosis of cardiac arrhythmias, local plastic surgery-based face recognition, heart disease prediction, one-shot malaria parasite recognition and kidney disease prediction, respectively.

**Pradeep N,
Sandeep Kautish,
Sheng-Lung Peng**

Big data in healthcare analytics

Foundations of healthcare informatics

1

B. Annappa[a], M.V. Manoj Kumar[b], and Likewin Thomas[c]

[a]*Department of Computer Science and Engineering, National Institute of Technology Karnataka, Mangalore, India,* [b]*Department of Information Science and Engineering, Nitte Meenakshi Institute of Technology, Bengaluru, India,* [c]*Department of Computer Science and Engineering, PES Institute of Technology and Management, Shivamogga, India*

1.1 Introduction

Healthcare informatics bridges the gap between healthcare and information engineering. It uses patient information and deals with managing the healthcare process. Healthcare informatics comprises multiple allies such as a computer, information, and the social, behavioral, and management sciences. The National Library of Medicine (NLM) defines healthcare as a multidisciplinary domain of planning, developing, adopting, and applying information technology-related inventions in the delivery, administration, and development of healthcare facilities (Haux, 2010).

Health informatics has become an inevitable part of modern healthcare. Due to the volumes of health data being generated, it is inevitable to deploy computers to manage patient data to offer a better healthcare process. Healthcare informatics aims at offering efficient, improved, time-critical, cost-effective, and user-friendly services (Wyatt & Liu, 2002; Young & Saxena, 2014). This chapter discusses various aspects of health informatics in detail, including what is health informatics, the range of applications, and delineating how the application of informatics in healthcare makes a difference in the patient care process as well as exploring the most significant and future aspects of healthcare informatics.

On the one hand, health informatics deals with devices, resources, and methods of patient care. On the other hand, it is related to storing, retrieving, and using the recorded data in optimizing the healthcare process. Healthcare informatics is designed to improve the efficiency of the patient care process while simultaneously generating data at the best quality (Mcdonald & Tierney, 1988).

It is evident from the normal application of healthcare informatics that it is used for information gathering as well as analyzing, interpreting, and utilizing the gained knowledge in guiding the process. It plays a vital role in making decisions and controlling/predicting healthcare progress. According to Haux, medical informatics is the discipline "concerned with the systematic processing of data, information, and knowledge in medicine and healthcare." This asserts that it is not just the application of computers in the healthcare process, which broadly involve aspects such as storing, retrieving, managing, analyzing, discovering, and synthesizing data/information/knowledge related to all aspects of healthcare (Saleem et al., 2015). A systematic overview of the staged application of computing devices in supporting the execution of the healthcare process is shown in Fig. 1.1.

Recording: This involves documenting the symptoms of illness, and this is used for communicating among doctors and everyone else in the healthcare process. Further, the same can be utilized for teaching and learning processes.

Storing and communicating: Once the data are recorded with appropriate measures, the further stages such as storing, processing, analyzing, and communicating the data and synthesized information are carried out.

Investigating/analyzing: This is utilizing the recorded data in a meaningful way to build information systems that facilitate the execution of the healthcare process.

An overview of the significant stages in healthcare informatics is illustrated in Fig. 1.2. There are several stages in healthcare informatics that can be grouped into categories as follows.

Opening medical data for all: Exposing medical data at the social level requires multiple checks such as anonymization, auditing, quality control, and standardization. Data made available to the public can be used for research studies/academic purposes to build alternative solutions (Musen, Middleton, & Greenes, 2014).

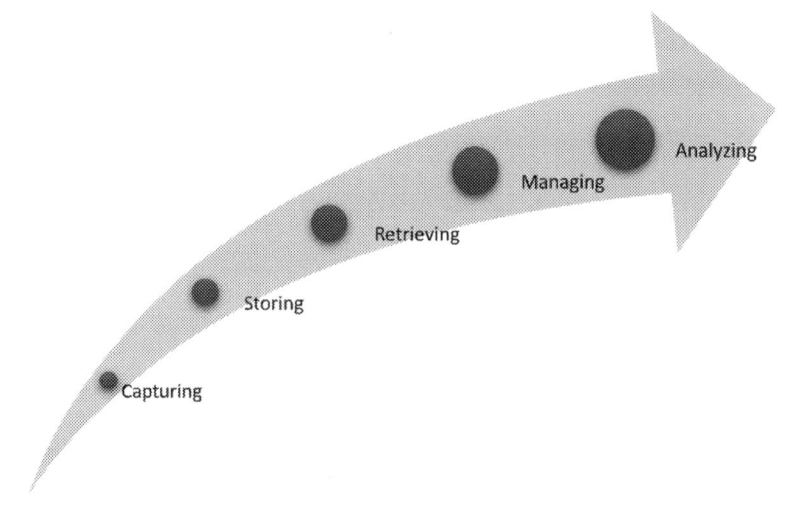

FIG. 1.1

Application of computers in the healthcare process.

No permission required.

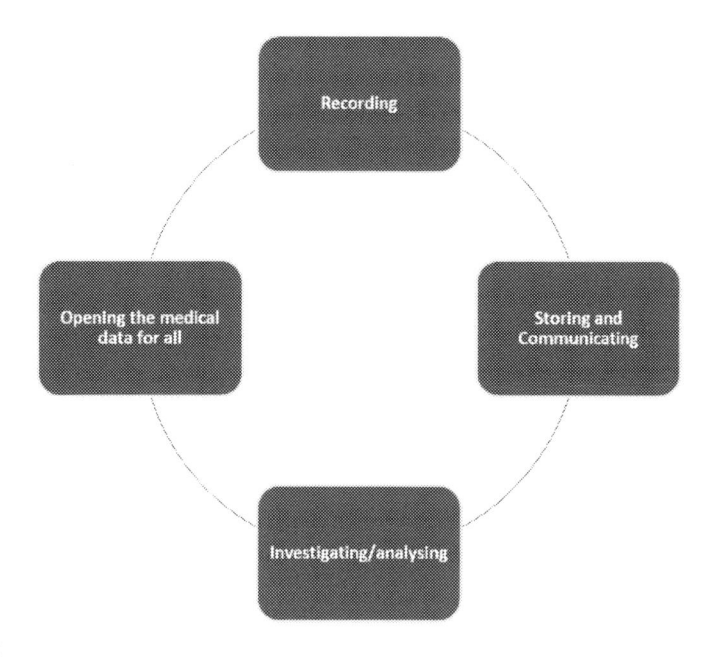

FIG. 1.2

Stages of healthcare informatics.

No permission required.

1.2 **Goals of healthcare informatics**

Based on understanding the four core elements (recording, storing and communicating, investigating, and opening medical data for the masses), the following core elements in health informatics are identified (VanLeit, 1995; Morak, Hayn, Kastner, Drobics, & Schreier, 2009):

(1) The way healthcare professionals think about patients.
(2) The way a diagnosis is made and evaluated as well as the way treatments are defined, selected, and evolved.
(3) The method of creating, shaping, and applying medical knowledge.
(4) Methods for creating healthcare systems and training healthcare professionals to operate these complex systems.

Patient care supported by healthcare informatics is always centered on providing solutions for problems related to processing recorded data, extracting information, and acquiring knowledge to improve the process (Doumont, 2002; Dumas, van der Aalst, & ter Hofstede, 2005; Halperin, Ma, Wolfson, & Nussinov, 2002). The knowledge acquired is further improved by repeated modification in the operating procedure of the healthcare process. Any healthcare information system has to be ready with solutions such as a structure for collecting, circulating, and implementing the clinical evidence. Reducing resource utilization at the organizational level while operating at a maximal benefit and offering the tools and means to reach the defined goals.

1.3 Focus of healthcare informatics

The pivotal focus of the healthcare process is the patient. Hence, a patient must be the central/focus and beneficiary of the healthcare process. Without the patient, there would be no need for healthcare informatics (Fowler & Heater, 1983).

The primary source of data and information is the patient. The lion's share of communication in the healthcare domain is related to patient data. The management, administration, and billing data are always centered on the kind of treatment given to the patient. An aspect such as medical data management primarily considers a patient-centered approach. Nursing activities and nursing data are focused on recording/analyzing patient data. Even at the organizational level, it is aimed at offering the best patient care at minimized/reduced (effective) costs (Eysenbach, 2000).

To further stress the fact that the patient is the center of the healthcare process, refer to Fig. 1.3, where a high-level view of various interacting entities with a patient is given.

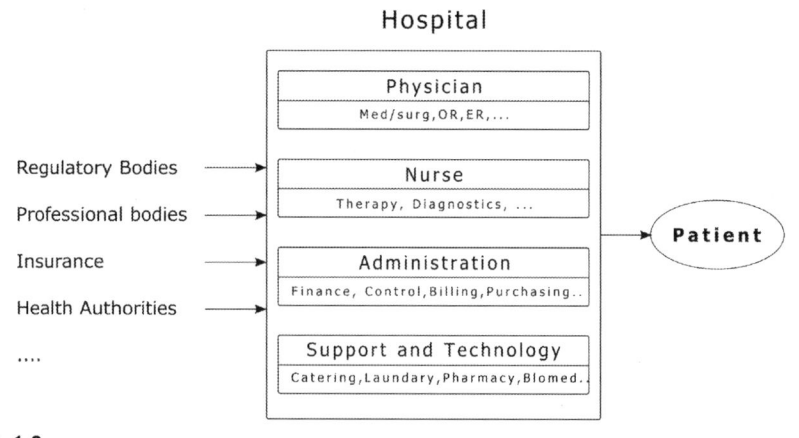

FIG. 1.3

The patient is the central focus of health informatics.

1.4 Applications of healthcare informatics

One could identify vast applications of health informatics in the ever-evolving field of patient care. Through varied applications, the process of patient care has greatly benefited. Some of the fundamental areas that greatly benefit are administration, accounting, auditing, and management. It has wide applications in structured documentation, therapy and diagnostics, and facilitating communication among the different entities in and between organizations.

It could be further analyzed to get the connections between different entity levels in an organization. Through this, one gets to know how the different entities are interacting to execute the healthcare process toward its completion. It facilitates in taking the decision through a decision support system. Health informatics has an array of applications in managing recorded information, which is most necessary for further evolving the process (Gunther & Alejandro, 2001; Wright et al., 2010).

These applications can be deployed at various levels such as during monitoring of the patient, at ventilators, and at imaging machinery. There is a possibility of the deployment of these systems beyond the point of patient care, such as an information system managing the patient record (a clinical electronic patient record). The deployment can take place at the level of the individual care unit/general wards (which assist the patient care process related to admission, medicine, maintenance, and discharge) (Lorenzi & Riley, 2013). The scientific community could greatly benefit by utilizing the patient care data recorded at various level of the healthcare process. All in all, the end beneficiary is society at large.

1.5 Medical information

Due to the reduction in storage and computation costs, hospitals are recording a massive quantity of data related to the healthcare process. The actual challenge now is to process this mammoth amount of data and to extract valuable information that could guide the efficient execution of the healthcare process.

Data is a fundamental driving force of future businesses. Any organization not ready for transforming its operation through insights obtained from data will be left behind (Yasnoff, O'Carroll, Koo, Linkins, & Kilbourne, 2000). This is true in every sector, but it applies more exclusively to healthcare organizations. Currently, healthcare information is growing at a pace of 48% year over year. Studies carried out at Sanford University, predict that the amount of healthcare data by 2020 will amount to 2314 exabytes (Talmon et al., 2009). It is very important for healthcare businesses to adapt their operation procedures to achieve actionable insights. The ability to discover and adapt to the operating environment will power the healthcare domain. The following are some instances that illustrate the amount of data being generated with respect to different healthcare activities (Yasnoff et al., 2000).

Abundant amounts of data are being recorded during the initial stages of the patient care process. Most of this information now can be captured with the help of process-aware information systems such as a patient care information system or a clinical information system. These can offer complete patient info at your fingertips. A study carried out by Friede, Blum, and McDonald (1995) and Yasnoff et al. (2000) gives the efficiency and usefulness of clinical information systems at the clinical level. While one can consider recording more than 2000 attributes of the patient, a detailed study on the number of attributes to be considered for different patient requirements is documented in Moen and Brennan (2005). It is normal to record at least 200 variables of any given patient, the same amount confronted by a physician during any given day at normal rounds (Houston & Ehrenberger, 2001). It has been experimentally proven that it is impossible for a good physician to comprehend a patient case if the number of attributes goes beyond seven variables. It is apparent that it is impossible to estimate the degree of relatedness between variables beyond some limit (Doumont, 2002). Any physician can increase his perception level, relatedness, and ability of accurate inference from data consisting of a large number of features (Dumas et al., 2005).

The previous paragraph discussed the information overload at the level of individual patients, that is, 2000 attributes overall-out of which 200 attributes are the most basic information of a patient. Consider, on any given day physician has to deal with numerous patients. It becomes impossible to extract the relatedness between attributes of data. This information overload becomes an issue. Besides all this, managing the hospital organization become more and more complicated every day (Houston & Ehrenberger, 2001). The literature related to this realm has been growing ever since the concept of healthcare informatics came to the forefront.

1.6 Clinical decision support systems

Clinical decision support systems originated due to information overload and the inability of decision making at the right time. It supports the decisions of healthcare personal in real time with evidence backed up by suitable process execution data. It helps in reducing the cost and time of executing the healthcare process. A clinical decision system is known for offering therapy guidelines directly. With the help of clinical decision support systems, one can enhance the quality of patient care with the help of high-value practices by not practicing what has very little value. Decision support systems offer the best healthcare under all circumstances. The end goal of utilizing clinical decision support systems can be achieved by the following practices (Houston & Ehrenberger, 2001).

- At the point of care, available therapy guidelines will reduce the intra- and interpersonal care variances (between person to person who operates different patients).
- It encourages the development of standard practices and promotes a uniform working culture.

- It facilitates the generation of open and standardized treatment procedures.
- It offers excellent documentation of the patient care process. The patient care data will be at the fingertips of administrators and management at any time.
- The use of clinical decision support system in hospitals helps in standardizing the working procedure and offers guidelines on what to do at which condition of process execution. It even helps to upgrade the patient care and therapy processes once any scientific advancements are done.

1.7 Developing clinical decision support systems

The basis of any clinical decision support system is the knowledge base, rules, and facts guiding the clinical process execution. The knowledge needs to be extracted from the domain experts and the recorded data. The fields of interest in data also play a vital role in extracting the knowledge base for a clinical decision support system. The knowledge base comprises components from management, ethics, medicine, and economics at larger portions. Depending on the ways in which the knowledge is acquired, the knowledge base for a clinical decision support system can broadly be classified as *traditional expert systems, evidence-based,* and *artificial intelligence and statistical inferential based systems* (Savel et al., 2012).

1.7.1 Traditional systems

Collecting information for the formation of a knowledge base relies heavily on the domain experts in the definite field of interest. The traditional approach has been very effective in generating some successful and powerful decision support knowledge rule bases—facilitating pathways. But the following are some of the shortcomings.

- Validating the knowledge base against clinical data is a challenging task in traditional systems.
- Validating the knowledge base generated using traditional systems in intensive care units is the most difficult/impossible task.
- Testing the knowledge base based on traditional systems is costly and time-consuming.

1.7.2 Evidence-based medicine

Evidence-based medicine combines available medical knowledge and scientific evidence. Evidence-based medicine does not offer its own scientific data, but it is built and validated on existing medical knowledge along with therapy knowledge and behaviors with respect to existing standards in the patient care process. The goal of evidence-based medicine is to set up databases from which one can access systematic reviews.

Evidence-based methods generate decision support system rule bases by combining good-quality clinical research evidence with reasoning, nursing expertise, and patient feedback. The evidence-based approach typically follows six staged approaches for laying out clinical pathways (Brender, Ammenwerth, Nykänen, & Talmon, 2006).

First, in evidence-based medicine, it is most important to identify and formulate the clinical problem. Second, the shortcomings of the existing approach must be thoroughly researched to decide on a new framework. Evaluating the alternative framework forms the third stage in designing knowledge-based medicine. As a fourth step, a clinical pathway and a supported rule base must be developed using the gathered evidence. Once the clinical pathway is generated, healthcare professionals must be thoroughly trained. This requires the continuous monitoring and controlling of healthcare professionals during the training and testing period of healthcare pathway deployment. After the healthcare pathway is successfully deployed, a thorough evaluation of its working at successive intervals and strategic interventions is required.

1.7.3 Artificial intelligence and statistical inference-based approaches

Evidence-based medicine has some shortcomings, such as it cannot use real patient data from actual repositories for medical knowledge discovery and validation. It is known that evidence-based medicine and traditional expert systems consume an excessive amount of time and are costly to implement. One real-life example of deploying the evidence-based approach at the MVM hospital, New Jersey, United States, took almost 30 years. This was an enormous amount of effort to develop and stabilize robust clinical procedures (Darkins et al., 2008).

The alternative approach for evidence-based medicine and traditional expert systems is to use artificial intelligence and statistical inference-based approaches. This method is a combined method for knowledge base discovery in large knowledge databases. This approach was originally proposed by Darkins et al. (2008), Doumont (2002), and Gunther and Alejandro (2001) using the optimization methods of artificial intelligence and confidence methods in the statistical inference discipline. At this point in time, these kinds of systems are thoroughly tested in healthcare information systems that deal with hemodynamic patients in critically ill situations.

This approach is dynamic and allows the real-time validation of rulesets in the healthcare information system. It offers an effective approach compared to any other method of knowledge base discovery. With the development in inferential statistics, we can expect that better strategies could be seen.

1.8 Healthcare information management

In modern organizations, data is the new oil, and one who knows how to utilize data to improve process execution gets enormously benefits. The hospitals and organizations that want to reap these benefits require a modern information processing facility/infrastructure. Centralized systems for supporting healthcare operations are slowly being converted to distributed applications. This offers a leap from isolated knowledge to structured team knowledge. The evolution is still possible if the organizations shift from ubiquitous computing to a client-server paradigm with the help of internet browsers as clients.

In the healthcare system, the term SOP primarily refers to standard operating procedures and their consequences on managing the overall organization at large. In order to increase efficiency, the optimization of standard operating procedures is normally carried out. This typically involves a reconstruction of the organizational/operational structures in the healthcare process. The healthcare process execution detail typically gets recorded in one of several process-aware information systems. This information plays a big role in modeling the actual process, giving clarity over what has been planned and how to process what is being executed in the system (Anderson, 2002; Hersh, Margolis, Quirós, & Otero, 2010).

If looked at from different perspectives, healthcare data can be classified into control, organizational, data, and time perspective. Here, we give a brief overview of each of these perspectives and their importance in healthcare informatics.

1.9 Control flow

Control flow refers to the steps followed to achieve a single task. Each task in the sequence is called an activity. A case is a collection of activities. Each activity in a case records various attributes such as the activity name, the time of execution, the resource that executed the activity, the cost for executing the activity, etc. The control flow of any hospital or healthcare process can be represented as a set of nodes connected by directed edges (Hersh et al., 2010).

The goal of a control flow model is to visualize which activities need to be executed and in what order. Based on the transitions between activities, execution can be serial, concurrent, optional, or repeated. Process mining offers a plethora of techniques for discovering the control flow model out of the event log. These discovery algorithms can represent control flow using various notations. This section aims to introduce a subset of the most widely used and important process modeling notations (Hersh et al., 2010).

There are various notations that pictorially represent the control flow of a process, including the Petri net (shown in Fig. 1.4), the transition net (shown in Fig. 1.5), the business process modeling notation (shown in Fig. 1.6), and another workflow

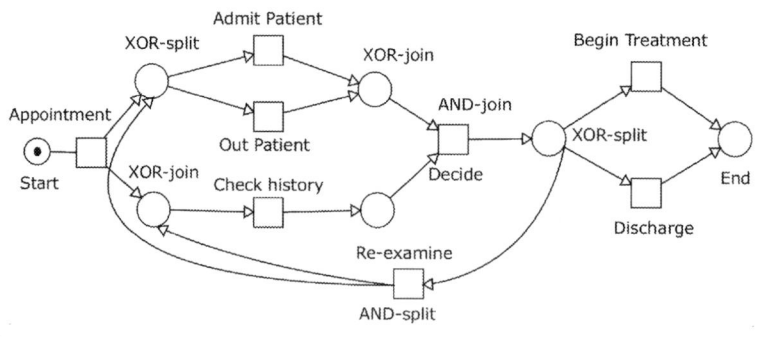

FIG. 1.4

Petri net.

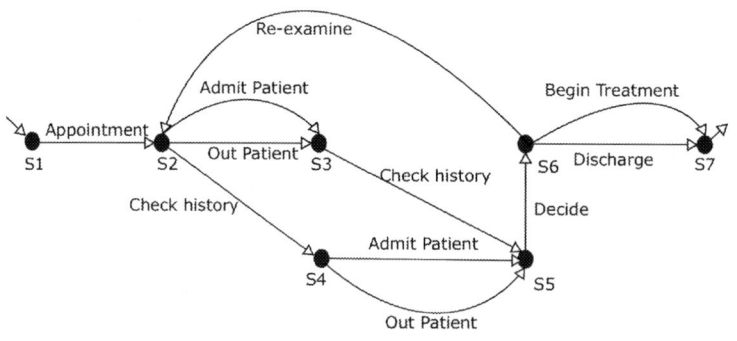

FIG. 1.5

Transition net.

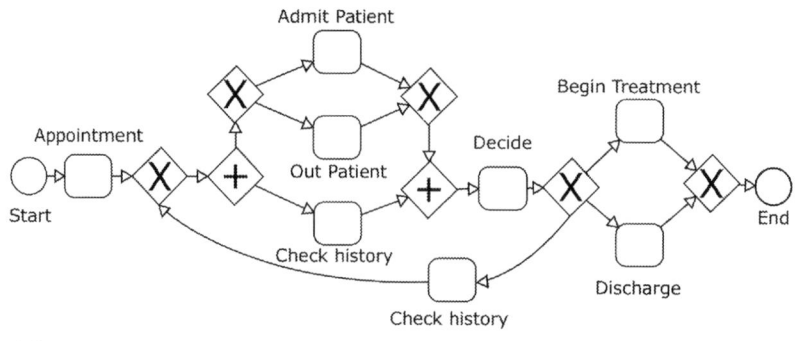

FIG. 1.6

Business process modeling notation.

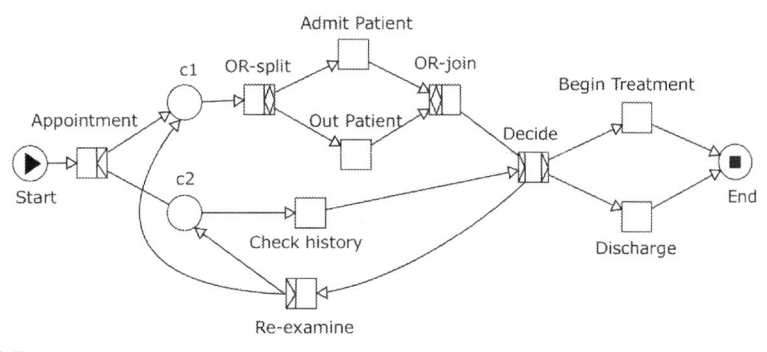

FIG. 1.7

Yet another workflow language.

No permission required.

language (shown in Fig. 1.7). The following diagram illustrates a high-level overview of the control flow of a hospital process. This comprises the sequence of steps followed during the admission process in a hospital.

1.10 Other perspectives

The time perspective is concerned with the time of execution of the activity. With the help of time, one can know the actual bottlenecks, service levels, and service times that could evaluate the quality of service delivery (Hebda, Czar, & Mascara, 2005).

The organizational perspective deals with extracting the connections between various organizational entities. Organizational entities can be an interaction between abstraction. It can be at the level of different roles such as the manager, CEO, CFO, etc. It can even be a relationship among individual entities in the organization.

Data perspective deals with the kind of information exchanges among different entities in the organization. Information can be in terms of primitive or complex data structures.

1.11 Conclusion

This chapter presented an overview of how the application of a computer/information system can leverage the functionality of the healthcare process. Due to the large amount of data being generated in the daily activities of the healthcare process in hospitals, the use of computers makes the life of the healthcare professional easier. Health information systems offer efficient, improved, and time and cost-efficient solutions for executing the healthcare process.

This chapter also gave an overview of how the recorded data in healthcare information systems can be used to extract models briefing different perspectives. Overall, by utilizing health informatics, end-to-end patient care in existing hospitals can be leveraged to greater heights.

References

Anderson, J. G. (2002). Evaluation in health informatics: Social network analysis. *Computers in Biology and Medicine*, *32*(3), 179–193. https://doi.org/10.1016/S0010-4825(02)00014-8.

Brender, J., Ammenwerth, E., Nykänen, P., & Talmon, J. (2006). Factors influencing success and failure of health informatics systems: A pilot Delphi study. *Methods of Information in Medicine*, *45*(1), 125–136. Schattauer GmbH https://doi.org/10.1055/s-0038-1634049.

Darkins, A., Ryan, P., Kobb, R., Foster, L., Edmonson, E., Wakefield, B., & Lancaster, A. E. (2008). Care coordination/home telehealth: The systematic implementation of health informatics, home telehealth, and disease management to support the care of veteran patients with chronic conditions. *Telemedicine and e-Health*, *14*(10), 1118–1126. https://doi.org/10.1089/tmj.2008.0021.

Doumont, J. L. (2002). Magical numbers: The seven-plus-or-minus-two myth. *IEEE Transactions on Professional Communication*, *45*(2), 123–127. https://doi.org/10.1109/TPC.2002.1003695.

Dumas, M., van der Aalst, W. M. P., & ter Hofstede, A. H. M. (2005). Process-aware information systems: Bridging people and software through process technology. In *Process-aware information systems: Bridging people and software through process technology* (pp. 1–409). John Wiley and Sons. https://doi.org/10.1002/0471741442.

Eysenbach, G. (2000). Consumer health informatics. *British Medical Journal*, *320*(7251), 1713–1716.

Fowler, G. A., & Heater, B. (1983). Guidelines for clinical evaluation. *The Journal of Nursing Education*, *22*(9), 402–404.

Friede, A., Blum, H. L., & McDonald, M. (1995). Public health informatics: How information-age technology can strengthen public health. *Annual Review of Public Health*, *16*, 239–252. https://doi.org/10.1146/annurev.pu.16.050195.001323.

Gunther, E., & Alejandro, J. (2001). Evidence-based patient choice and consumer health informatics in the Internet age. *Journal of Medical Internet Research*, e19. https://doi.org/10.2196/jmir.3.2.e19.

Halperin, I., Ma, B., Wolfson, H., & Nussinov, R. (2002). Principles of docking: An overview of search algorithms and a guide to scoring functions. *Proteins, Structure, Function and Genetics*, *47*(4), 409–443. https://doi.org/10.1002/prot.10115.

Haux, R. (2010). Medical informatics: Past, present, future. *International Journal of Medical Informatics*, *79*(9), 599–610. https://doi.org/10.1016/j.ijmedinf.2010.06.003.

Hebda, T., Czar, & Mascara, C. (2005). *Handbook of informatics for nurses and health care professionals* (pp. 120–121). Pearson Prentice Hall.

Hersh, W., Margolis, A., Quirós, F., & Otero, P. (2010). Building a health informatics workforce in developing countries. *Health Affairs*, *29*(2), 275–278. https://doi.org/10.1377/hlthaff.2009.0883.

Houston, T. K., & Ehrenberger, H. E. (2001). The potential of consumer health informatics. *Seminars in Oncology Nursing*, *17*(1), 41–47. https://doi.org/10.1053/sonu.2001.20418.

Lorenzi, N. M., & Riley, R. T. (2013). *Organizational aspects of health informatics: managing technological change*. Springer Science & Business Media.

Mcdonald, C. J., & Tierney, W. M. (1988). Computer-stored medical records: Their future role in medical practice. *JAMA: The Journal of the American Medical Association, 259*(23), 3433–3440. https://doi.org/10.1001/jama.1988.03720230043028.

Moen, A., & Brennan, P. F. (2005). Health@Home: The work of health information management in the household (HIMH): Implications for consumer health informatics (CHI) innovations. *Journal of the American Medical Informatics Association, 12*(6), 648–656. https://doi.org/10.1197/jamia.M1758.

Morak, J., Hayn, D., Kastner, P., Drobics, M., & Schreier, G. (2009). Near field communication technology as the key for data acquisition in clinical research. In *Proceedings—2009 1st international workshop on near field communication, NFC 2009* (pp. 15–19). https://doi.org/10.1109/NFC.2009.12.

Musen, M. A., Middleton, B., & Greenes, R. A. (2014). Clinical decision-support systems. In *Biomedical informatics: Computer applications in health care and biomedicine (4th ed., pp. 643–674)*. London: Springer. https://doi.org/10.1007/978-1-4471-4474-8_22.

Saleem, J. J., Plew, W. R., Speir, R. C., Herout, J., Wilck, N. R., Ryan, D. M., ... Phillips, T. (2015). understanding barriers and facilitators to the use of clinical information systems for intensive care units and anesthesia record keeping: A rapid ethnography. *International Journal of Medical Informatics, 84*(7), 500–511. https://doi.org/10.1016/j.ijmedinf.2015.03.006.

Savel, T. G., Foldy, S., & Control, C. for D. (2012). The role of public health informatics in enhancing public health surveillance. *Morbidity and Mortality Weekly Report. Surveillance Summaries (Washington, D.C. : 2002), 61*, 20–24.

Talmon, J., Ammenwerth, E., Brender, J., de Keizer, N., Nykänen, P., & Rigby, M. (2009). STARE-HI-statement on reporting of evaluation studies in health informatics. *International Journal of Medical Informatics, 78*(1), 1–9. https://doi.org/10.1016/j.ijmedinf.2008.09.002.

VanLeit, B. (1995). Using the case method to develop clinical reasoning skills in problem-based learning. *American Journal of Occupational Therapy*, 349–353. https://doi.org/10.5014/ajot.49.4.349.

Wright, G., Ammenwerth, E., Demiris, G., Hasman, A., Haux, R., Hersh, W., ... Martin-Sanchez, F. (2010). Recommendations of the international medical informatics association (IMIA) on education in biomedical and health informatics. *Methods of Information in Medicine, 49*(2), 105–120. https://doi.org/10.3414/ME5119.

Wyatt, J. C., & Liu, J. L. Y. (2002). Basic concepts in medical informatics. *Journal of Epidemiology and Community Health, 56*(11), 808–812. https://doi.org/10.1136/jech.56.11.808.

Yasnoff, W. A., O'Carroll, P. W., Koo, D., Linkins, R. W., & Kilbourne, E. M. (2000). Public health informatics: Improving and transforming public health in the information age. *Journal of Public Health Management and Practice, 6*(6), 67–75. https://doi.org/10.1097/00124784-200006060-00010.

Young, P. J., & Saxena, M. (2014). Fever management in intensive care patients with infections. *Critical Care, 18*(2). https://doi.org/10.1186/cc13773.

Smart healthcare systems using big data

2

Chinmay Chakraborty[a] and Megha Rathi[b]

[a]Dept. of Electronics & Communication Engineering, Birla Institute of Technology, Mesra, Ranchi, Jharkhand, India, [b]Dept. of Computer Science & Engineering, JIIT, Noida, Uttar Pradesh, India

2.1 Introduction

2.1.1 Background and driving forces

Big data plays a major role in Internet of Things (IoT) applications by providing good insight and improving efficiency. The big data-enabled IoT consists of applications, platforms, and challenges. The platforms of big data are cloud computing, fog computing, edge computing, MapReduce, and the columnar database used to analyze the IoT data effectively. Big data handles terabytes of petabytes of data. The sensor-generated data are processed to a cloud storage platform. Technologies such as ZigBee, 6LoWPAN, LoRaWAN, Wi-Fi, and Bluetooth are used for data transmission (Mandal, Mondal, Banerjee, Chakraborty, & Biswas, 2020).

The sensor cloud framework offers sensor and sensor network virtualization, automatic prevision and data management, on-demand services, monitor usage, operations, health status, and secured data collection and processing. The sensor cloud framework provides four important services: acquisition, visualization, monitoring, and analysis. Big data plays an important role in healthcare; it generates patient health data and sends that into the cloud using IoT. These healthcare data are mapped to the digital health industry. These connected data enhance healthcare resources. The different sources of big data are machine-to-machine communication, databases, cloud services, RFID, and wireless sensor networks. Big data analytics covers predictive analysis, diagnostic analysis, descriptive analysis, preemptive analysis, reports, visualization, statistical and predictive modeling, optimization, information extraction, and augmentation.

Nowadays, big data is gaining popularity in every subfield of computer science and other engineering fields. Big data, as the name suggests, contains an enormous amount of data. For complex and sensitive health problems, big data not only provides support in dealing with technical problems but also provides a solution through insights from the huge dataset. The term "big data" describes a collection of data that is large in size and exponentially grows with time. For handling such type complex and large datasets,

Demystifying Big Data, Machine Learning, and Deep Learning for Healthcare Analytics.
https://doi.org/10.1016/B978-0-12-821633-0.00009-X

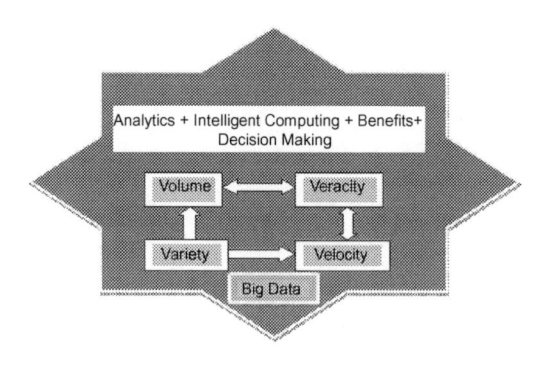

FIG. 2.1

Context of big data.

No permission required.

other data-oriented techniques have not proven to be efficient so for handling complex and huge data set big data technologies are very efficient from the perspective of storage and processing of data. Ideally, big data is comprised of a variety of data that assists in delivering the right information to the concerned person or authority in the right amount and at the correct time for further decision making (Wu, Zhu, Wu, & Ding, 2013). Big data can be utilized by making use of a scalable, flexible, and current data structure, coupled with the latest machine-learning techniques. Big data can be examined at two levels, as shown in Fig. 2.1. At a rudimentary level, it's simply a collection of data that can be analyzed for the profitability of organizations. On the other level, it is exceptional data that presents distinctive data along with its profitability. Big data is quite distinctive from other traditional data in every viewpoint such as memory requirement, time, and functionality. In terms of size, big data is a thousand times greater than traditional data. In terms of data production and transmission speed, it is also a thousand times greater than other data management tools. The following are a few big data generation sources: text, images, audio, video, web data, web session data, various organizations, machine data, and many more.

As shown in Fig. 2.1, various contexts of big data may be defined as:

- **Volume**: A very large dataset. Big data is usually measured in petabytes or exabytes. One exabyte = 1 million terabytes.
- **Velocity**: Sources of big data are also very large and data grow exponentially. Data are generated by the web, various machines, organizations, etc., and transmitted at the speed of light.
- **Variety**: The form, function, and source define the three major aspects of big data. Data are in different formats such as images, graphs, web data, videos, and many more. The function provides the required metadata to process data. For example, social network data is processed differently than medical data. The last aspect deals with the reality of data generation, such as data collected from sensors or any machine.

Table 2.1 Features and comparison of big data with traditional data.

Characteristic	Big data	Other data
Objective	To communicate, administer, and analyze	Manage commercialized applications
Generation source	Media, sensors, web logs, organizations, machines	Documents and transactions related to business
Data structure	Unstructured	Fully structured
Tools	Hadoop, MapReduce, NoSQL, and Spark	Business systems
Cost	Very high	Low to medium
Volume	Petabytes, exabytes	Gigabytes, terabytes
Variety	Text, audio, video, web logs, graphs, medical records	Alphanumeric
Veracity	Reliability and quality depends on data generation source	Clean and reliable
Velocity	Real time (growing rapidly)	Controlled level of data generation
Storage structure	Distributed systems, cloud	Storage area network
Data manipulation	Parallel processing	Ordinary data processing

- **Veracity**: The quality, reliability, and correctness of all such aspects relate to veracity.

Big data is a crucial advanced technological domain that impacts everyone and is very helpful in the health domain. Big data requires data to be gathered, structured, and analyzed fruitfully so that all aspects of big data are managed. Traditional data is like a lake and big data is like a fast-flowing river. The data can be considered from various sources such as imaging data, health records, genetic data, unstructured data, and reports. Table 2.1 presents a comparison of big data with traditional data (Sagiroglu & Sinanc, 2013).

2.2 Big data analytics in healthcare

The healthcare sector can experience fruitful results through big data analytics. Big data analytics plays a significant role in health science. Big data is a huge collection of data generated from various sources that gets integrated with other technologies such as machine learning and intelligent computing (Gandomi & Haider, 2015). Hadoop is used in a big data platform for data management. The Hadoop map-reduce framework is used to process large datasets across different clusters. The Hadoop-distributed file system is applied to manage the large volume of data. Applied to the medical domain, it makes use of medical data for the prevention, diagnosis, curing, and treatment of various diseases. The four major sectors of the healthcare system are patient-centric care, real-time monitoring, predictive disease analysis, and enhancement of treatment methods. Big data provides early prediction of diseases. Big data analytics considers descriptive, diagnostic, predictive, and prescriptive methods for supporting smarter and cost-effective decisions. Listed below are some applications of big data in Healthcare.

2.2.1 Disease prediction

Disease prediction is a main application of big data. Medical records are huge in size and symptoms of every disease differ from each other. The prediction of the right disease at the right time is essential for a life-threatening disease such as cancer. Big data along with associated technologies helps in achieving this objective.

2.2.2 Electronic health records

Electronic health records (EHR) is another widespread utilization of big data tools and Techniques in the health sector. Personalized records include attributes such as personal information, medical history, pathological tests, allergies, sensitive disease, etc. Medical records are transferred via a secure and safe medium and every medical record is editable by a doctor. All the edits in the original file are saved with no security danger while data consistency and integrity are also managed with big data analytics. EHRs can also be used to send warnings and reminders to a patient.

2.2.3 Real-time monitoring

Real-time monitoring is a vital objective of an advanced health model. In hospitals, medical technological support systems provide effective online solutions on the spot for any complex medical problem; they also assist doctors in prescriptive decisions on any sensitive/severe patient. For tracking critical patients from home, big data analytics are required. Big data has the potential to achieve this objective by extracting data through sensors and wearables. Real-time analytics can analyze and generate insights from all available information. The large volume of data can be generated during real-time scenarios.

2.2.4 Medical strategic planning

The use of big data in medicine permits strategic arrangements for better health awareness. Health practitioners can collect and analyze normal routine checkup results from divergent locations, demographics, and groups to find factors for the development of certain disease types and also for upgrading the treatment strategies.

2.2.5 **Telemedicine**

Telemedicine is another advantageous application of big data in the health industry (Chakraborty, Gupta, & Ghosh, 2013). With the invention of smartphones and the IoT, one can utilize mobile phones for online consultation, drug suggestions, diagnosis, real-time patient monitoring, tracking health milestones, and many other things. Different biomedical images such as chronic wound images can be processed to a central hub using telemedicine platforms for remote monitoring (Chakraborty, Gupta, & Ghosh, 2014).

2.2.6 **Drug suggestions**

Symptom-based drug recommendation is also possible with the use of big data analytics. Vital areas in drug industry research and development include drug suggestions, alerts when drugs could have side effects with each other, alerts regarding the negative effects of the continuous use of any drug, and finding the cheapest drugs. Drug suggestions and recommendations make the online treatment process effective.

2.2.7 **Medical imaging**

Big data has the potential to store and gather different data formats. Millions of images are produced during medical scans. Effectively storing and analyzing all these images is essential for treatment. Also, hospitals to need to save them for a long time for novel findings. Big data analytics could alter the way images are captured and stored. Algorithms are designed to read and store pixel-based data, then convert such data into a form that is easily readable and understandable by doctors.

2.3 **Related work**

Big data is of no use unless it provides great insights from medical data to improve the quality of services offered by the healthcare sector. Big data analytics have proven to be very advantageous in the medical domain. Some of the latest related studies in the application of big data in healthcare include.

In the work of Chen, Hao, Hwang, Wang, and Wang (2017), the objective was to predict disease outbreaks. Machine-learning techniques were implemented for the accurate prediction of disease outbreaks from infrequent disease groups. Experiments were conducted on real data gathered from a hospital in China from 2013 to 2015. The model accuracy is reduced with incorrect and incomplete data during the learning phase. To handle inaccurate and incomplete data, a latent factor model is used. The authors developed a new convolution neural network (CNN) that works on both structured and unstructured medical data. The proposed algorithm achieves 94.8% accuracy.

In a significant contribution from Saravana Kumar, Eswari, Sampath, and Lavanya (2015), predictive analysis predicted the diabetes type and complications associated with its treatment. Advanced medical systems generate massive amounts medical records. Big data is required to store and analyze these records. As we know that most of the unstructured data is big in nature so transformation is required to make it consistent for further use. In developing countries such as India, a disease such as diabetes is a noncommunicable hazardous disease. In this research work, big data tools and techniques are utilized to predict the prevalent diabetes type and the hurdles associated with its treatment. In yet another novel work Yuvaraj and SriPreethaa (2019) presented the importance of machine-learning techniques in Hadoop-based data clusters for the prediction of disease. In this work, the authors used the diabetes dataset, and the results are very promising in reflecting the importance of big data analytics for storage and effective retrieval. A dataset was collected from the Pima Indians Diabetes Database from the National Institute of Diabetes and Digestive Diseases (Pima Indians Diabetes Database, 2016). The research work of Cichosz, Johansen, and Hejlesen (2016) provided insights into the management of diabetes-related complications. Diabetes is one of the most common diseases worldwide and also a rich source of medical data. These abundant medical records of diabetes patients are effective for designing protocols and policies for complications related to diabetes. Such knowledge can further be utilized in awareness, treatment, and health administration for improving patient care. Regression models are used for prediction; the proposed model has proven to be very effective for handling diabetes data.

The main emphasis of the proposed research work by Rajamhoana et al. (2018) is to review the latest studies for the classification and prediction of heart disease. Advanced computational techniques such as machine learning and deep learning were applied to a heart disease dataset collected from the UCI ML repository. The authors implemented various machine-learning and deep-learning algorithms to validate the performance on the heart dataset. It has been found that an artificial neural network (ANN) performs better than other existing approaches.

In a potential contribution from Manogaran et al. (2018), a data-processing architecture was developed for cancer diagnosis and prognosis. The hidden Markov model and Gaussian mixture (GM) clustering were used to build the cancer prediction framework. Alterations in some DNA sequences can lead to a cancerous tumor in the human body. Additionally, DNA copy number alteration is also associated with various life-threatening diseases such as cancer. To diagnosis cancer, DNA copy alteration is required, and it is the one vital attribute for cancer detection. In this work, a framework using a hidden Markov model and GM clustering was developed to find the DNA copy change across the genome. The proposed model has proven to be very notable in the research area of cancer prediction.

Ow and Ow and Kuznetsov (2016) emphasized developing strategies for cancer precision. The application of big data analytics in precision medicine is of great use.

The authors validate the performance of their earlier developed algorithm known as prognostic signature vector matching (PSVM) via the optimization of prognostic variable weights. This article is the extension of their research work in which they developed a novel model for disease classification. Also, the proposed approach was compared with other approaches such as K nearest neighbor, support vector machine, random forest, neural networks, and logistic regression to validate the results. The application of big data for drug discovery is presented in the study by Szlezák, Evers, Wang, and Pérez (2014). According to this study, big data plays a crucial role in drug discovery and the development of new drugs. Big data is the latest trend and it dramatically alters pharmaceutical businesses, biology, drugs, and healthcare. With the help of big data tools and technologies, one can discover innovative and smart ways for drug development and profitability. In another contribution of big data analytics, Chen and Butte (2016) provided information on the discovery of new drugs from big datasets. The massive data of pharmaceutical organizations, medicine salts, and herbs motivates health practitioners and scientists to utilize and integrate diverse datasets for the development of new and effective drugs. Big data helps in the analysis of these massive medical datasets to gain knowledge from disease datasets and develop drugs for the respective disease. It aids in providing new and more effective drugs in a cost-effective market. This work highlights the state of the art of integrating big data to point out the latest targets, drug symptoms, and side effects for precision medical care.

In the study from Al Mayahi, Al-Badi, and Tarhini (2018), the intent was to analyze possible applications of big data analytics in the medical sector to exponentially increase outcomes of healthcare services. This article implemented the naive Bayes (NB) and J48 algorithms to predict kidney disease and achieve accuracies of 95% and 100%, respectively. A summary of other recent studies of big data is in Table 2.2 shown below.

2.4 Big data for biomedicine

Big data is impacting the health domain, especially biomedicine. The main advantages of utilizing big data in personalized medicine are upgrading disease treatment, developing personalized treatment frameworks, precision medicine, and symptom-based disease prediction. The primary objective of applying big data in biomedicine is to remarkably improve patient care. Presented below is a summary of papers of big data in biomedicine.

2.5 Proposed solutions for smart healthcare model

As big data is growing in healthcare, a cost-effective smart healthcare solution can be proposed that will try to remove barriers to using big data in healthcare. It will also assist doctors in making rapid decisions as well as enhancing individual care. Patients with complicated medical stories who are distressed from numerous situations will benefit with this model. A series of pathological tests is required to validate

Table 2.2 Summary of recent related works.

Study	Purpose	Approach	Results	Limitations
Wang, Kung, and Byrd (2018)	Find gaps in the implementation of big data	Presents overview for data munging, integration, development, and summarization	Correct and complete the data storage structure. Cloud-based data storage structure	Requirement of data analytical techniques for unstructured data
Saini, Baliyan, and Bassi (2017)	Prediction of heart disease	Ensemble classification technique (bagging + boosting)	Accuracy is improved with the implementation of hybrid machine learning approach for heart disease prediction	Other ensemble approaches need to be explored for further validation of results
Maji and Arora (2019)	Heart disease prediction using data-mining approach	Hybrid approach (decision tree, C4.5 integrated with ANN)	Ensemble model produces better accuracy	Proposed approach needs to be validated on large datasets also with increased experimental runs
Jain and Singh (2018)	Feature selection and classification techniques for disease detection	Filter, wrapper, embedded, and hybrid feature selection; traditional, adaptive, and parallel classification	Out of all feature selection methods, the filter method when embedded with the classification model enhances accuracy	For improving the accuracy of hybrid data mining models, its essential to remove noisy and redundant features from the dataset
Vlachostergiou, Tagaris, Stafylopatis, and Kollias (2018)	Parkinson's disease prediction	Multitasking learning (MTL) layer	MTL outperforms other machine learning models, and provide 83% accurate results while predicting Parkinson's disease	Other attributes such as brain scan results, disease duration, treatment duration, and drug dose can be taken into account for more models
	Real-time online consultation with doctor	Web-based patient care tracking framework	Sensor-based framework is developed for online tracking and consultation with the doctor, which is also cost effective	Variety of sensors must be utilized for data storage on cloud and retrieval of data

Narayanan, Paul, and Joseph (2017)	Big data application areas are needed for health domain	Rough set theory and neural network	Advanced neural network-based model for future research	Risks of big data need to be considered
Tse, Chow, Ly, Tong, and Tam (2018)	To analyze the importance of big data in the medical domain	A qualitative methodology of studying the importance of big data analytics	A data ascendancy strategy should be prepared	Balance between medical data and security concerns

any disease, but with the help of an IoT-enabled framework, one can detect disease quickly. Just enter certain symptoms, and based on those symptoms, the disease will be predicted.

For accurate prediction, data-mining, machine-learning, and deep-learning models are required. Ensemble learning techniques and hybrid models (artificial intelligence + intelligent computing + IoT) are also proven to be efficient for disease detection. Fig. 2.2 presents the characteristics of smart healthcare models that consist of effective decision making, drug suggestion and discovery, cost and time savings, real-time monitoring, online consultation and diagnosis, and improved patient care. Parallel processing is required to create smart and innovative models, especially for the healthcare sector. The parallel processing system will work on divergent small jobs and is suited for large datasets such as big data with ease, high throughput, and productivity. MapReduce is one of the known parallel programming frameworks for processing large datasets. The main advantage of processing medical data with MapReduce is that minimal shifting of data on distributed systems is required for data processing. Smart health frameworks require data to be distributed on different nodes at different demographic locations, so big data architecture is required for parallel processing of data. Along with parallel processing, analytical techniques are also required for effective decision making. Techniques such as artificial intelligence, machine learning, data and web mining, data analytics, IoT, sensor networks, deep learning, computer vision, algorithms, and many more are required to develop high-quality hybrid health frameworks.

The deep-learning scheme analyzes very large and complex datasets (medical images, videos, unstructured information, texts). An architecture of big data is needed to generate such integrated high-quality solutions for complex healthcare problems. Fig. 2.3 presents the architecture of big data for handling, storing, and retrieving large distributed medical records.

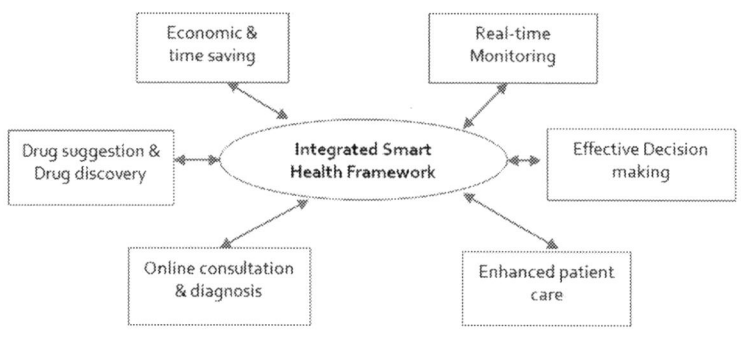

FIG. 2.2

Smart health model characteristics.

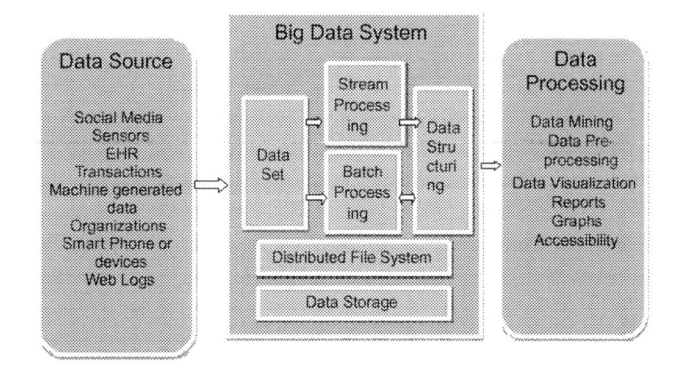

FIG. 2.3

Big data health architecture.

2.6 **Role of sensor technology for eHealth**

Sensor technology plays an important role in advanced healthcare technology. Wireless sensors improve healthcare performances by providing functionalities in a ubiquitous setting that permits sensors to associate with the Internet anywhere at anytime. Healthcare sectors demand reliable, real-time, and precise diagnostic outcomes delivered by sensors remotely from homes, clinics, and hospitals. Sensors and devices are still collecting necessary information and supplementary data. Numerous sensors are used in clinics, including for blood sugar, respiratory rate, heart rate, pulse, blood pressure, and temperature. Healthcare sensors mostly deal with patient monitoring, the environment, and diagnosis. Biosensors are used to analyze the biological and physiochemical detectors. Biosensors monitor blood glucose, cancer screening, pathogen detection, drug discovery, and routine analytical check-ups. The smart sensors cover the most effective solutions such as multisensing, computation, self-calibration, and cost-effectiveness. Another major promising technology, wireless medical sensor networks, is used for early diagnosis and effective cures. In wireless sensor networks, the smart dust concept is applied for computation, power, communication, and self-contained sensing purposes. Smart or intelligent sensors enhance healthcare through the adoption of smartphones, microelectromechanical items, nanotechnology, and wireless sensor networks. The microelectromechanical sensor includes optical projection, fluid-flow control, pressure measurement, and motion sensing. Tactile sensors give tactile sensing capabilities and also measure physical interaction information from the environment. Smart sensors are used in supply chain management. Smart living tools associated with diverse sensors and assistive smartphone medical devices help to collect everyday data from the elderly. IoT sensors will be the most promising technology to resolve the health issue for homecare. Important characteristics of sensor include low cost, small size, wireless/wired connection, self-validation/identification, less power, robust, self-healing/diagnosis/calibration, and data preprocessing to perform IoT. Device-to-device communication becomes powerful, where the devices are acquiring healthcare data from sensors under the Internet of Medical Things platform (IoMT) (Chinmay & Rodrigues, 2020). Structural health monitoring is applied for regular monitoring of stress, strain, temperature, damage detection, and inspection with less labor involved on a real-time/online basis (Chang, 1999).

Big data from different data sources is put into a cloud storage framework for analyzing the stored data. The various IoT data sources are medical devices, car sensors, satellites, video games, road cameras, refrigerators, etc. Here, large volumes of data from millions of connected objects are shared. This generated data offer smart IoT services. The IoT generates massive amounts of data for vehicle tracking, weather-monitoring stations, wearable fitness bands, retail inventory, and health.

The important features for big data sensing methods are: (a) elastic data sensing scalability, (b) data collection and management, (c) on-demand service, (d) security, (e) billing, (f) multitenanted access and service, and (g) automated data-sensing provision. Big data enhances the usage and sharing of sensor networks, minimizes the sensor network management costs, supports global connectivity, reduces upgrading and maintenance costs, generates various business models, and provides sensor domain-specific applications. Sensors are able to transmit the data, so IoT can be recommended as the intelligence of things for medical domain. From India's perspective, big data is generated from various sources such as genomic data, lab and imaging outcomes, insurance claim providers, clinic providers, mHealth applications and devices, manual survey results, electronic medical records, retail pharmacies, and public healthcare data. Wearable sensor devices are constantly gathering structured and unstructured data and monitoring patients. The useful information that can be extracted from large datasets is also a big challenge. Cloud computing is applied to manage the data. Fog computing can be used before the cloud for processing the data. The cloud framework makes fast decisions in real-time scenarios.

2.7 Major applications and challenges

The big data-enabled IoT provides various applications, including smart homes, smart cars, smart grids, smart agriculture, smart lighting, smart appliances, intrusion detection, gas detectors, smart parking, smart transportation, air pollution monitoring, forest fire detection, river flood detection, smart payments, shipment monitoring, and healthcare. The main challenges of big data-enabled IoT are the real-time analysis, noises in data, the large volume of data, storage, diverse data sources, variety, high redundancy, quality of service, and security. The data can be lost at the sensing devices due to miscalibrations, device compatibility, environmental complications, minimum sensing resources, capability, etc. Machine learning and big data approaches can be applied to the acquired data for the development of predictive diagnostic tools. But some challenges involve security, scalability, usability, privacy, methods, and standards. IoT-big data is the most promising technology that supports different on-demand services where big data is collected from several sensors. In healthcare, big data handling is a difficult task for real-time remote monitoring. The patient's allocation based on prioritization (surgical room, surgery, transplantation) is an issue of the telemedicine system. This chapter focuses on the role of big data for healthcare and provides a discussion on big data sensing, challenges, and services. The applications of big data include different sectors such as public, healthcare, industry, education, learning, insurance, transportation, natural resources, banking, fraud detection, entertainment, etc. The beneficiaries of healthcare data analytics are clinicians, governments, insurance companies, pharmaceutical companies, and advertising and marketing agencies.

Healthcare analytics enhance the quality of life/care by optimizing treatment costs, reducing waiting time, envisaging outbreaks of epidemics, etc. The various data such as sensing data, clinical data, electronic health data, omics data, and public data can be processed through a big data warehouse for analytics purposes using diagnostic, predictive, prescriptive, and descriptive models while also enhancing the smarter solutions. The various big data applications in biomedical engineering include medical health records, real-time alerts using robust clinical decision support, improved patient participation, building up strong strategic planning using medical data, curing vital diseases, predictive health data analytics, fraud reduction, improved security, telemedicine, personal health records, the correlation between big data with medical imaging, avoiding unnecessary patient visits, the intelligence of the healthcare unit, patient history management, and predictions of future patient conditions.

The major issues based on our interpretation are as follows: lack of standardization, lack of security, interconnectivity among various sensor networks, sensor resource virtualization, on-demand big data service, data collection and storage requirements, technology, policies, quality of service, big data monitoring, and analytics services. Another major challenge of big data in healthcare is legal issues for personal information handling. Big data needs large processing power and massive storage units.

Amit, Chinmay, Anand, and Debabrata (2019) discussed the recent trends in IoT and big data analytics for healthcare technologies. The authors highlighted with examples of advanced medical imaging, telemedicine, wearable devices, IoT, biotechnological advances, and big data analytics and management. Akash, Chakraborty, and Bharat (2019) presented an IoT-enabled cloud platform for sharing the relevant information regarding seizures for improving e-health service by minimal costs, maximized accuracy, and faster solutions. Chakraborty (2019) discussed the use of mobile health for remote chronic wound monitoring. The major requirements for developing specialized techniques in big data analytics are: (a) multimedia data—EHR consists of very large amounts unstructured data from real-time sensors and also deals with heterogeneous data, (b) complex background knowledge—health data generated from complex metadata and optimal decisions required for analyzing the data, (c) highly qualified users—doctors, clinics, researchers, lab technicians, and bioinformaticians are highly professional. The expert-driven systems support clinicians to monitor the analytics process, (d) complex decision—technology enables patients, doctors, and clinics using smartphones, wearable devices, and sensors, (e) privacy—healthcare data are very sensitive and need to be protected during transmission from the patient hub to the server (Chakraborty, 2019).

2.8 Conclusion and future scope

We have interpreted that big data analytics provide valuable insights into the health sector and provide assistance in analyzing factors influencing extraordinary health services. The urgent need is to merge big data analytics with the medical sector. Big data works on the principle of huge amounts of data; one can forecast future events using large medical datasets. To leverage the gap between structured and unstructured data sources, researchers are using big data tools and techniques. Nowadays, big data is expanding in every field from industry to academia to provide reliable future projections and analysis. Medical datasets are growing exponentially, and this has forced researchers to develop intelligent computational approaches for effective decision making in healthcare. Expert systems, frameworks, and models are designed and developed to upgrade health services. We have to retain financial assistance, time, and strength for effectively deploying big data in health organizations to assist patients in tasks such as real-time consultation/diagnosis, drug suggestion, health preventive schemes, online diagnosis of severe patients, and many more. Lastly, we have concluded that the application of big data in the medical domain can upgrade it to an advanced level.

Further, big data can also be used for medical strategic planning, for curing cancer, finding factors that influence cancer, early diagnosis of cancer, real-time alert system, maintaining electronic health records for further investigation and analysis, improved health predictions, telemedicine, and many more. The deep-learning models will perform well to predict possible diseases. Good security schemes will be required for smart or intelligent sensor data processing.

References

Akash, G., Chakraborty, C., & Bharat, G. (2019). Sensing and monitoring of epileptical seizure under IoT platform. In *IGI: Smart medical data sensing and IoT systems design in healthcare* (pp. 201–223).

Al Mayahi, S., Al-Badi, A., & Tarhini, A. (2018). Exploring the potential benefits of big data analytics in providing smart healthcare. In *Lecture notes of the Institute for Computer Sciences, Social-Informatics and Telecommunications Engineering, LNICST: Vol. 200* (pp. 247–258). Springer Verlag. https://doi.org/10.1007/978-3-319-95450-9_21.

Amit, B., Chinmay, C., Anand, K., & Debabrata. (2019). Emerging trends in IoT and big data analytics for biomedical and health care technologies. In *Handbook of data science approaches for biomedical engineering.*

Chakraborty, C. (2019). Mobile health (m-health) for tele-wound monitoring. In *Vol. 1. Mobile health applications for quality healthcare delivery* (1st ed., pp. 98–116). IGI. https://doi.org/10.4018/978-1-5225-8021-8.ch005.

Chakraborty, C., Gupta, B., & Ghosh, S. K. (2013). A review on telemedicine-based WBAN framework for patient monitoring. *Telemedicine and e-Health, 19*(8), 619–626. https://doi.org/10.1089/tmj.2012.0215.

Chakraborty, C., Gupta, B., & Ghosh, S. K. (2014). Mobile metadata assisted community database of chronic wound images. *Wound Medicine, 6*, 34–42. https://doi.org/10.1016/j.wndm.2014.09.002.

Chang, F.-K. (1999). *A summary report of the 2nd workshop on structural health monitoring held at Stanford University on*.

Chen, B., & Butte, A. J. (2016). Leveraging big data to transform target selection and drug discovery. *Clinical Pharmacology and Therapeutics, 99*(3), 285–297. https://doi.org/10.1002/cpt.318.

Chen, M., Hao, Y., Hwang, K., Wang, L., & Wang, L. (2017). Disease prediction by machine learning over big data from healthcare communities. *IEEE Access, 5*, 8869–8879. https://doi.org/10.1109/ACCESS.2017.2694446.

Chinmay, C., & Rodrigues, J. J. C. P. (2020). A comprehensive review on device-to-device communication paradigm: Trends, challenges and applications. *Wireless Personal Communications*, 185–207. https://doi.org/10.1007/s11277-020-07358-3.

Cichosz, S. L., Johansen, M. D., & Hejlesen, O. (2016). Toward big data analytics: Review of predictive models in management of diabetes and its complications. *Journal of Diabetes Science and Technology, 10*(1), 27–34. https://doi.org/10.1177/1932296815611680.

Gandomi, A., & Haider, M. (2015). Beyond the hype: Big data concepts, methods, and analytics. *International Journal of Information Management, 35*(2), 137–144. https://doi.org/10.1016/j.ijinfomgt.2014.10.007.

Jain, D., & Singh, V. (2018). Feature selection and classification systems for chronic disease prediction: A review. *Egyptian Informatics Journal, 19*(3), 179–189.

Maji, S., & Arora, S. (2019). Decision tree algorithms for prediction of heart disease. In *Information and communication technology for competitive strategies* (pp. 447–454). Singapore, Chicago: Springer.

Mandal, R., Mondal, M. K., Banerjee, S., Chakraborty, C., & Biswas, U. (2020). Data de-duplication approaches-concepts, strategies and challenges. *A survey and critical analysis on energy generation from Datacenter* (pp. 203–230). Elsevier. https://doi.org/10.1016/B978-0-12-823395-5.00005-7.

Manogaran, G., Vijayakumar, V., Varatharajan, R., Malarvizhi Kumar, P., Sundarasekar, R., & Hsu, C. H. (2018). Machine learning based big data processing framework for cancer diagnosis using hidden Markov model and GM clustering. *Wireless Personal Communications, 102*(3), 2099–2116. https://doi.org/10.1007/s11277-017-5044-z.

Narayanan, U., Paul, V., & Joseph, S. (2017). Different analytical techniques for big data analysis: A review. In *2017 International conference on energy, communication, data analytics and soft computing (ICECDS)* (pp. 372–382). Chicago: IEEE.

Ow, G. S., & Kuznetsov, V. A. (2016). Big genomics and clinical data analytics strategies for precision cancer prognosis. *Scientific Reports, 6*. https://doi.org/10.1038/srep36493.

Pima Indians Diabetes Database. (2016). https://www.kaggle.com/uciml/pima-indians-diabetes-database.

Rajamhoana, S. P., Devi, C. A., Umamaheswari, K., Kiruba, R., Karunya, K., & Deepika, R. (2018). Analysis of neural networks based heart disease prediction system. In *Proceedings—2018 11th international conference on human system interaction, HSI 2018 (pp. 233–239)*Institute of Electrical and Electronics Engineers Inc. https://doi.org/10.1109/HSI.2018.8431153.

Sagiroglu, S., & Sinanc, D. (2013). Big data: A review. In *2013 International conference on collaboration technologies and systems (CTS)* (pp. 42–47).

Saini, M., Baliyan, N., & Bassi, V. (2017). Prediction of heart disease severity with hybrid data mining. In *2017 2nd International conference on telecommunication and networks (TEL-NET)* (pp. 1–6). IEEE.

Saravana Kumar, N. M., Eswari, T., Sampath, P., & Lavanya, S. (2015). Predictive methodology for diabetic data analysis in big data. *Procedia Computer Science*, *50*, 203–208. Elsevier B.V. https://doi.org/10.1016/j.procs.2015.04.069.

Szlezák, N., Evers, M., Wang, J., & Pérez, L. (2014). The role of big data and advanced analytics in drug discovery, development, and commercialization. *Clinical Pharmacology and Therapeutics*, *95*(5), 492–495. https://doi.org/10.1038/clpt.2014.29.

Tse, D., Chow, C. K., Ly, T. P., Tong, C. Y., & Tam, K. W. (2018). The challenges of big data governance in healthcare. In *2018 17th IEEE International conference on trust, security and privacy in computing and communications/12th IEEE International conference on big data science and engineering (TrustCom/BigDataSE)* (pp. 1632–1636). IEEE.

Vlachostergiou, A., Tagaris, A., Stafylopatis, A., & Kollias, S. (2018). Multi-task learning for predicting Parkinson's disease based on medical imaging information. In *2018 25th IEEE International conference on image processing (ICIP)* (pp. 2052–2056). Chicago: IEEE.

Wang, Y., Kung, L., & Byrd, T. A. (2018). Big data analytics: Understanding its capabilities and potential benefits for healthcare organizations. *Technological Forecasting and Social Change*, *126*, 3–13.

Wu, X., Zhu, X., Wu, G. Q., & Ding, W. (2013). Data mining with big data. *IEEE Transactions on Knowledge and Data Engineering*, *26*(1), 97–107.

Yuvaraj, N., & SriPreethaa, K. S. (2019). Diabetes prediction in healthcare systems using machine learning algorithms on Hadoop cluster. *Cluster Computing*, 1–9. https://doi.org/10.1007/s10586-017-1532-x.

Big data-based frameworks for healthcare systems

3

Aboobucker Ilmudeen

Department of Management and Information Technology, Faculty of Management and Commerce, South Eastern University of Sri Lanka, Oluvil, Sri Lanka

3.1 Introduction

Currently, the domain of big data has rapidly grown up where massive volumes of data are produced from various sources. As a result, big data is getting ever more popular by extending into several forms such as eHealth, mHealth, and the Internet of medical things (Firouzi et al., 2018). Big data and healthcare systems have been closely interrelated with the help of advanced and sophisticated technologies. Big data has ample potential to offer various healthcare functionalities such as disease observation, medical decision support, and patient health monitoring. Therefore, effective big data-based healthcare systems are required to observe patient symptoms and identify clinical decisions by medical officers and physicians. Similarly, healthcare systems normally keep large amounts of data produced by patient clinical records, compliance and governing desires, and patient care. The conventional data-processing techniques, frameworks, or algorithms are unable to deal with large volumes of big data (Hossain & Muhammad, 2016).

Today, there is a paradigm change in healthcare that has seen a shift from cures to early detection and prevention of diseases (Kim & Seu, 2014). In healthcare, big data has the power of analytical capability to extract the hidden patterns, unseen links, insights, predictions, and various trends from big volumes of data from varied data sources. Accordingly, recent developments in healthcare such as patient electronic health records as well as the integration of smart health, mHealth, and eHealth devices have helped the expansion of innovative healthcare systems that enable medical precision and tailored healthcare solutions. Recent advancements in the Internet of medical things and cloud servers can regularly collect transmitted data that can be handled and processed by machine learning and big data analytic techniques (Syed, Jabeen, Manimala, & Alsaeedi, 2019). Further, healthcare datasets are complex in nature and are strongly intertwined. Thus, activities such as simplifying the difficulty in data, recognizing the interlinks among numerous healthcare factors, and the choice of target features for healthcare analytics need extremely refined and developed domain-centric methods and practices (Palanisamy & Thirunavukarasu, 2019).

Healthcare big data includes various organized, semiorganized, and unorganized data sources that cannot be fully handled by the traditional algorithms, frameworks, tools, and techniques (Hossain & Muhammad, 2016). Similarly, most existing big data tools for storing, processing, retrieving, and analyzing the heterogeneous large volume of data are insufficient (Inoubli, Aridhi, Mezni, Maddouri, & Mephu Nguifo, 2018). In recent years, advancements in sophisticated healthcare systems or applications have received serious attention by academia and practitioners to predict diseases, clinical decisions, and patient care (Hossain & Muhammad, 2016). Hence, this necessitates a big data-based framework that simplifies collecting, storing, mining, classifying, modeling, and processing huge amounts of heterogeneous data (Hossain & Muhammad, 2016).

While the tools, techniques, frameworks, and platforms of big data have been widely used in different fields, the implementation and delivery of innovative healthcare services would be a suitable direction for research in the healthcare sector. Similarly, the paradigm of big data and healthcare systems rest on the basic architecture and deployment of suitable tools that are proved in the prior research attempts. Big data analytics are well recognized due to their distinctive capabilities as they have analytical power, design elements, superior methodological and practical solutions, and flexibility (Desarkar & Das, 2017). Despite the advantages offered by healthcare systems, there are various issues that hinder the efficient implementation of healthcare systems. Hence, this requires a closer collaboration between the hardware and software components, including developers, designers, architects, and medical professionals for efficient healthcare system development.

This chapter first gives an outline of the most widespread and generally used frameworks, architectures, and models in big data. It focuses on identifying key features such as the architecture, the programming platform, the design elements, the big data processing models, and the components that depict the physiognomies of big data frameworks in detail. This chapter lists prior big data framework-related studies based on their key features. In addition, this chapter includes past studies that applied healthcare systems in practical applications that evaluate the performance of each existing framework and present a comparative illustration by their technical, design, and architecture principles. Finally, this chapter concludes by proposing a big data-based conceptual framework in healthcare systems with their design elements.

3.2 The role of big data in healthcare systems and industry

With the recent developments and related technologies in big data, its impact in the healthcare industry has identified various data sources such as telematics, sensors and wearable devices, and social media platforms (Palanisamy & Thirunavukarasu, 2019). Today, connecting IoT devices and big data creates unique openings for delivering healthcare services to users by integrating machine learning, cloud computing, and data mining (Mahmud, Iqbal, & Doctor, 2016). When it comes to healthcare, large volumes of data are being stored by drug and pharmaceutical manufacturing companies. These data are highly complicated in nature. Sometimes, they cannot be linked with other information by the practitioners, but they have some potential insights for better decision making. Modern analytic techniques can extract hidden insights into the genetic and ecological roots of diseases from these complex medical datasets.

Today, big data analytics have become important and vital. Problems such as security, privacy, compliance procedures, and establishing standards to advance big data technologies will draw much attention. Along this line of thinking, arrangements such as more effective platforms for data processing, smart data collection techniques, smart and accurate computation, visualization, and storage techniques need to be developed to extract the value. Pramanik, Lau, Demirkan, and Azad (2017) identified three paradigm shift trends in their smart health and smart city study: (1) in healthcare, traditional health to ubiquitous health, and then smart health; (2) in the city, the traditional city to a digital city, and then to a smart city; and (3) in data, the database has shifted to data mining that enables big data.

Data sources in healthcare can be generally categorized into the following: First, *structured data* that refers to data that follow a well-defined data type, structure, and format, including the classified terminologies of different diseases, information about disease symptoms and diagnosis, laboratory results, electronic health records, information about patients such as admission histories, and clinical and drug details. The second is semistructured data with a self-describing nature that is organized in a minimal structure. For example, data generated from the IoT and sensors for patient health conditions, doctor-to-patient email, social media, and the web. The third is unstructured data that describes no natural structure such as medical prescriptions written by physicians/doctors using human languages, clinical records, biomedical descriptions, discharge records, claims, informal texts, and so forth.

3.3 Big data frameworks for healthcare systems

The success of the healthcare system rests on the primary architecture and the deployment of suitable frameworks that have been identified as novel research areas in the big data field (Palanisamy & Thirunavukarasu, 2019). Numerous healthcare-related big data frameworks have been identified in recent times to handle huge amounts of varied records from various databases to produce meaningful insights and trends. In the past, big data frameworks have been categorized based on features such as the models in programming languages, the compatibility of languages, the data source types, the iterative data analyzing ability, the adaptability of the framework with the present machine learning libraries, and the fault acceptance strategy (Inoubli et al., 2018). Many past studies have attempted to evaluate the big data frameworks (e.g., Inoubli et al., 2018; Mahmud et al., 2016; Palanisamy & Thirunavukarasu, 2019; Sicari, Rizzardi, Grieco, Piro, & Coen-Porisini, 2017). The section below discusses various prior studies for healthcare systems that applied big data frameworks (see more in Table 3.1).

Table 3.1 Big Data frameworks-related studies in healthcare systems.

Author	Name of the framework	Techniques/ platform features	Data sources	Analytical capability	Application domain	Key highlights
Syed et al. (2019)	Smart healthcare framework using Internet of medical things and big data analytics	Parallel process in Hadoop MapReduce used	Multiple wearable sensors	Multinomial naïve Bayes classifier that matches MapReduce	Smart healthcare for ambient assisted living	Smart healthcare facilitates to remotely observe health status of elderly people
Chawla and Davis (2013)	Patient-centered big data-driven personalized healthcare framework	Computing and analytics framework used to integrate big data	Electronic medical records, patient experiences, histories	Collaborative filtering to predict diseases	Patient-centered healthcare	Data driven computational system for patient-centric healthcare
Raghupathi and Raghupathi (2014)	Conceptual and architectural big data framework	Outlines methodological and architectural design of big data	Multiple locations, physically dissimilar data sources in many formats.	Analytical queries and generating reports	Healthcare design domain	Proposes a state-of- the-art model for designing healthcare big data framework
Fang, Pouyanfar, Yang, Chen, and Iyengar (2016)	Health informatics processing pipeline	Machine learning techniques and algorithms compared	Electronic health records, public health, genomic, behavioral data	Feature selection Machine learning (classification, regression, clustering)	Decision support system via computational health informatics for practitioners	Data capturing, storing, sharing, analyzing, searching, and decision support through computational health informatics

Lin, Dou, Zhou, and Liu (2015)	Cloud-based framework for self-caring service	Lucene-based distributed search cluster and Hadoop cluster are used	Patient profile, data, and clinical data	Hadoop cluster is used for highly concurrent and scalable medical record retrieval, data analysis, and privacy protection	Home diagnosis—self-caring service	Home self-caring based on historical medical records and disease-symptom
del Carmen Legaz-García, Martínez-Costa, Menárguez-Tortosa, and Fernández-Breis (2016)	Semantic web-based framework for interoperability and exploitation of clinical archetypes	Semantic web technologies for interoperability and exploitation of archetypes, EHR data, and ontologies	Electronic health records (EHR), ontologies	OWL-based ontology	Classification based on clinical criteria	Integration of Semantic web resources with EHR
Sicari et al. (2017)	Policy enforcement framework IoT-based smart health	Security and quality threats in dynamic large-scale smart hearth environments. Cross-domain policies have been defined using XML	RFID and instruments generated data, patient, environmental data	Policy-based access control mechanism for availing healthcare resources	Smart health applications to prevent security threats in large scale heterogeneous health environment	Implementing policy framework for smart healthcare
Hossain and Muhammad (2016)	Voice pathology assessment big data framework	Machine-learning algorithms	Speech signals	Classifiers such as support vector machine, extreme learning machine, and a Gaussian mixture model	The audio features classified as normal or pathological	A framework to handle healthcare big data
Mahmud et al. (2016)	Cloud-based data analytics and visualization framework—health-shocks prediction	Amazon web services linked with geographical information systems and fuzzy rule summarization technique	Healthcare data focused on socio, economic, cultural, and geographical conditions	Generated predictive model using fuzzy rule summarization	Public households to increase healthcare facilities	Cloud enabled geographical information system

Continued

Table 3.1 Big Data frameworks-related studies in healthcare systems—con't

Author	Name of the framework	Techniques/ platform features	Data sources	Analytical capability	Application domain	Key highlights
Rahman, Bhuiyan, and Ahamed (2017)	RFID-based framework to preserve two privacy issues in healthcare system	Authentication and access control are ensured for RFID tag application in healthcare domain	RFID tags	Techniques to preserve privacy in RFID	Protecting healthcare domain services	Better privacy mechanism to protect RFID applications in healthcare domain
Pramanik et al. (2017)	Big data-aided framework for smart healthcare system	State-of-the-art design and architecture of smart healthcare services	Electronic health record, patient diagnosis and biometric data, social media, and surveillance data	Smart healthcare services at smart cities via advanced healthcare systems	Smart integration and technologies to provide state-of-the-art healthcare services	Combining big data and healthcare designed smart services for the smart city
Forkan, Khalil, Ibaida, and Tari (2015)	Cloud-centric bigdata framework for personalized patient care through context-aware computing system	Knowledge discovery-based context-aware framework	Profile data, patient medical records, activity logs, vital signs, and context cum environmental sensor data	Mine trends and patterns with associated probabilities used to learn proper abnormal conditions	Classification to identify real abnormal conditions of patients having variations in blood pressure	Personalized healthcare services through context aware decision making approach

Accordingly, Rahman et al. (2017) suggested an RFID-based model to illustrate two privacy problems in RFID-related healthcare system: (1) authentication protocols for identification and monitoring purposes, and (2) access control to limit illegal access to protected data. Their research reveals the security and privacy aspects in the technical design of RFID systems in the healthcare domain. In reality, the privacy and sensitivity of medical data in the healthcare sector must be strictly controlled with security procedures and data quality requirements. Pramanik et al. (2017) proposed an intelligent healthcare model based on big data with a three-dimensional structure of a paradigm shift and three technical branches of big data healthcare systems that addressed the possible challenges and opportunities in executing this system in the healthcare business context. Similarly, Raghupathi and Raghupathi (2014) suggested a big data framework considering the theoretical and methodological aspects in which the proposed conceptual architecture of big data analytics consists of big data sources, the transformation of big data, existing platforms and tools, and the applications of big data analytics.

Moreover, Sicari et al. (2017) suggested a policy execution structure to address security threats that are expected during the development of IoT-based applications for smart health. Their modeling is appropriate for heterogeneous IoT-based applications and their architecture in this smart healthcare context. Syed et al. (2019) developed a smart healthcare framework that can observe the health of aged people by using the Internet of medical things; this is then analyzed by machine-learning algorithms. This proposed system employed a multinomial naïve Bayes classifier that supports the MapReduce paradigm as well as a system for the faster analysis of data and better disease decision making with better treatment recommendations in which elderly health conditions could be remotely monitored. Youssef (2014) introduced a big data analytics-based healthcare information systems framework in mobile cloud computing environments. Their proposed framework offers features such as interoperability, greater adaptability, readiness, and the ability to transfer healthcare records among medical officers, patients, and physicians to enable finding valuable insights for effective decision making at the right time.

3.4 Overview of big data techniques and technologies supporting healthcare systems

3.4.1 Cloud computing and architecture

The advent of mobile devices permits the services offered by smart devices that can simplify the remote monitoring of healthcare services at any time wherever with the help of cloud computing. The features of cloud computing including storage capacity, server management, bandwidth, and network efficiency have made cloud computing as one of the foremost reason that the big data become so pervasive (Fang et al., 2016). Similarly, cloud computing provides data analytics, insights, and the mining of critical data, resulting in increased speeds in many analytical decisions in healthcare systems (Mahmud et al., 2016). Cloud computing has been identified as one of the most broadly adopted and deployment architectures by the healthcare application developers (Mahmud et al., 2016). The Internet of Things (IoT)-supported cloud computing can be flexible to extend the development of novel applications in the healthcare sector over smart platforms (Kumar, Lokesh, Varatharajan, Chandra Babu, & Parthasarathy, 2018). Hence, for real-time disease analysis, healthcare systems frequently use cloud computing services to save the information between the data originator layer/sensor device and the data analyzer layer/end-user applications (Singh, Bansal, Sandhu, & Sidhu, 2018).

3.4.2 Fog computing and architecture

Fog computing is an addition to cloud computing that has been recognized as an emerging paradigm with shorter delays and improved network efficiency for healthcare systems (Manogaran et al., 2018). Today, fog computing is a sophisticated model for the effective handling of healthcare records that are generated by various IoT sensor devices. Although cloud and fog computing have become paradigms in delivering on-demand services in healthcare, cloud computing cannot be a good option as it has time delays for systems that require real-time responses (Tuli et al., 2020). The important advantage of fog computing is increasing the scalability and storage capacity while improving the efficiency for collecting, storing, processing, and analyzing data (Manogaran et al., 2018).

Fog computing means varied physically dispersed devices that are generally connected to a network in order to share computing resources and to offer storage capability, flexible connection, and calculating power (Mutlag, Abd Ghani, Arunkumar, Mohammed, & Mohd, 2019). Fog computing seems to be a viable system for healthcare services, as it shortens the time, virtual supervision, and illness analysis (Vijayakumar, Malathi, Subramaniyaswamy, Saravanan, & Logesh, 2018). Fog computing components can be connected by several components such as network access points, network gateways, and routers together with network protocols. In general, fog computing has many benefits such as reduced latency, improved bandwidth, reliability, energy-efficient operation, safety, and flexibility (Kraemer, Braten, Tamkittikhun, & Palma, 2017). It can offer immediate outcomes and provide sophisticated healthcare, much-improved diagnostics, and treatments (Singh et al., 2018). The foremost notion of fog computing is to improve effectiveness and flexibility while reducing the volume of data that can be transferred to the cloud-based components for analyzing, managing, and processing big data.

The figure below illustrates the design of fog computing for a healthcare network. It consists of three layers: the smart terminal layer, the fog computing layer, and the cloud computing layer. The first layer consists of different smart intelligent devices and tools used to collect healthcare data from the patient. The collected data are transferred to the fog computing layer for additional processing. The second layer is the fog computing layer that functions as a middle layer between the smart terminal and cloud computing layers. This layer involves preprocessing, handling, analyzing, distributing, storing, controlling, encrypting, and distributing the data. The cloud computing layer involves data processing, transferring and providing healthcare data to relevant entities and users. Fig. 3.1 shows the architecture of fog computing in healthcare.

3.4.3 Internet of things (IoT)

The IoT includes different tools and technologies such as sensor devices, controllers, and wired and wireless services that enable connections among physical entities and virtual representations (Manogaran et al., 2017). The IoT and cloud computing have emerged as an integrated platform that plays a vital role in healthcare services to connected users by minimizing the cost and time related to data collection. In this

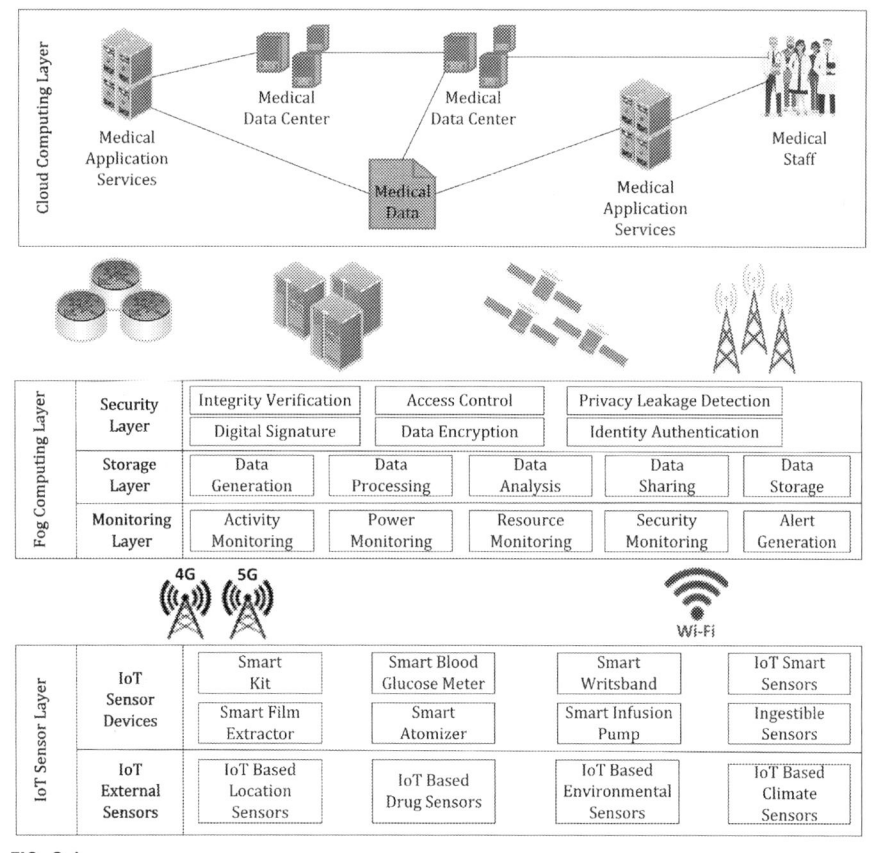

FIG. 3.1

Fog computing architecture for healthcare networks.

No Permission Required.

healthcare context, IoT-based applications can be applied to monitor patient health conditions, ubiquitous healthcare management, telemedicine, detecting clinical issues and diseases, maintenance, and logistics arrangements in healthcare systems (Sicari et al., 2017). The efficient implementation of the IoT for gathering, processing, and supplying data produced by medical devices, sensors, wearables, and humans is important for modern healthcare systems. Fig. 3.2 shows the role of IoT in healthcare.

3.4.4 Internet of medical things (IoMT)

The IoMT is a set of intelligent applications and devices (see Fig. 3.3) that interacts over a network for smart healthcare services (Toor et al., 2020). The growing applications of IoT and IoMT electronic devices have made medical treatment more systematic; also, healthcare records can be well organized (Jin, Yu, Zhang, Pan, & Guizani, 2019). The recent emerging advancements in healthcare equipment, the use of IoT devices, state-of-the-art wireless, and computing power have enabled the extensive applications of the IoMT (Kotronis et al., 2019). Therefore, the practice of data handling in healthcare is improved. particularly by employing techniques such as machine and deep learning, data mining, artificial intelligence, and algorithmic modeling. The IoMT connects with diverse objects such as end users, sensor devices, and network nodes that operate simultaneously for supreme disease analysis and diagnosis, as a result it increase to reduce the cost, finest analysis, superior supervision of clinical procedures and more precise cures (Guntur et al., 2019). Equally, IoMT devices have been extensively employed to detect symptoms and predict weather and the climate as well as for virtual observation and the superior connection of healthcare devices. Fig. 3.3 reveals the infrastructure components in IoMT.

3.4.5 Machine learning (ML)

ML is the perfect method to exploit the unseen insights and patterns from large datasets with the least support from humans (Desarkar & Das, 2017). ML contains a range of techniques such as data mining, pattern recognition, predictive analytics, and other modeling. The healthcare industry is highly interested in utilizing ML techniques in practical knowledge bases by executing predictive and prescriptive analytics to support intelligent clinical services. Scholars define ML as a kind of artificial intelligence that enables machines to learn without being clearly programmed; it is used to increase future outcomes based on past results (Manogaran et al., 2017). In healthcare, ML techniques are used for predicting disease severity and reasoning, decision support for medical surgery or therapy, extracting healthcare knowledge, analyzing various health data, and drug discovery (Pramanik et al., 2017). Various ML techniques used for mining large datasets are decision trees, support vector machines, neural networks, dimensionality reduction, etc. (Desarkar & Das, 2017).

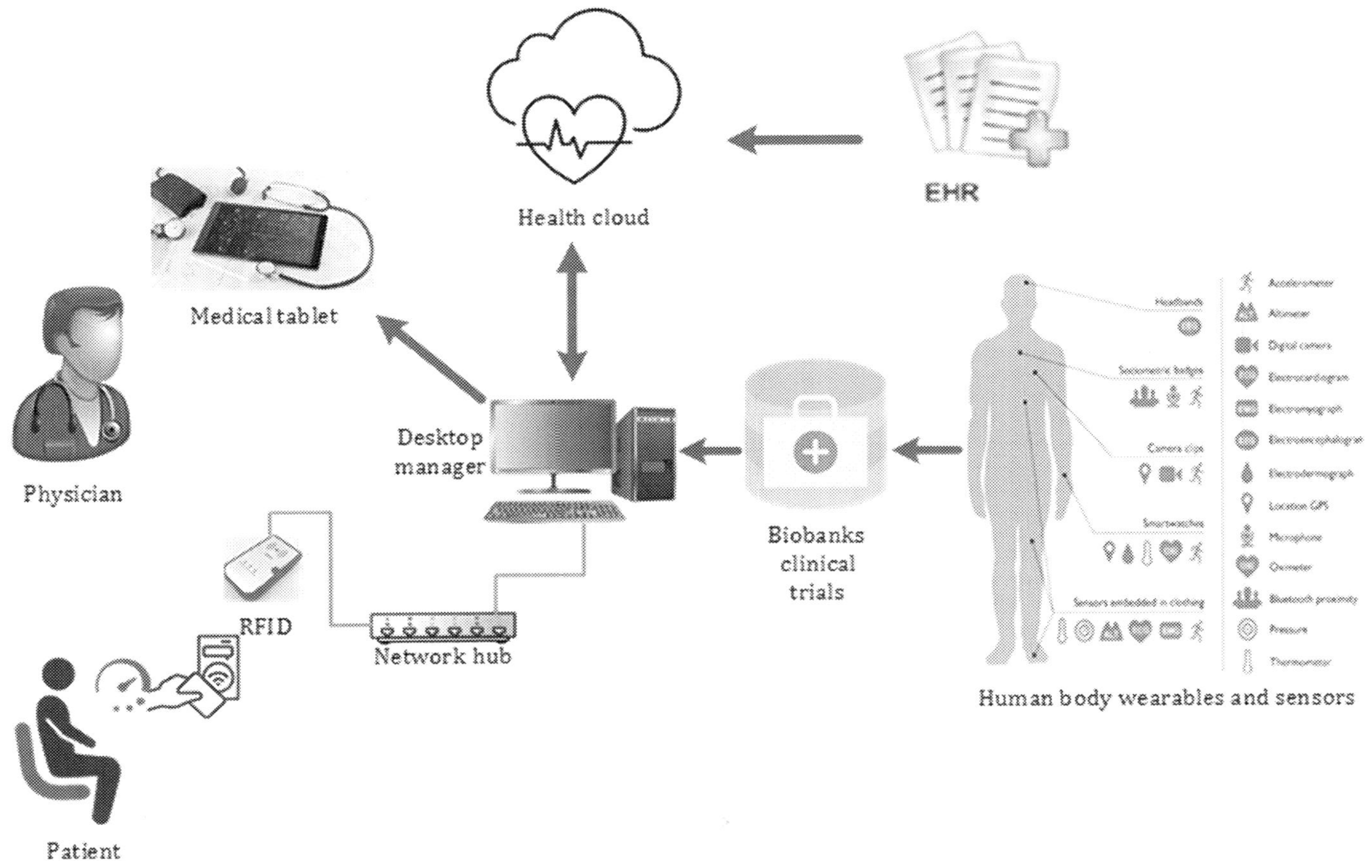
Health cloud
EHR
Medical tablet
Physician
Desktop manager
Biobanks clinical trials
Human body wearables and sensors
RFID
Network hub
Patient

FIG. 3.2

Role of IoT in healthcare.

No Permission Required.

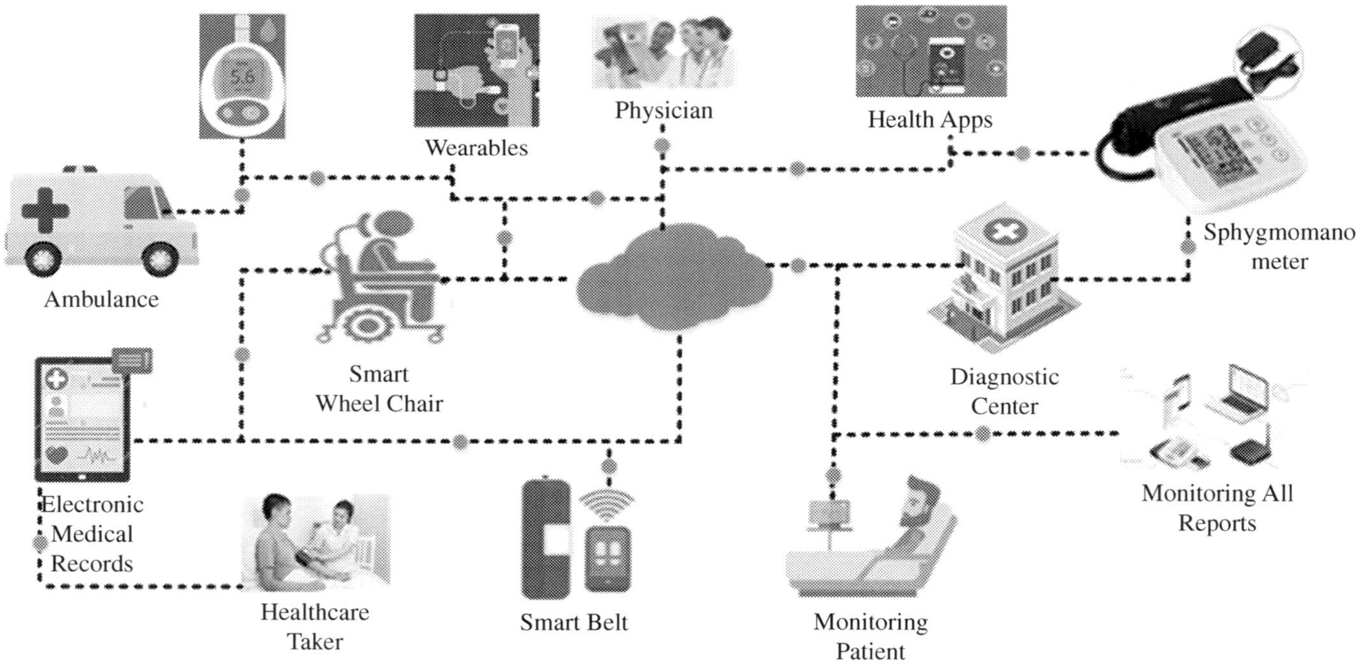

FIG. 3.3

Internet of Medical Things (IoMT) infrastructure.

No Permission Required.

3.4.6 **Deep learning**

Deep learning (DL) is a subset of ML that involves extracting high-level, complex abstractions as data representations with the support of a hierarchical learning process (Desarkar & Das, 2017). DL aims to predict and classify healthcare data with extremely high accuracy. Its key functionality is the ability to explore large unorganized datasets, which is highly essential as they are unclassified (Desarkar & Das, 2017).

3.4.7 **Intelligent computational techniques and data mining**

The traditional methods for handling medical records are inadequate because they are unable to treat large volumes of multifaceted records (Fang et al., 2016). Big data is used with the sole reason of analytics in which knowledge and significant insights are mined from big data. There are various sources that can be used to collect large volumes of big data such as media, the cloud, the web, IoT sensors, and databases (Rakesh Kumar et al., 2019).

Pramanik et al. (2017) identified three comprehensive technical branches—intelligent agents, text mining, and machine learning—as modern healthcare technologies in their Smart Health in a Smart City study. Intelligent agents in healthcare are defined as entities that gather instructions and interrelate with environments in which they recognize physical and virtual settings by using various sensing devices to perform the given tasks (Pramanik et al., 2017). The execution of intelligent agents in the medical field includes retrieving health information from big data; disease diagnostic decision support systems; planning and scheduling tasks for doctors, nurses, and patients; medical information sharing; medical image processing; automation; simulation; bioinformatics; medical data management; and health decision support systems (Pramanik et al., 2017).

In healthcare, text mining enables the mining of important insights from textual data in which some analytical structures can spontaneously code unorganized records with text mining grouping, research activities on the biomedical field, and knowledge management and discovery (Pramanik et al., 2017). In healthcare, various data mining procedures can be used such as classification, association rule mining, regression, clustering, detection, analysis, decision trees, and visualization to extract effective details (Rakesh Kumar et al., 2019) (Mahmud et al., 2016). Data mining is a technique that extract patterns and connections that can generate knowledge or insights from databases or large datasets (Rakesh Kumar et al., 2019). The powerful amalgamation of healthcare informatics and data mining by employing big data analytics techniques will increase the quality of healthcare services with effective decision making (Kanti, 2017).

ML is applied to extract patterns and models, whereas data mining is a mixture of statistics and ML that is mostly involved with large datasets; it analyzes huge, complex, and unstructured/structured datasets (Fang et al., 2016). Researchers have identified the fuzzy logic systems as a perfect option to design healthcare systems as they can handle the uncertainties, inaccuracies, complications, and comprehensiveness of the data. Further, fuzzy systems offer apparent and convenient rule-oriented models that can be a methodology for designing predictive models and cataloging by employing imprecise cognitive models of uncertain data and information (Mahmud et al., 2016).

3.5 Overview of big data platform and tools for healthcare systems

3.5.1 Hadoop architecture

In healthcare, several techniques, tools, and structures are established to process big data. Among these, Hadoop is the distributed data processing structure used to save and handle huge volumes of data by employing the MapReduce technique. Hadoop is the famous open-source distributed processing platform for big data analytics that can function as the data analytics and data organizer in the Apache environment (Pramanik et al., 2017). There are various big data-related tools employed to handle huge volumes of records with different technical designs, such as PIG, Oozie, MapReduce, Cassandra, HBase, Mahout, Avro, etc. These tools and platforms can support the Hadoop distributed computing platform. An important functionality of Hadoop is computation and data partitioning across different nodes and the processing of application execution in parallel near to their dataset (Desarkar & Das, 2017). Hadoop has the ability to share divided datasets to various hosts and can perform dual roles such as an analytics tool and a data organizer (Raghupathi & Raghupathi, 2014). Hadoop is based on a *master-slave* architecture; thus it possesses greater fault-tolerance ability as the master node will be ready when a slave node is down, then it reassigns the work to the remaining slave nodes (Kanti, 2017). Fig. 3.4 below describes the Hadoop architecture.

3.5.2 Apache hadoop

Hadoop contains two key components: the *Hadoop Distributed File System* (HDFS) for data storage and *Hadoop MapReduce* for the simultaneous analysis of huge data records. HDFS offers an extendable dispersed file management method for keeping huge datasets over scattered notes in a more consistent and well-organized way (Inoubli et al., 2018).

3.5.3 Apache spark

Spark and Flink are both from Apache-hosted big data analytics frameworks that simplify the creation of multistep data pipelines by using straightly acyclic graph patterns (Marcu, Costan, Antoniu, & Pérez-Hernández, 2016). Spark depends on

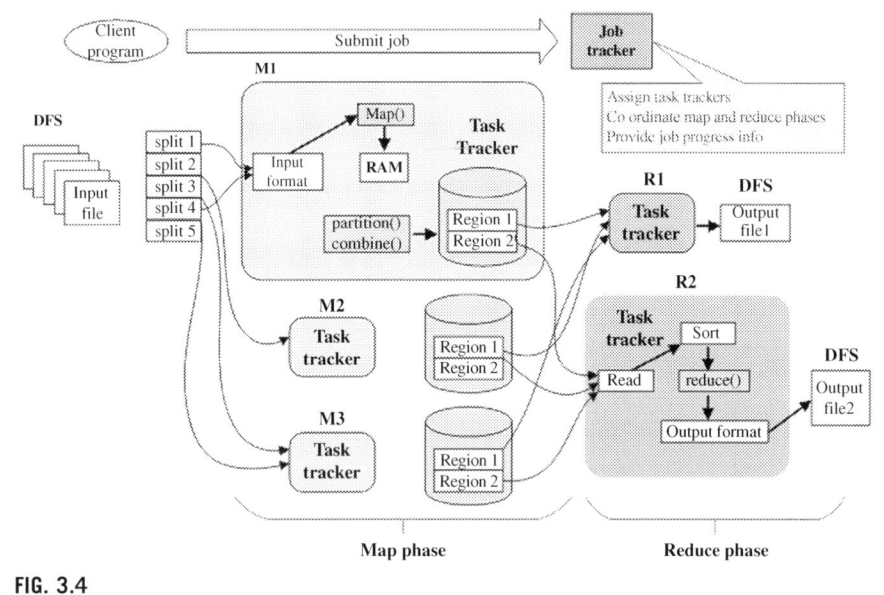

FIG. 3.4

Hadoop MapReduce architecture.

No Permission Required.

two key ideas: resilient distributed datasets (RDD) that possess the data objects in system memory, and transformations that are executed on the datasets in parallel (Spangenberg, Roth, & Franczyk, 2015). Apache Spark is a prevailing processing framework that gives an ease-of-use tool for the effective analytics of heterogeneous data; it has several benefits over other big data frameworks such as Hadoop and Storm (Inoubli et al., 2018). Spark is assembled on top of read-only, resilient collections of objects partitioned across multiple nodes that possess origin information and can be restructured in case of downs by fractional r-computation from ancestor RDDs (Marcu et al., 2016).

3.5.4 Apache storm

Storm is a fault-tolerant framework for processing huge structured and unstructured datasets in real time; this is appropriate for simultaneous data handling, ML, parallel, and computation of iteration (Inoubli et al., 2018). The big data analytics frameworks handle *heterogeneous dataflows* and heterogeneous data-analyzing semantics such as batch analyzing in Hadoop and constant stream processing in Storm (Ranjan, 2014)

3.6 Proposed big data-based conceptual framework for healthcare systems

Sensors and wearable devices regularly generate large volumes of data that contain structured and unstructured data. Table 3.2 lists a comparison of big data-related healthcare system features in prior and present studies. Following a comprehensive review, a big data-based conceptual framework is proposed in healthcare systems that contains IoT sensors, data sources, big data types, a big data analytics platform and tools, analytics data output, patient health monitoring, and recommendation system. This is shown in Fig. 3.5.

3.6.1 Proposed system functionalities

3.6.1.1 Data sources

This section describes the functionalities of the proposed system. The healthcare-related data will be collected from embedded sensors that are placed on a patient's body. Similarly, a variety of healthcare data is available on social media and websites that can also be collected for this healthcare system. The collected data would be transmitted to the database server for further processing in the cloud environment. Meanwhile, the patient's signs and health status records can be transferred to the fog computing environment through a data-processing node. Once the transferred data are matched with the existing data in the stored database, a particular patient's health status report will be produced that describes the current nature with the required preliminary treatment/needed medical therapy to be taken. The collected healthcare data can be in any form such as structured data, unstructured data, and semistructured data from the patient body/social media. To handle this data, there are a number of big data analytics methods, techniques, and platform tools. These techniques and tools will be used for analysis based on the data types. The output of these analyses will be in forms such as queries, reports, OLAP, and extracted data mining.

3.6.1.2 Patient healthcare-related data

Modern developments, especially in IoT and sensors, have made a patient's monitoring more advanced, and that will provide superior healthcare to sick people. Healthcare records may include a variety of symptoms and patient health issue data. For instance, it may include body symptoms such as fevers, headaches, pain in bones and joints, skin eruptions, body pains, red eyes, nausea, vomiting, and myalgia. To collect these symptoms, IoT sensors or physical tests will be used. If it is an IoT device, it will be embedded in the patient's body to identify or detect conditions. The symptoms detected by the IoT sensors will be transmitted to the fog computing environment, which is linked through cloud computing for storage in single or multiple data storage components.

Table 3.2 Comparison of big data-based healthcare system features.

Author	Disease/system	Platform/techniques features								Performance feature
		Cloud computing	Fog computing	IoT	Machine learning	Deep learning	Data mining	Artificial Intelligence	Big data	
Tuli et al. (2020)	Heart disease	✓	✓	✓	✓	✓	✓		✓	
Forkan et al. (2015)	Chronic disease	✓					✓	✓	✓	
Hossain and Muhammad (2016)	Voice pathology assessment framework				✓	✓	✓		✓	Accuracy and time efficiency
Kim and Seu (2014)	Ubiquitous healthcare systems for blood pressure and motion data based on vital signs						✓		✓	Supports performance and the reliability of healthcare system
del Carmen Legaz-García et al. (2016)	Semantic web-based framework for clinical models and HER data for colorectal cancer						✓		✓	Framework for EHR and semantic web technologies for the interoperability and exploitation of archetypes
Lin et al. (2015)	Cloud-based framework for self-caring service	✓	✓	✓					✓	Home diagnosis provides disease precaution knowledge particularly for elders and individuals with chronic diseases

Continued

Table 3.2 Comparison of big data-based healthcare system features—con't

Author	Disease/system	Platform/techniques features								Performance feature
		Cloud computing	Fog computing	IoT	Machine learning	Deep learning	Data mining	Artificial Intelligence	Big data	
Mahmud et al. (2016)	Cloud-based data analytics and visualization framework for health-shocks prediction	✓			✓				✓	The precision of the model for sorting health-shock shows 89% performance in predicting health-shocks
Manogaran et al. (2018)	MapReduce-based prediction model to forecast heart diseases	✓	✓	✓					✓	Throughput, accuracy, sensitivity, and f-measures measured to confirm the efficiency
Ilmudeen (2021)	Designing IoT-based system to predict and analyze dengue	✓	✓	✓					✓	The proposed design and method offers to predict dengue
This proposed big data-based conceptual framework	Proposed framework for healthcare systems	✓	✓	✓	✓	✓	✓		✓	Conceptual design illustrates various aspects in the context of healthcare systems.

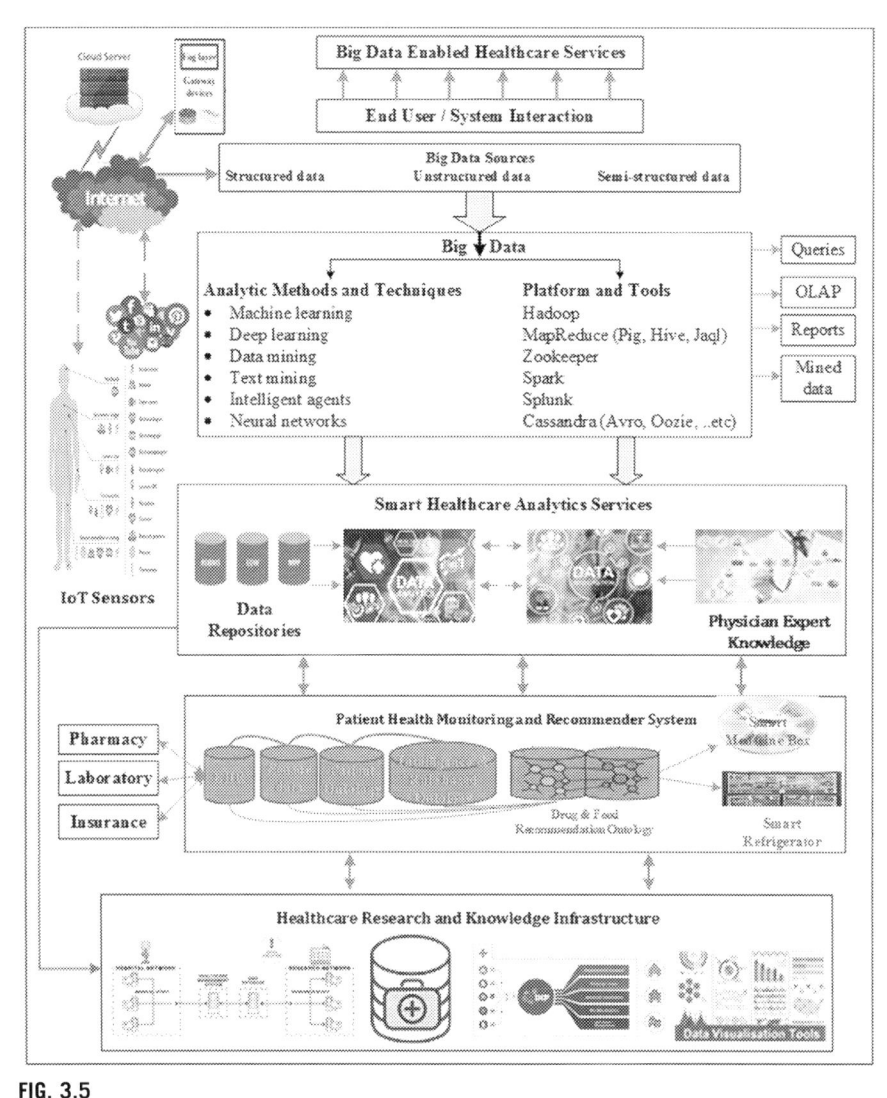

FIG. 3.5

Proposed big data systems.

No Permission Required.

3.6.1.3 Cloud and fog computing components

In this proposed system, cloud and fog computing environments have been employed. As the healthcare data seems to be highly sensitive and private, there is a need for the protected transfer of these medical records. Hence, to secure the data, a data encryption mechanism will be employed. The cloud layer saves data from the Internet for the concurrent retrieval of data by devices linked with this ecosystem. The data that require further sorting or handling will be transmitted to the cloud layer. The cloud layer is responsible for the storing, handling, and processing of records that cannot be handled by the fog computing layer. The fog computing component gathers data from the Internet for further analysis. This fog computing component is responsible for data storage, processing, and transmission as well as communication-related activities. Therefore, the patient's health condition and the associated symptoms and signs will be collected, transferred, handled, and analyzed at this fog component. Further, this fog computing component is used for real-time data handling and processing from IoT devices. The sophisticated cloud and fog computing applications have received greater attention in recent healthcare system developments.

3.6.1.4 Big data analytics methods, techniques, and platform tools

The output from the big data analytics techniques and platforms will be transferred to a smart healthcare analytics services that includes data repositories, analytics, and integration as well as expert physician knowledge. The healthcare-related big data will be systematically analyzed with modern state-of-the-art techniques. Each component in this category will collaborate effectively. The components will offer updated valuable insights, disease analysis, medical therapies, drug suggestions, healthcare clinical management, healthcare scheduling, and many other services. Employing ML techniques and inference algorithms, an IoT healthcare system can aptly learn from sensors and a patient's medical history to describe the present and projected future health conditions of the patient, and even sound alarms to the medical officers and patient if needed (Firouzi et al., 2018). Accordingly, the data collected by the IoT devices can be processed and analyzed via the MapReduce platform that can handle large volumes of data by employing ML techniques to decide a subject's actions over time.

3.6.1.5 Patient healthcare monitoring and recommendation system

This component database includes electronic health records (EHR), sensor data, patient ontology, and intelligence rule-based ontology. The data will be gathered for these databases from various healthcare units/departments that are located across the country such as clinical and administrative departments including pharmacies, radiology, medical laboratories, billing, insurance, etc. The above departments are linked via the EHR unit in the cloud that allows data sharing and interoperability. Further, the EHR consists of patient records that are linked from different units (pharmacy, lab, insurance, etc.) such as city, state, or region and are stored in the cloud. The second component is the food and drug recommendation ontology. The recent developments in Internet-based technologies have made greater alternatives for a recommendation system to support patients for their routine lifestyle (Subramaniyaswamy et al., 2019). The sophisticated collection from smart devices, agents, techniques, and linking components makes this recommendation system. Further, the needed data for patient drugs and food will be gathered and preprocessed for recommendations by looking at the patient status. This contains a collection of data for drugs such as the dose, chemical structure, properties, etc. With regard to meals, the data such as nutrition value and ingredients will be considered for the recommendation. Further, for a patient's diet, favorite details such as the spice level and options such as vegetarian or nonvegetarian are also combined with this recommendation system.

6.1.6 Healthcare research and knowledge infrastructure

This layer includes various healthcare-related research, innovation, and knowledge infrastructure components. The purpose of these components is to build a system that offers state-of-the-art healthcare services and support for patients, researchers, hospital management, medical professionals, healthcare agencies, government organizations, and policy makers. The databases and knowledge repositories in this layer keep the symptoms, diseases, drugs, drug composition and dose level, clinical records, clinical procedures, disease prediction, prevention mechanisms, medical practices, etc., to explore new insights and trends for future use and knowledge discoveries. Researchers and data scientists will use advanced simulations, data visualization, data mining, classification, and modeling to support the knowledge demands of medical professionals and policy makers.

3.7 Conclusion

Today's digital and information era has messed with big volume of data that are created by heterogeneous sources in many forms. The size of big data has increased exponentially. The process involved in collecting, storing, extracting, analyzing, and optimizing these data has also become very significant in healthcare systems. Hence, the mixture of big data and healthcare systems can accelerate potential benefits in the healthcare sector. This chapter discusses various big data-related methods, techniques, platforms, and architectures in the context of big data frameworks in healthcare systems. In addition, this chapter proposed a theoretically developed conceptual big data framework that illustrates various aspects in big data for a healthcare system. The key contributions of this chapter include a systematic review of big data-related frameworks in healthcare systems; various big data- and analytic-based methods, techniques, and platforms; and a proposal for a modern conceptually designed healthcare system framework.

References

Chawla, N. V., & Davis, D. A. (2013). Bringing big data to personalized healthcare: A patient-centered framework. *Journal of General Internal Medicine, 28*(3), 660–665.

Desarkar, A., & Das, A. (2017). Big-data analytics, machine learning algorithms and scalable/parallel/distributed algorithms. In *Internet of things and big data technologies for next generation healthcare* (pp. 159–197). Springer.

del Carmen Legaz-García, M., Martínez-Costa, C., Menárguez-Tortosa, M., & Fernández-Breis, J. T. (2016). A semantic web based framework for the interoperability and exploitation of clinical models and EHR data. *Knowledge-Based Systems, 105*, 175–189.

Fang, R., Pouyanfar, S., Yang, Y., Chen, S. C., & Iyengar, S. S. (2016). Computational health informatics in the big data age: A survey. *ACM Computing Surveys*, *49*(1). https://doi.org/10.1145/2932707.

Firouzi, F., Rahmani, A. M., Mankodiya, K., Badaroglu, M., Merrett, G. V., Wong, P., & Farahani, B. (2018). Internet-of-Things and big data for smarter healthcare: From device to architecture, applications and analytics. *Future Generation Computer Systems*, *78*, 583–586. https://doi.org/10.1016/j.future.2017.09.016.

Forkan, A. R. M., Khalil, I., Ibaida, A., & Tari, Z. (2015). BDCaM: Big data for context-aware monitoring—A personalized knowledge discovery framework for assisted healthcare. *IEEE Transactions on Cloud Computing*, *5*(4), 628–641.

Guntur, S. R., Gorrepati, R. R., Dirisala, V. R., Dey, N., Borra, S., Ashour, A. S., & Shi, F. (2019). *Robotics in Healthcare: An Internet of Medical Robotic Things (IoMRT) Perspective* (pp. 293–318). Academic Press. https://doi.org/10.1016/B978-0-12-816086-2.00012-6. Chapter 12.

Hossain, M. S., & Muhammad, G. (2016). Healthcare big data voice pathology assessment framework. *IEEE Access*, *4*, 7806–7815. https://doi.org/10.1109/ACCESS.2016.2626316.

Inoubli, W., Aridhi, S., Mezni, H., Maddouri, M., & Mephu Nguifo, E. (2018). An experimental survey on big data frameworks. *Future Generation Computer Systems*, *86*, 546–564. https://doi.org/10.1016/j.future.2018.04.032.

Ilmudeen, A. (2021). Design and development of IoT-based decision support system for dengue analysis and prediction: Case study on Sri Lankan context. In *Healthcare paradigms in the internet of things ecosystem* (pp. 363–380). Academic Press.

Jin, Y., Yu, H., Zhang, Y., Pan, N., & Guizani, M. (2019). Predictive analysis in outpatients assisted by the Internet of Medical Things. *Future Generation Computer Systems*, *98*, 219–226. https://doi.org/10.1016/j.future.2019.01.019.

Kanti, S. B. (2017). Big data for secure healthcare system: a conceptual design. *Complex & Intelligent Systems*, 133–151. https://doi.org/10.1007/s40747-017-0040-1.

Kim, T. W., & Seu, J. H. (2014). Big data framework for u-Healthcare system. *Life Science Journal*, *11*(7), 112–116. http://www.lifesciencesite.com/lsj/life1107s/020_24428life1107s14_112_116.pdf.

Kotronis, C., Routis, I., Politi, E., Nikolaidou, M., Dimitrakopoulos, G., Anagnostopoulos, D., … Djelouat, H. (2019). Evaluating internet of medical things (IoMT)-based systems from a human-centric perspective. *Internet of Things*, *8*, 100125. https://doi.org/10.1016/j.iot.2019.100125.

Kraemer, F. A., Braten, A. E., Tamkittikhun, N., & Palma, D. (2017). Fog computing in healthcare—A review and discussion. *IEEE Access*, *5*, 9206–9222. https://doi.org/10.1109/ACCESS.2017.2704100.

Kumar, P. M., Lokesh, S., Varatharajan, R., Chandra Babu, G., & Parthasarathy, P. (2018). Cloud and IoT based disease prediction and diagnosis system for healthcare using Fuzzy neural classifier. *Future Generation Computer Systems*, *86*, 527–534. https://doi.org/10.1016/j.future.2018.04.036.

Lin, W., Dou, W., Zhou, Z., & Liu, C. (2015). A cloud-based framework for home-diagnosis service over big medical data. *Journal of Systems and Software*, *102*, 192–206.

Mahmud, S., Iqbal, R., & Doctor, F. (2016). Cloud enabled data analytics and visualization framework for health-shocks prediction. *Future Generation Computer Systems*, *65*, 169–181. https://doi.org/10.1016/j.future.2015.10.014.

Manogaran, G., Lopez, D., Thota, C., Abbas, K. M., Pyne, S., & Sundarasekar, R. (2017). Big data analytics in healthcare internet of things. In *Understanding complex systems* (pp. 263–284). Springer Verlag. https://doi.org/10.1007/978-3-319-55774-8_10. Issue 9783319557731.

Manogaran, G., Varatharajan, R., Lopez, D., Kumar, P. M., Sundarasekar, R., & Thota, C. (2018). A new architecture of Internet of Things and big data ecosystem for secured smart healthcare monitoring and alerting system. *Future Generation Computer Systems*, *82*, 375–387. https://doi.org/10.1016/j.future.2017.10.045.

Marcu, O. C., Costan, A., Antoniu, G., & Pérez-Hernández, M. S. (2016). Spark versus flink: Understanding performance in big data analytics frameworks. In *Proceedings—IEEE international conference on cluster computing, ICCC* (pp. 433–442). Institute of Electrical and Electronics Engineers Inc. https://doi.org/10.1109/CLUSTER.2016.22.

Mutlag, A. A., Abd Ghani, M. K., Arunkumar, N., Mohammed, M. A., & Mohd, O. (2019). Enabling technologies for fog computing in healthcare IoT systems. *Future Generation Computer Systems*, *90*, 62–78. https://doi.org/10.1016/j.future.2018.07.049.

Palanisamy, V., & Thirunavukarasu, R. (2019). Implications of big data analytics in developing healthcare frameworks—A review. *Journal of King Saud University—Computer and Information Sciences*, *31*(4), 415–425. https://doi.org/10.1016/j.jksuci.2017.12.007.

Pramanik, M. I., Lau, R. Y. K., Demirkan, H., & Azad, M. A. K. (2017). Smart health: Big data enabled health paradigm within smart cities. *Expert Systems with Applications*, *87*, 370–383. https://doi.org/10.1016/j.eswa.2017.06.027.

Raghupathi, W., & Raghupathi, V. (2014). Big data analytics in healthcare: promise and potential. *Health Information Science and Systems*, *2*(1), 3. https://doi.org/10.1186/2047-2501-2-3.

Rahman, F., Bhuiyan, M. Z. A., & Ahamed, S. I. (2017). A privacy preserving framework for RFID based healthcare systems. *Future Generation Computer Systems*, *72*, 339–352. https://doi.org/10.1016/j.future.2016.06.001.

Rakesh Kumar, S., Gayathri, N., Muthuramalingam, S., Balamurugan, B., Ramesh, C., Nallakaruppan, M. K., … Kumar, R. (2019). *Medical Big Data Mining and Processing in e-Healthcare* (pp. 323–339). Academic Press. https://doi.org/10.1016/B978-0-12-817356-5.00016-4. Chapter 13.

Ranjan, R. (2014). Modeling and simulation in performance optimization of big data processing frameworks. *IEEE Cloud Computing*, *1*(4), 14–19. https://doi.org/10.1109/MCC.2014.84.

Sicari, S., Rizzardi, A., Grieco, L. A., Piro, G., & Coen-Porisini, A. (2017). A policy enforcement framework for Internet of Things applications in the smart health. *Smart Health*, *3–4*, 39–74. https://doi.org/10.1016/j.smhl.2017.06.001.

Singh, S., Bansal, A., Sandhu, R., & Sidhu, J. (2018). Fog computing and IoT based healthcare support service for dengue fever. *International Journal of Pervasive Computing and Communications*, *14*(2), 197–207. https://doi.org/10.1108/IJPCC-D-18-00012.

Spangenberg, N., Roth, M., & Franczyk, B. (2015). Evaluating new approaches of big data analytics frameworks. In *Vol. 208*. *Lecture notes in business information processing* (pp. 28–37). Springer Verlag. https://doi.org/10.1007/978-3-319-19027-3_3.

Subramaniyaswamy, V., Manogaran, G., Logesh, R., Vijayakumar, V., Chilamkurti, N., Malathi, D., & Senthilselvan, N. (2019). An ontology-driven personalized food recommendation in IoT-based healthcare system. *Journal of Supercomputing*, *75*(6), 3184–3216. https://doi.org/10.1007/s11227-018-2331-8.

Syed, L., Jabeen, S., Manimala, S., & Alsaeedi, A. (2019). Smart healthcare framework for ambient assisted living using IoMT and big data analytics techniques. *Future Generation Computer Systems*, *101*, 136–151. https://doi.org/10.1016/j.future.2019.06.004.

Toor, A. A., Usman, M., Younas, F., Fong, A. C. M., Khan, S. A., & Fong, S. (2020). Mining massive e-health data streams for IoMT enabled healthcare systems. *Sensors (Switzerland)*, *20*(7). https://doi.org/10.3390/s20072131.

Tuli, S., Basumatary, N., Gill, S. S., Kahani, M., Arya, R. C., Wander, G. S., & Buyya, R. (2020). HealthFog: An ensemble deep learning based smart healthcare system for automatic diagnosis of heart diseases in integrated IoT and fog computing environments. *Future Generation Computer Systems*, *104*, 187–200. https://doi.org/10.1016/j.future.2019.10.043.

Vijayakumar, V., Malathi, D., Subramaniyaswamy, V., Saravanan, P., & Logesh, R. (2018). Fog computing-based intelligent healthcare system for the detection and prevention of mosquito-borne diseases. *Computers in Human Behavior*, *100*.

Youssef, A. E. (2014). A framework for secure healthcare systems based on big data analytics in mobile cloud computing environments. *International Journal of Ambient Systems and Applications*, *2*, 1–11.

Predictive analysis and modeling in healthcare systems

4

M.V. Manoj Kumar[a], Pradeep N[b], and H.A. Sanjay[a]

[a]*Department of Information Science and Engineering, Nitte Meenakshi Institute of Technology, Bengaluru, India,* [b]*Computer Science and Engineering, Bapuji Institute of Engineering and Technology, Davangere, Karnataka, India*

4.1 Introduction

Every second, new data are being created. The rapid development of the digital universe has brought us into the era of "big data." For example:

- eBay, the online auction and web shopping company, processes around 50 petabytes of information daily (Brandtzæg, 2013).
- Facebook, the popular social networking platform, processes more than 500 terabytes of data daily (Brandtzæg, 2013).

Now, less than 0.5% of all data is ever analyzed and used (Antonio, 2013). The immediate need is to extract and use information from the recorded data. The knowledge gained by analyzing the raw data could play an invaluable role in numerous realms. One such aspect where data plays a key role (in understanding, implementing, and improving) is healthcare and healthcare-related organizations.

The increased level of competition compels individual healthcare units to perform in the best way possible. For the top 1000 healthcare organizations, just a 10% increase in data accessibility will result in a 65% increase in research output and innovation (de Medeiros, van Dongen, van der Aalst, & Weijters, 2004). This justifies the potential behind properly utilizing the data. The availability of data and the capability of analyzing it could create a better healthcare process. Depending on the context of the application and the problems at hand, a variety of analysis techniques would give interesting and useful information.

Most healthcare organizations are increasingly moving toward digitizing (or automating) their day-to-day processes with the help of information systems. Event logs offer a wealth of information for today's organizations, but they are rarely exploited to provide meaningful insights into business processes.

By 2020, the prediction was that each person would generate about 7 megabytes of new information every second (de Medeiros et al., 2004). A cloudburst of digital data is produced daily and will continue to increase exponentially in the coming days. Data volume with respect to healthcare is no exception, as extremely large amounts of data are being generated in healthcare. This plays an important role in modern patient care. The following are some noteworthy facts about the increasing volume of healthcare data pointed out by Belle et al. (2015).

During the beginning of the decade 2011–2020, healthcare data volume was 30% of digital data worldwide.

In 2013, the amount of healthcare data was 153 exabytes, which amounted to roughly 40% of the total digital data worldwide. That same volume of data is likely to reach more than 2000 exabytes by 2020. With the ever-increasing quantity of data, we need to recognize the significant part it plays in improving the results and overall efficiency in the healthcare domain. In this chapter, we explore the process modeling dimension of healthcare using process mining (van der Aalst, 2011). This uses the data recorded in information systems to extract the executed version of the healthcare process (the sequence of steps carried out to complete the task), which gives insights into how the process was executed versus how it was planned.

There are several healthcare information systems that produce event logs (Dumas, van der Aalst, & ter Hofstede, 2005; Lanz, Kreher, Reichert, & Dadam, 2010). Some of the widely used PAISs are (Lawrence, Bouzeghoub, Fabret, & Matulovic-broqué, 1997), enterprise resource planning (Reichert, Rinderle-Ma, & Dadam, 2009), FLOWer (Van Der Aalst, Weske, & Grünbauer, 2005), PDM (Lanz et al., 2010), customer relationship management (Schmitt, 2010), IBM WebSphere (Bennett, Wells, & Freelon, 2011), ChipSoft (van Cann, Jansen, & Brinkkemper, 2013), electronic medical records and electronic health records (Jha et al., 2009), practice management system (Pedersen, Schneider, & Santell, 2001), patient portals (Irizarry, De Vito Dabbs, & Curran, 2015), etc.

Added to these information systems, nowadays most embedded systems are also capable of generating event logs. An example is the "CUSTOMerCARE Remote Services Network" of Philips Healthcare (Vijayaraghavan, Haex-Director, & Both, 2006), where events occurring within an x-ray machine (for example moving the table, setting the deflector, etc.) are recorded and analyzed. These examples show that systems record (parts of) the actual behavior, and thus implicitly or explicitly create event logs.

This chapter briefs the top-down and bottom-up approaches for process modeling and execution; the significance of process mining and its basic techniques; the format of event logs and their attributes; perspectives of information system processes; some of the most frequently used control flow modeling notations; and the predictive modeling of control flow perspective using a heuristic miner at different precision levels. The chapter ends with a brief conclusion and exercise questions.

4.2 Process configuration and modeling in healthcare systems

The main factors for increased interest in healthcare process modeling are two-fold. First, more and more healthcare-related event data are getting recorded, and this enables one to make informed decisions. Second, there is a need to optimize the process and make the healthcare workflow efficient.

Traditional systems in healthcare follow the top-down approach (shown in Fig. 4.1), where the high-level process model is designed at first. Then, the system is configured for controlling and managing the execution of the healthcare process. The problem with the traditional approach for healthcare process modeling is shown in Fig. 4.2. In this approach, the planned model (existing model) could be completely different from the one that is followed.

The problem of the top-down approach could be overcome by the bottom-up approach shown in Fig. 4.3. Here, the first step is to record the event log (PAISs assisting the execution of healthcare processes record all steps of execution). Analyzing these data with process mining gives the actual process model. It offers

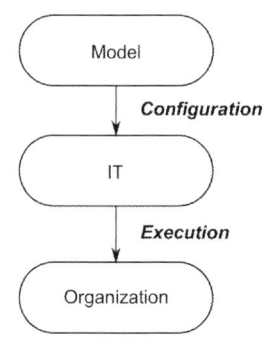

FIG. 4.1

Traditional healthcare process management.

No Permission Required.

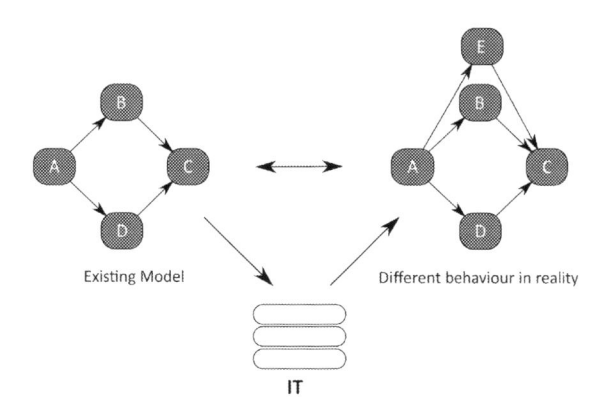

FIG. 4.2

Problem with traditional healthcare process configuration and management.

No Permission Required.

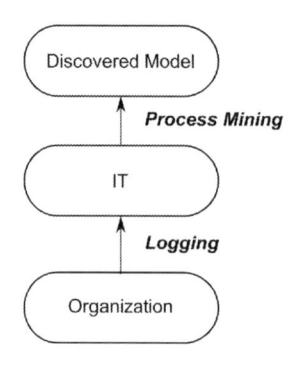

FIG. 4.3

Bottom-up healthcare process management.

real x-ray vision into the process. We follow the bottom-up approach to extract the process model from the healthcare process and further predict the causal relationship structure of the process on the real-life event log; this is discussed in further sections.

4.3 Basic techniques of process modeling and prediction

The basic techniques in process modeling using the bottom-up approach involve discovery, conformance checking, and enhancement.

The basic idea of process modeling is shown in Fig. 4.4. Fig. 4.5 describes the basic types of in-process model construction and related operations. Discovery techniques take the event log as input and generate various models related to different perspectives such as control flow, organization, etc.

4.3.1 Process discovery

Conformance checking takes an event log and the generated model from the discovery step and uncovers any discrepancies between how the process was planned and how it was executed. For enhancement, the process model is extended with additional perspectives, which makes the process model more informative.

Process discovery methods are most useful to provide insights into what occurs. Discovery techniques produce control-flow, data, organizational, time, and case models. Numerous methods exist in process mining to elicit process models of various notations.

Control-flow perspective depicts the causal relationships between the activities of the process. It discovers the visual process graph in terms of nodes and edges. It gives the ordering of the execution of an activity. For example, Section 4.4 briefs modeling the process control-flow perspective in more than 10 different modeling formats such as Petri net, BPMN, EPN, etc.

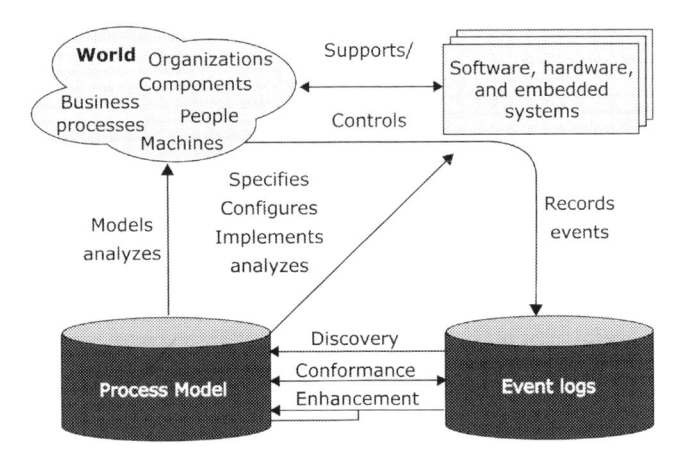

FIG. 4.4

The basic idea of process mining.

From Process Mining Manifesto. (c. 2011). https://www.win.tue.nl/ieeetfpm/doku.php?id=shared:process_mining_manifesto.

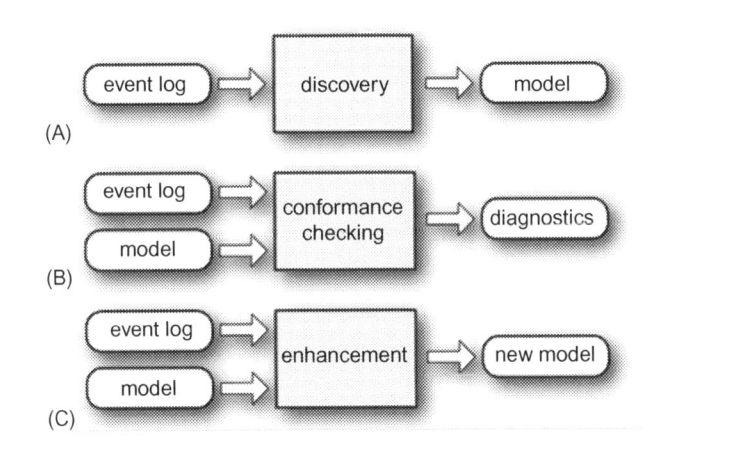

FIG. 4.5

Basic techniques in process model construction.

From Process Mining Manifesto. (c. 2011). https://www.win.tue.nl/ieeetfpm/doku.php?id=shared:process_mining_manifesto.

Organizational perspective structures the organization and the interaction between different organizational entities in it. It even deals with extracting the social network from the set of human entities in the organization and the kinds of interactions between them. It can be further constructed to extract the interactions between different hierarchies in the organization.

Case perspective (Jans, Van Der Werf, Lybaert, & Vanhoof, 2011) focuses on the assets of a case. A case typically consists of an execution path in the process. For example, if a case represents an individual patient with different ailments, each patient is a case and he/she should be treated with different approaches/departments.

4.3.2 Enhancement

The idea of enhancement (shown in Fig. 4.5) is to improve the discovered process model to make it more informative. There are various types of enhancements that one can perform, including repairing a process model to better reflect reality and the extension of process models with information extracted from event logs. For example, a control-flow model can be enhanced and made more readable by overlaying additional information such as a timestamp (van der Aalst et al., 2007).

4.4 Event log

The starting point of process mining is the observed behavior of process executions stored in event logs (Kaymak, Mans, Van De Steeg, & Dierks, 2012). An overview of recording the process execution data is shown in Fig. 4.6. We now formalize the various notations related to the event log.

4.4.1 Event and attributes

Because the event logs of information systems provide factual data about the underlying processes, they are a precious source of information (Mans, Schonenberg, Song, Van Der Aalst, & Bakker, 2008; Weijters, van Der Aalst, & De Medeiros,

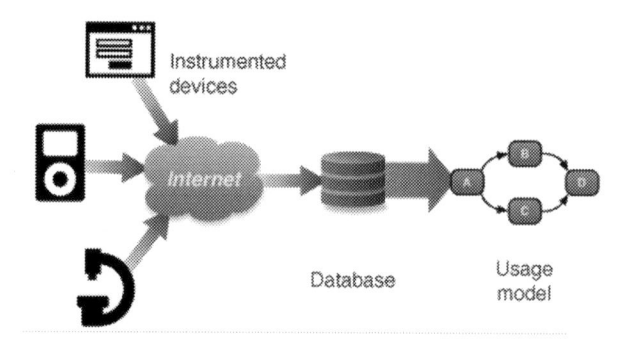

FIG. 4.6

Process of recording event log and model generation from recorded data.

From Process of recording event log and model generation from recorded data. (c. 2015). https://fluxicon.com/technology/.

| Case id | Event id | Properties | | | | |
		Timestamp	Activity	Resource	Cost	...
	2342	10-10-2012:01.20	Appointment	Anne	200	...
	2343	11-10-2012:07.24	Admit patient	Jhon	100	...
xx12	2344	12-10-2012:08.24	Check history	Thomas	400	...
	2345	13-10-2012:12.24	Decide	Clar	50	...
	2346	13-10-2012:13.25	Begin treatment	Ram	10	...
	3347	14-10-2012:01.29	Appointment	Anne	200	...
	3348	14-10-2012:01.34	Out patient	Wil	20	...
xx13	3349	14-11-2012:16.34	Check history	Thomas	300	...
	3350	15-11-2012:08.22	Decide	Clar	50	...
	3351	15-11-2012:09.17	Discharge	Ram	100	...

FIG. 4.7

Event log of the hospital admission process.

2006). Fig. 4.7 shows an example of an event log of the hospital patient admission process. This event log will be used for discussion and illustration in this chapter.

Let ε be the set of all event identifiers and AN be the set of all attributes. For an attribute $s \in AN$, let X_X be a universal set (consists of a set of all possible values of x). Given ε and an attribute $x \in AN$, $\#x(e)$ is the value for all attributes x not defined in e and $\#x$.

$$\#activity(e) \in A$$

signifies the activity associated with event e.

$$\#resource(e) \in R$$

indicates the resource performing the event e.

$$\#time(e) \in R$$

indicates the timestamp of event e.

$$\#trans(e) \in TT$$

indicates the transaction type associated with event e.

4.4.2 Case, trace, and event log

The event log on a control perspective is normally defined over a set of activity names. A sequence of activity is referred to as a trace (a set of traces is given in Fig. 4.8). A simple event log is defined as a bag of traces (Rebuge & Ferreira, 2012; Rojas, Munoz-Gama, Sepúlveda, & Capurro, 2016)

$\mathcal{L}$ =[⟨ Appointment, Admit patient, Check history, Decide, begin Treatment ⟩. ⟨ Appointment, Out patient, Check history, Decide, Discharge ⟩. ⟨ Appointment, Admit patient, Check history, Decide, Re-examine, Out patient, Check history, Decide, Discharge ⟩. ⟨ Appointment, Check history, Admit patient, Decide, Discharge ⟩. ⟨ Appointment, Out patient, Check history, Decide, Re-initiate, Admit patient, Check history, Decide, Begin treatment ⟩. ..]

FIG. 4.8

Control flow traces.

No Permission Required.

$\mathcal{L}$ =[⟨ Anne, Jhon, Thomas, Clar, Ram ⟩, ⟨ Anne, Wil, Thomas, Clar, Ram ⟩, ⟨ Anne, John, Thomas, Clar, Steve, Wil, Thomas, Clar, Ram ⟩, ⟨ Anne, Thomas, John, Clar, Ram ⟩, ⟨ Anne, Wil, Thomas, Clar, Steven, John, Thomas, Clar Ram ⟩, ..]

FIG. 4.9

Resource perspective.

No Permission Required.

Projection using a resource classifier can be used in mining organizational models, social networks, etc. (Ghasemi & Amyot, 2016). By choosing a resource classifier $\#_{resource}$ (a sample outcome is shown in Fig. 4.9), the sample event log results as,

4.4.3 Structure of an event log

To better understand the structure of an event log, a tree diagram of a hospital patient admission event log is given in Fig. 4.10. Using the tree structure, we can describe the structure of the event log as follows:

- *Case ID* is used to distinctly identify an instance of the process. For example, xx12, xx13, etc., represent the case IDs in the hospital admission event log given in Fig. 4.7.
- *Event ID* assigns a distinct identifier for every event related to a specific case. For example, event 2346 of case xx12 and event of 3347 of case xx13.
- *Activity* assigns a readable name for every event of a case. For example, event 2346 of case xx12 and event of 3347 of case xx13 point to an activity named Appointment. But the activity Appointment will be carried out separately for both cases.
- *Resources* identify the individuals or machines who are assigned and responsible for executing a specific activity. For example, Ram is assigned as a resource for executing the activity Discharge related to all cases.

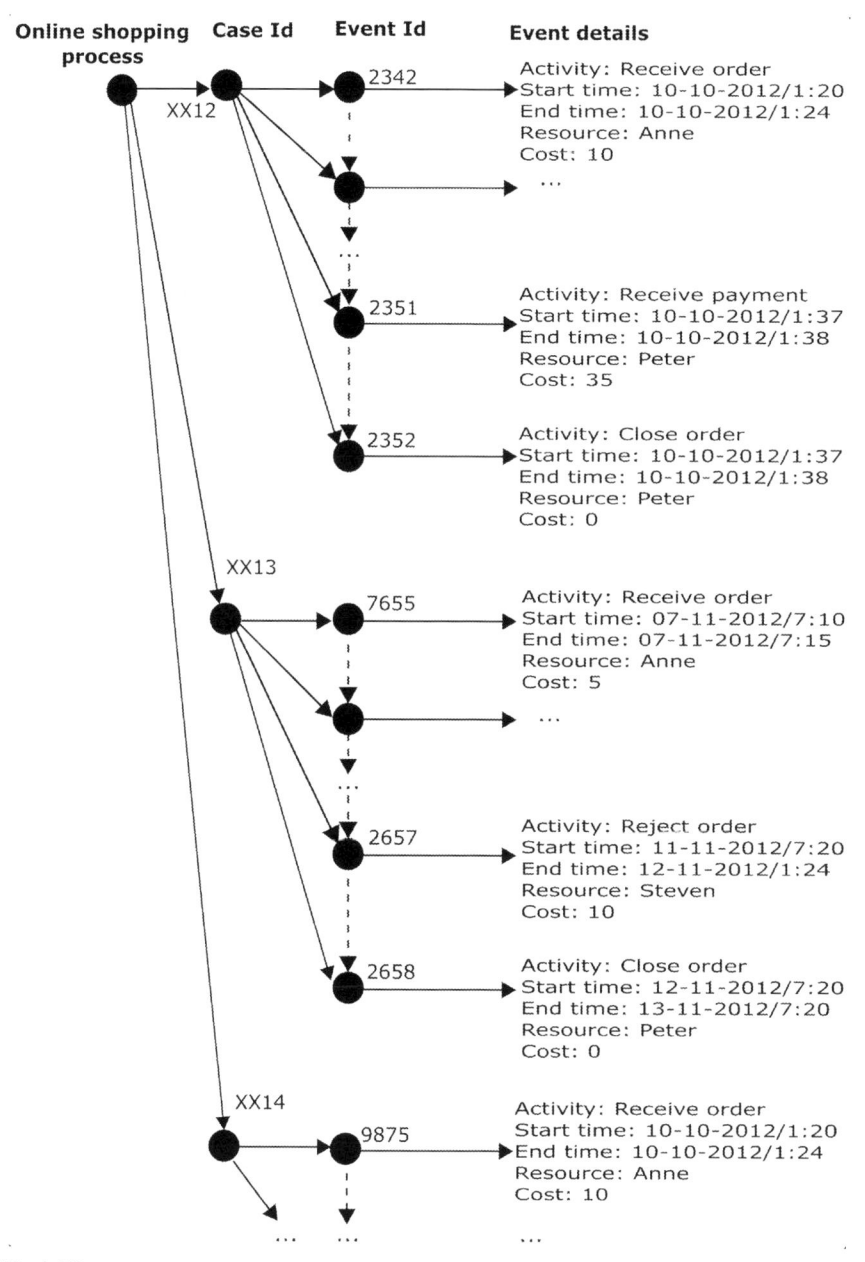

Online shopping process   **Case Id**   **Event Id**   **Event details**

XX12   2342
Activity: Receive order
Start time: 10-10-2012/1:20
End time: 10-10-2012/1:24
Resource: Anne
Cost: 10

...

2351
Activity: Receive payment
Start time: 10-10-2012/1:37
End time: 10-10-2012/1:38
Resource: Peter
Cost: 35

2352
Activity: Close order
Start time: 10-10-2012/1:37
End time: 10-10-2012/1:38
Resource: Peter
Cost: 0

XX13   7655
Activity: Receive order
Start time: 07-11-2012/7:10
End time: 07-11-2012/7:15
Resource: Anne
Cost: 5

...

2657
Activity: Reject order
Start time: 11-11-2012/7:20
End time: 12-11-2012/1:24
Resource: Steven
Cost: 10

2658
Activity: Close order
Start time: 12-11-2012/7:20
End time: 13-11-2012/7:20
Resource: Peter
Cost: 0

XX14   9875
Activity: Receive order
Start time: 10-10-2012/1:20
End time: 10-10-2012/1:24
Resource: Anne
Cost: 10

...

FIG. 4.10

The tree structure of the process log.

- *Timestamps* record the duration.
- The expenditure incurred while executing a specific activity is recorded in the *cost* field.
- *Data* objects related to the process are recorded in the data field. Typical data objects are messages, files, documents, guards, videos, voice, and conditions.

The goal of predictive modeling is to reduce the cost, keep the process under execution in control, improve the quality of the process, speed up the process, recommend alternatives, and improve the satisfaction of all stakeholders.

4.5 Control perspective of hospital process using various modeling notations

The control flow notations connect activities (i.e., functions, tasks, transitions) in the process through constructs such as conditions, places, events, connectors, and gateways. Control flow, perspective models, the causal relationship between different steps in the process execution. Based on the transitions between activities, execution can be serial, concurrent, optional, or repeated. Process mining offers a plethora of techniques for discovering the control-flow model from the event log. These discovery algorithms can represent control flow using various notations. This section aims to introduce a subset of the most widely used and important process modeling notations.

4.5.1 Transition systems

A transition system is the most rudimentary control flow modeling notation. It consists of states and transitions. It is represented using the triplet TS$=\{S, A, T\}$, where S represents a set of states, T is a set of transitions, and A is a set of activities.

Transition systems are simple but tend to fail when they are used for representing concurrent systems due to the state explosion problem (Valmari, 1996). A hospital admission process modeled in the transition system notation is given in Fig. 4.11.

4.5.2 Petri net

The α algorithm (Van Der Aalst, Weijters, & Maruster, 2004) available in process mining can generate the control flow of a process in the Petri net notation. The Petri net model of the hospital admission process is given in Fig. 4.12. A Petri net is a triplet $N = \{P, T, F\}$ where P is a set of places and T is a set of transitions such that $P \cap T = \phi$.

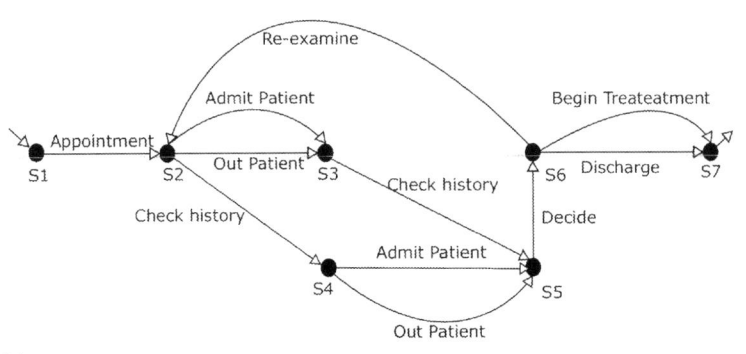

FIG. 4.11

Transition net.

No Permission Required.

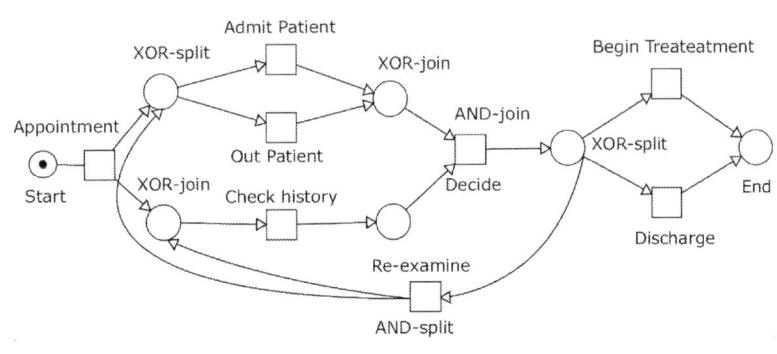

FIG. 4.12

Petri net.

No Permission Required.

4.5.3 **Workflow nets**

A workflow net is an extension of Petri nets with a specific start and end place. A workflow net is represented with attributes as $N = \{P, T, F, A, L\}$, where N is the workflow net, P is the output place, and F is the final place. A sample workflow net is shown in Fig. 4.13.

4.5.4 **Yet another workflow language (YAWL)**

YAWL is an amalgamation of the workflow system and modeling language. The development of YAWL was greatly driven by the workflow patterns initiative (Weijters & Van der Aalst, 2003) based on a regular investigation of the notations used by existing control flow notations.

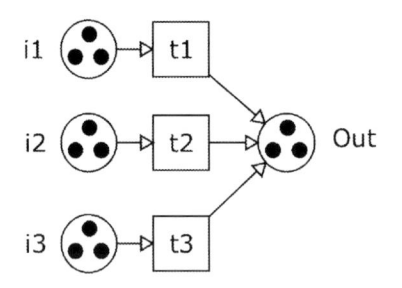

FIG. 4.13

Workflow net.

YAWL is based on the philosophy of facilitating a range of patterns without increasing the modeling language complexity. Fig. 4.14 shows the hospital process built using a YAWL modeling notation.

4.5.5 Business process modeling notation (BPMN)

In recent times, BPMN (White, 2004) has become a widely used notation to model operational processes. It preserves the concept of tasks from YAWL. The BPMN model of the hospital admission process is shown in Fig. 4.15.

4.5.6 Event-driven process chains (EPC)

EPCs (Scheer, Thomas, & Adam, 2005) fundamentally offer a subset of features from BPMN and YAWL with their own graphical notations. Activities in EPCs are called functions. Functions consist of one input and output arc. An EPC of a hospital process is shown in Fig. 4.16.

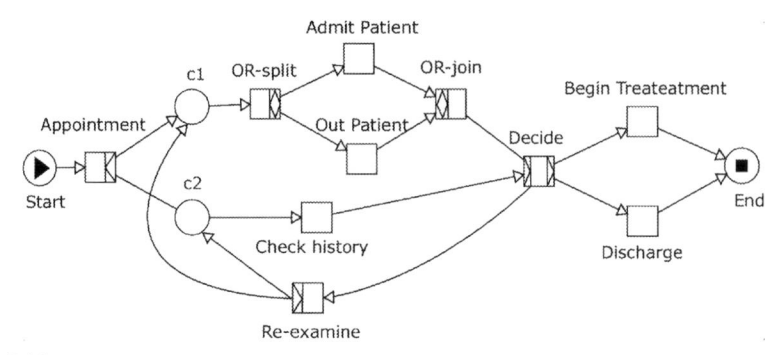

FIG. 4.14

YAWL.

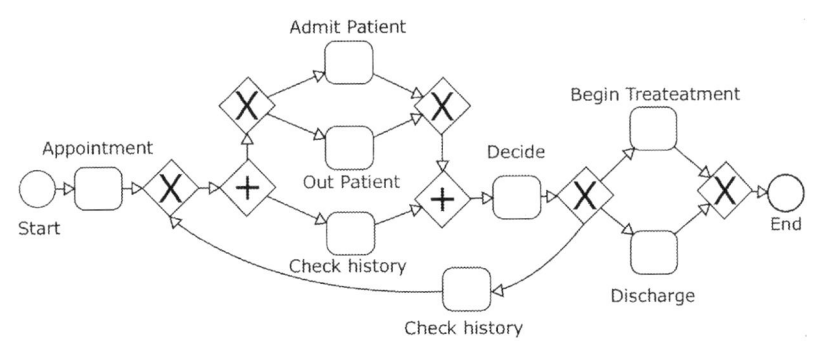

FIG. 4.15

BPMN.

No Permission Required.

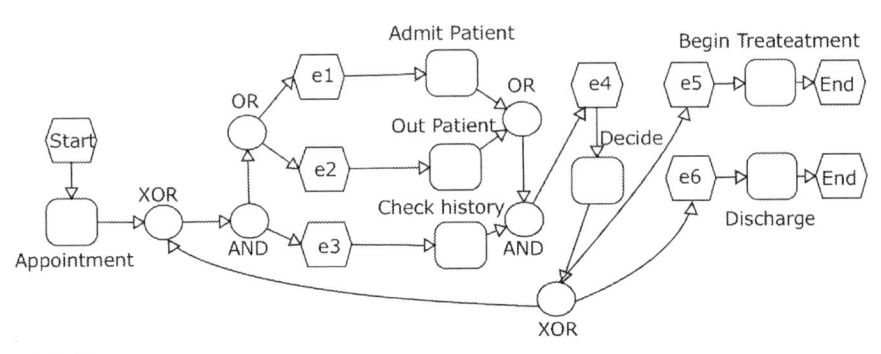

FIG. 4.16

EPC.

No Permission Required.

4.5.7 Causal nets

A causal net is a graph where arcs signify causal dependencies and nodes signify activities. Each activity has a set of possible input and output bindings that guides the routing logic in the causal net. Causal nets are highly suitable for control flow-related tasks of process mining. A causal net representation of the hospital admission process is shown in Fig. 4.17.

4.6 Predictive modeling control flow of a process using fuzzy miner

This section presents a case study to illustrate the modeling of a real-life healthcare process. We present a detailed analysis of the event logs of the hospital treatment

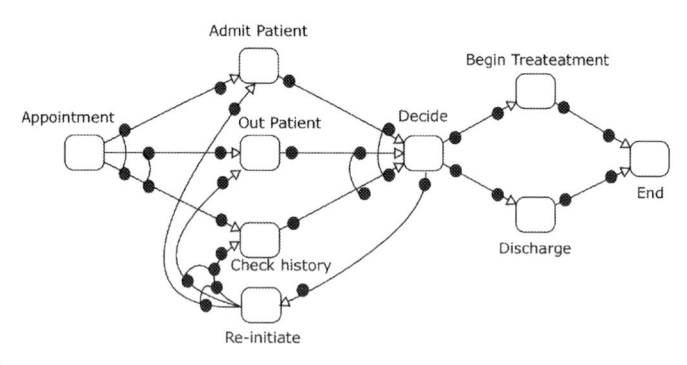

FIG. 4.17

Causal net.

process execution and the hospital billing process (Günther & Van Der Aalst, 2007). We have taken the data from the 4TU.Research Data center. The fuzzy miner will be used to predictively elucidate the control flow model of the process. Process is predictively reconstructed by vomiting the insignificant activities and paths in the process, this can be done using the fuzzy mining technique.

4.6.1 Hospital process

The "hospital billing" event log comprises events that are associated with billing for medical facilities. Each trace of the event log accounts for the activities performed to bill for a package of medical services.

Activities found in the billing process are depicted in Table 4.1. This table shows the relative frequency of each significant activity. All activities in the process have been depicted in the table, irrespective of execution frequency and shows the typical attributes and complexity of the dataset. There are 1 lakh cases, totaling 451,539 events over 18 different activities. Other typical attributes of the event log are shown in Table 4.2 and Table 4.3.

The predictive model of the hospital billing process with 100% paths and 100% activities is given in Fig. 4.18. The control flow model resembles a spaghetti structure and is cumbersome. It cannot be comprehended nor used for any sort of analysis. Further, it is predictively simplified by decreasing the number of activities and paths in it. For predictively generating the simple control flow models, the activities and paths are filtered out at some threshold value decided by the domain expert.

A simplified control flow model of the hospital billing process is shown in Fig. 4.19. This model considers all 18 activities and the most significant paths out of several insignificant ones. The model is much cleaner and simpler compared to the one given in Fig. 4.18, which considers all activities and paths in the process. The predictive model shown in the figure can be easily comprehended and can identify discrepancies in the planned and executed processes.

Table 4.1 Activities in the process and relative frequency.

Activity	Frequency	Relative frequency
New	101,289	~22
Fin	74,738	~16
Release	70,926	~15
Code ok	68,006	~15
Billed	67,448	~14
Change Dian	45,451	~10
Delete	8225	~1
Reopen	4669	~1
Code NOK	3620	~0.8
Storno	2973	~0.66
Reject	2016	~0.45
Set Status	705	~0.16
Empty	449	~0.1
Manual	372	~0.08
Join Pat	358	~0.08
Code error	75	~0.02
Change end	38	~0.01
Zdbc behan	1	~0

No Permission Required.

Table 4.2 Overview of the hospital billing process.

Events	451,359
Cases	10,000
Activities	18
Median case duration	14.6 weeks
Mean case duration	18.2 weeks
Start	12/12/2012
End	01/19/2016

No Permission Required.

Table 4.3 Overview of the hospital billing process.

Activities	18
Minimal frequency	1
Median frequency	3296
Mean frequency	25,075
Maximal frequency	101,289
Frequency Std. duration	35,074.67

No Permission Required.

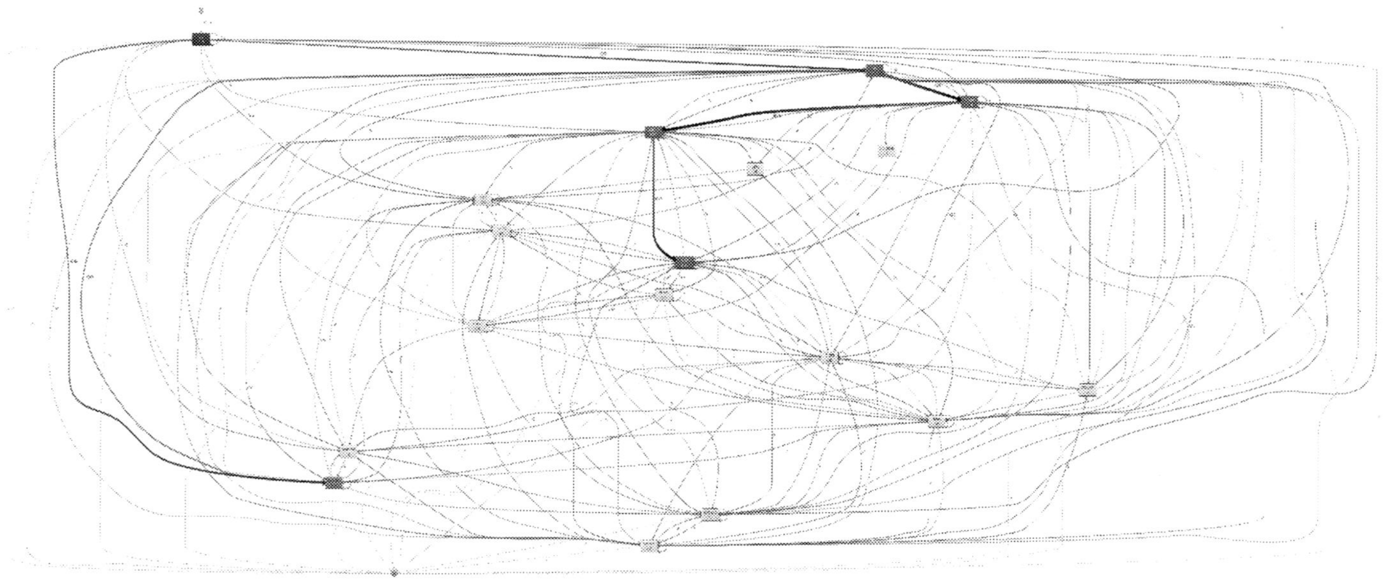

FIG. 4.18

100% activity and 100% paths in the process.

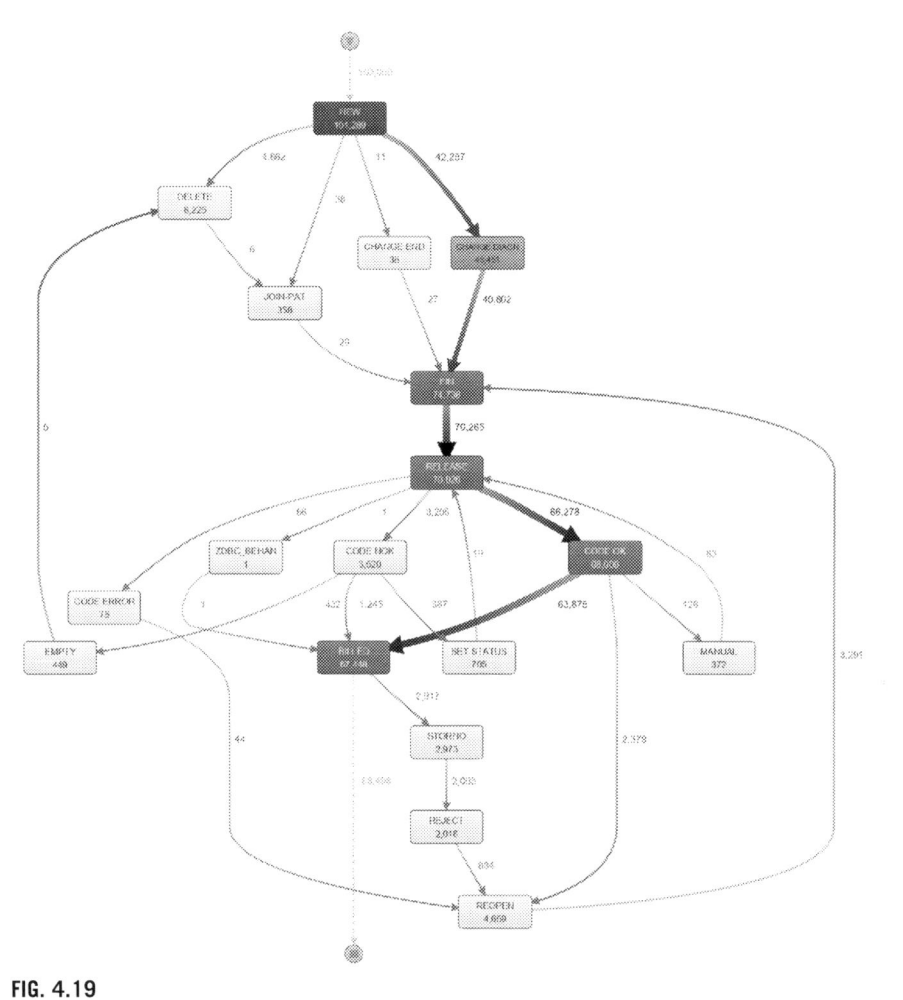

FIG. 4.19

100% activity with significant paths.

Another predictive model of the hospital billing process that considers all significant activities (out of 18) and paths is shown in Fig. 4.20. The much-simplified predictive version of the control flow of the hospital billing process is shown in Fig. 4.21. This control flow model considers both significant activities and paths. Paths and activities in the process that are insignificant are vomited while the rest of the activities/paths that are significant are considered. The predictive control flow model given in Fig. 4.21 is the simplest control flow model that one can produce, and it could be used to guide the information system process.

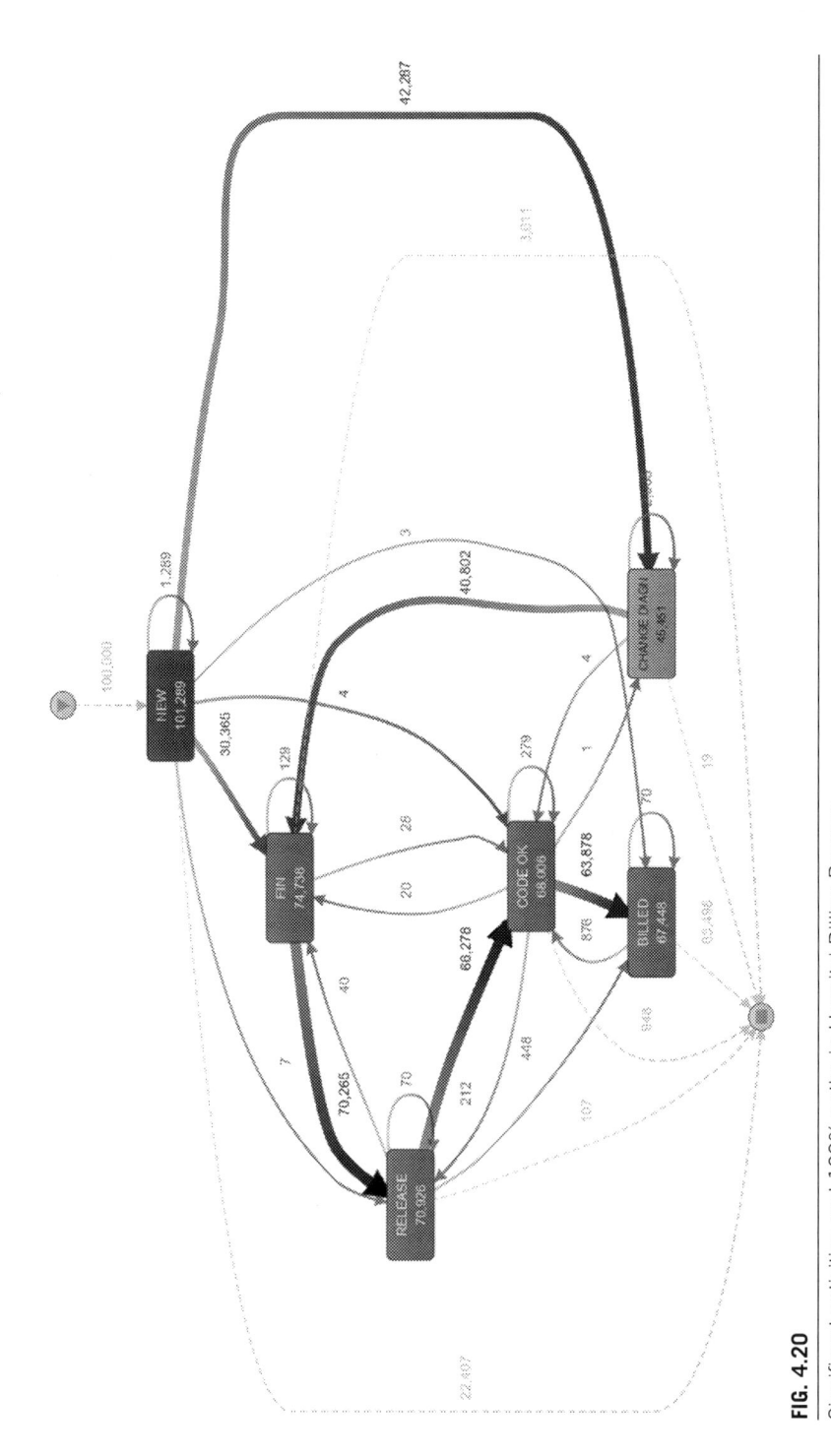

FIG. 4.20

Significant activities and 100% paths in Hospital Billing Process.

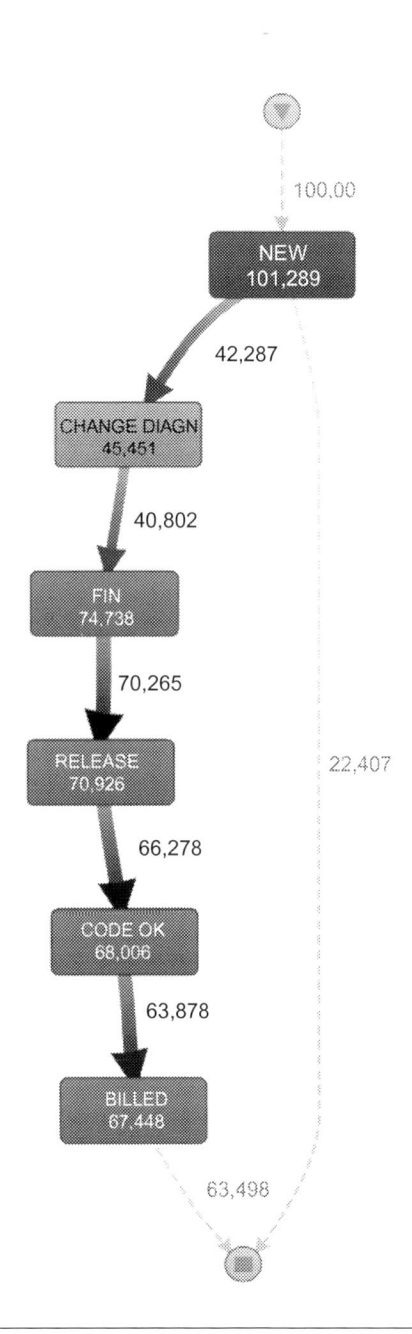

FIG. 4.21

Significant activities and paths of hospital process.

4.6.2 **Hospital treatment process**

This section briefs the predictive modeling of the hospital treatment process. The event log of the process has been obtained from the 4TU.Research Data center (Delft, 2010). The download link is given in the footnote section.

Fig. 4.22 shows some of the most significant activities and the relative frequencies. Due to the sheer number of activities in the process, we have only considered the activities that have been executed 1% of the total number of total activity executions.

The event log information has been given in Fig. 4.23 and Fig. 4.24. The hospital treatment process comprises 1.5 Lakh+ events over 625 activities making 1143

Activity	Frequency	Relative frequency	
aannamae laboratoriumonderzoek	15,353	10.22 %	
ligdagen - alle spec.beh.kinderg.-reval.	10,897	7.25 %	
190205 klasse 3b a205	9,351	6.22 %	
ordertarief	9,008	5.99 %	
190101 bovenreg.toesl. a101	6,241	4.15 %	
vervolgconsult poliklinisch	5,239	3.49 %	
kalium potentiometrisch	4,328	2.88 %	
natrium vlamfotometrisch	4,304	2.86 %	
hemoglobine foto-elektrisch	4,275	2.84 %	
creat' ine	3,955	2.63 %	
leukocyten tellen elektronisch	3,968	1.97 %	
trombocyten tellen - elektronisch	2,724	1.81 %	
differentiele telling automatisch	2,370	1.58 %	
administratief tarief - eerste pol	2,171	1.44 %	
190021 klinische opname a002	2,118	1.41 %	
calcium	2,042	1.36 %	
glucose	1,950	1.3 %	
kruisproef volledig -drie methoden-	1,705	1.13 %	
haemoglobine foto-electrisch - spoed	1,676	1.12 %	

FIG. 4.22

Significant activities in the process that form the control-flow perspective.

No Permission Required.

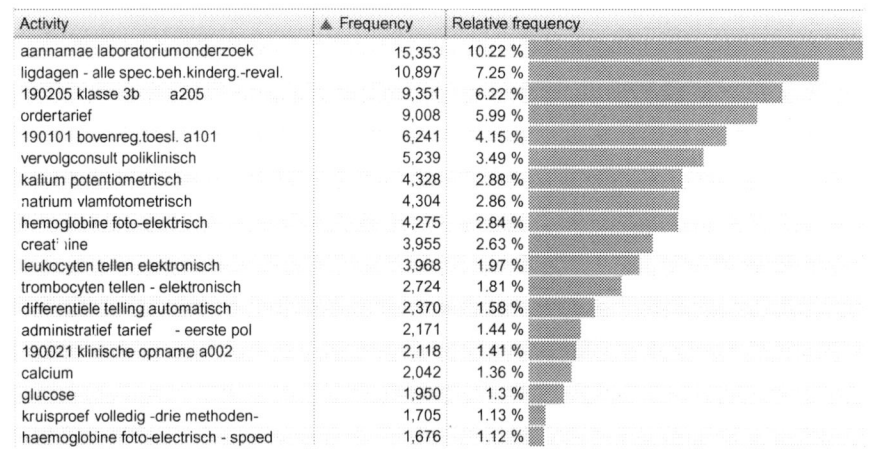

Events	
Cases	1,143
Activities	624
Median case duration	47.6 wks
Mean case duration	12.7 mths
Start	03.01.2005 04:30:00
End	20.03.2008 04:30:00

FIG. 4.23

Overview of the hospital treatment process.

No Permission Required.

Activities	624
Minimal frequency	1
Median frequency	4
Mean frequency	240.85
Maximal frequency	15,353
Frequency std. deviation	1,074.53

FIG. 4.24

Overview of activity classes in the hospital process.

distinct cases. When compared to the hospital billing process, which had 18 activities, 625 activities is a large process.

The predictive control flow model of the hospital process is given in Fig. 4.25. It uses 100% of all activities and paths in the event log. The generated model is not comprehensive, and it cannot be used to guide the future execution of the same process due to its spaghetti structure. A much simpler version of the process could be produced by excising the insignificant activities and paths from the event log. The same threshold criteria are used here to predictively generate the simplified control flow models from the original control flow model. As usual, the threshold limit is set by a domain expert.

The simplified predictive model considering all activities and significant paths is given in Fig. 4.26. Fig. 4.27 shows the predictive control flow model constructed by considering only significant activities and all paths.

A much-simplified predictive model of the hospital process is shown in Fig. 4.28. It considers both significant activities and paths. All insignificant paths and activities are removed while constructing the predictive control flow model.

4.7 Open research problems

This chapter presented comprehensive methods for predictively modeling the control flow perspective of the operational process. There are several other perspectives from which the healthcare process can be modeled. Perspectives such as case, organization, social networking, and time play a vital role in predictively modeling the processes. The following are the some of the open research directions that, if solved, could greatly benefit the predictive modeling research domain of the healthcare process.

FIG. 4.25

100% activity and 100% paths.

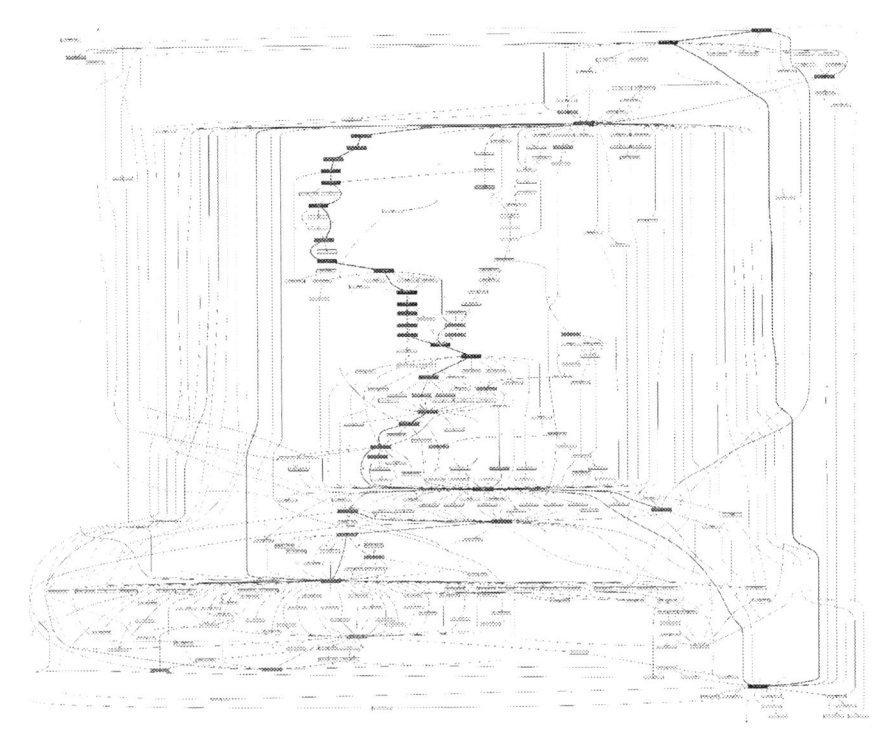

FIG. 4.26

100% activity with significant paths.

No Permission Required.

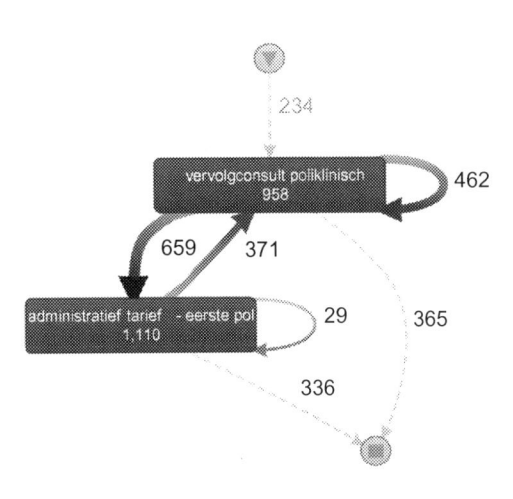

FIG. 4.27

Significant activities and 100% paths in Hospital Treatment Process.

No Permission Required.

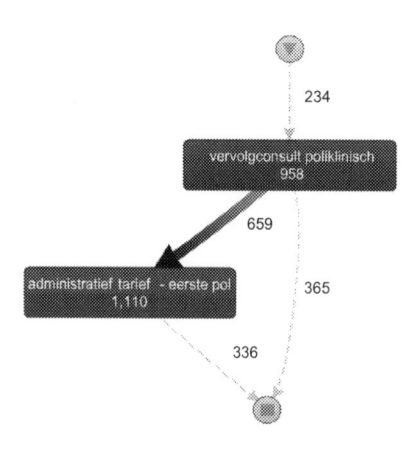

FIG. 4.28

Significant activity and paths hospital treatment process.

No Permission Required.

- The healthcare process tends to change its structure to cater to real-life requirements. Dynamic changes in the process are inevitable. This phenomenon of changing the process is called concept drift. Solving the problem of concept drift for predictively modeling the healthcare process is a challenging issue.
- Collating the event log from more than one event source and generating the consolidated predictive model is an open research problem. The event source could be from more than one process-aware information system. These could be homogeneous or heterogeneous systems.
- Mining the streams of event data to generate a predictive model is another issue in healthcare modeling. Streams of data from various sources tend to change with the evolving process. The algorithms used to construct the models from streaming data need to consider the change in the distribution of the underlying data. This dynamic change in data during process execution is called concept drift.
- Concept drift is a matter of interest in any research discipline that involves data analysis. It basically deals with the "process of handling the changing process." Discovering different types of concept drift such as sudden, recurring, incremental, gradual, and abrupt in various associated perspectives is a challenging problem.
- Finding the variants of the healthcare process grouped by different case and execution patterns.

The area of healthcare predictive modeling is rich with problems to be solved. There are ample opportunities to make the modeling and execution of the healthcare process more efficient, dynamic, and informative.

4.8 Conclusion

This chapter gave a brief overview of how to predictively model the control-flow perspective of the operational healthcare process. Data from process-aware information systems recorded in the form of an event log are considered for generating a predictive model. Predictive modeling has been tested on two real-life healthcare event logs obtained from a credible source. The predictive process models of both processes are depicted at the various configurations of activities and paths in them Typically, a process constructed by considering the significant activity and/or significant path yields more-informative models compared to the models that consider all available information in the event log. This is evident from the results in Section 4.6 of this chapter.

Considering the volume of healthcare data being generated, predictive analytics is a promising field that could practically enhance and optimize the quality of the healthcare process.

References

Antonio, R. (2013). *Business report: The data made me do it*. The next frontier for big data is the individual.

Belle, A., Thiagarajan, R., Soroushmehr, S. M. R., Navidi, F., Beard, D. A., & Najarian, K. (2015). Big data analytics in healthcare. *BioMed Research International, 2015*. https://doi.org/10.1155/2015/370194.

Bennett, W. L., Wells, C., & Freelon, D. (2011). Communicating civic engagement: Contrasting models of citizenship in the youth web sphere. *Journal of Communication, 61*(5), 835–856. https://doi.org/10.1111/j.1460-2466.2011.01588.x.

Brandtzæg, P. B. (2013). Big data, for better or worse: 90% of world's data generated over last two years. *Science Daily, 3*.

de Medeiros, A. K. A., van Dongen, B. F., van der Aalst, W. M. P., & Weijters, A. J. M. M. (2004). *Process mining: Extending the α-algorithm to mine short loops*. Springer.

Dumas, M., van der Aalst, W. M. P., & ter Hofstede, A. H. M. (2005). *Process-aware information systems: Bridging people and software through process technology* (pp. 1–409). John Wiley and Sons. https://doi.org/10.1002/0471741442.

Ghasemi, M., & Amyot, D. (2016). Process mining in healthcare: A systematised literature review. *International Journal of Electronic Healthcare, 9*(1), 60–88. https://doi.org/10.1504/IJEH.2016.078745.

Günther, C. W., & Van Der Aalst, W. M. P. (2007). Fuzzy mining—Adaptive process simplification based on multi-perspective metrics. In *Vol. 4714. Lecture notes in computer science (including subseries lecture notes in artificial intelligence and lecture notes in bioinformatics)* (pp. 328–343).

Irizarry, T., De Vito Dabbs, A., & Curran, C. R. (2015). Patient portals and patient engagement: A state of the science review. *Journal of Medical Internet Research, 17*(6), e148. https://doi.org/10.2196/jmir.4255.

Jans, M., Van Der Werf, J. M., Lybaert, N., & Vanhoof, K. (2011). A business process mining application for internal transaction fraud mitigation. *Expert Systems with Applications, 38*(10), 13351–13359. https://doi.org/10.1016/j.eswa.2011.04.159.

Jha, A. K., Desroches, C. M., Campbell, E. G., Donelan, K., Rao, S. R., Ferris, T. G., … Blumenthal, D. (2009). Use of electronic health records in U.S. Hospitals. *New England Journal of Medicine*, *360*(16), 1628–1638. https://doi.org/10.1056/NEJMsa0900592.

Kaymak, U., Mans, R., Van De Steeg, T., & Dierks, M. (2012). On process mining in health care. In *Conference proceedings—IEEE international conference on systems, man and cybernetics* (pp. 1859–1864). https://doi.org/10.1109/ICSMC.2012.6378009.

Lanz, A., Kreher, U., Reichert, M., & Dadam, P. (2010). Enabling process support for advanced applications with the aristaflow BPM suite. In *Vol. 615. CEUR workshop proceedings* (pp. 17–22).

Lawrence, P., Bouzeghoub, M., Fabret, F., & Matulovic-broqué, M. (1997). Workflow handbook. In *Proceedings of the international workshop on design and management of data warehouses.*

Mans, R. S., Schonenberg, M. H., Song, M., Van Der Aalst, W. M. P., & Bakker, P. J. M. (2008). Application of process mining in healthcare—A case study in a Dutch Hospital. In *Vol. 25. Communications in computer and information science* (pp. 425–438). https://doi.org/10.1007/978-3-540-92219-3_32.

Pedersen, C. A., Schneider, P. J., & Santell, J. P. (2001). ASHP national survey of pharmacy practice in hospital settings: Prescribing and transcribing—2001. *American Journal of Health-System Pharmacy*, *58*(23), 2251–2266. https://academic.oup.com/ajhp/issue.

Rebuge, A., & Ferreira, D. R. (2012). Business process analysis in healthcare environments: A methodology based on process mining. *Information Systems*, *37*(2), 99–116. https://doi.org/10.1016/j.is.2011.01.003.

Reichert, M., Rinderle-Ma, S., & Dadam, P. (2009). Flexibility in process-aware information systems. In *Vol. 5460. Lecture notes in computer science (including subseries lecture notes in artificial intelligence and lecture notes in bioinformatics)* (pp. 115–135). https://doi.org/10.1007/978-3-642-00899-3_7.

Rojas, E., Munoz-Gama, J., Sepúlveda, M., & Capurro, D. (2016). Process mining in healthcare: A literature review. *Journal of Biomedical Informatics*, *61*, 224–236. https://doi.org/10.1016/j.jbi.2016.04.007.

Scheer, A. W., Thomas, O., & Adam, O. (2005). Process modeling using event-driven process chains. In *Process-aware information systems: Bridging people and software through process technology* (pp. 119–145). John Wiley & Sons, Inc. https://doi.org/10.1002/0471741442.ch6.

Schmitt, B. H. (2010). *Customer experience management: A revolutionary approach to connecting with your customers.* John Wiley & Sons.

Valmari, A. (1996). The state explosion problem. In *Advanced course on petri nets* (pp. 429–528).

van Cann, R., Jansen, S., & Brinkkemper, S. (2013). ChipSoft. In *Software business start-up memories* (pp. 130–136).

van der Aalst, W. (2011). *Process mining: Discovery, conformance and enhancement of business processes. Vol. 2.* Springer-Verlag.

Van Der Aalst, W., Weijters, T., & Maruster, L. (2004). Workflow mining: Discovering process models from event logs. *IEEE Transactions on Knowledge and Data Engineering*, *16*(9), 1128–1142. https://doi.org/10.1109/TKDE.2004.47.

van der Aalst, W. M. P., Reijers, H. A., Weijters, A. J. M. M., van Dongen, B. F., Alves de Medeiros, A. K., Song, M., & Verbeek, H. M. W. (2007). Business process mining: An industrial application. *Information Systems*, *32*(5), 713–732. https://doi.org/10.1016/j.is.2006.05.003.

Van Der Aalst, W. M. P., Weske, M., & Grünbauer, D. (2005). Case handling: A new paradigm for business process support. *Data and Knowledge Engineering*, *53*(2), 129–162. https://doi.org/10.1016/j.datak.2004.07.003.

Vijayaraghavan, E. V., Haex-Director, M. G., & Both, M. P. (2006). *A conceptual data design for an installed base and configuration management for the Philips Healthcare IT products*.

Weijters, A. J. M. M., van Der Aalst, W. M., & De Medeiros, A. A. (2006). Process mining with the heuristics miner-algorithm. *Technische Universiteit Eindhoven Technology Reports*, *166*, 1–34.

Weijters, A. J. M. M., & Van der Aalst, W. M. P. (2003). Rediscovering workflow models from event-based data using little thumb. *Integrated Computer-Aided Engineering*, *10*(2), 151–162.

White, S. A. (2004). *Introduction to BPMN. Vol. 2*. IBM Cooperation.

Challenges and opportunities of big data integration in patient-centric healthcare analytics using mobile networks

5

M. Karthiga[a], S. Sankarananth[b], S. Sountharrajan[c], B. Sathis Kumar[d], and S.S. Nandhini[a]

[a]*Bannari Amman Institute of Technology, Erode, Tamil Nadu, India,* [b]*Excel College of Engineering and Technology, Namakkal, Tamil Nadu, India,* [c]*VIT Bhopal University, Bhopal, Madhya Pradesh, India,* [d]*VIT University, Chennai, India*

5.1 Introduction

Healthcare big data is comprised of huge sectors of electronic health data that are very tough to organize and master through the traditional means of software and hardware. The traditional methods of organizing such enormous data through the available tools are also very tricky. Healthcare big data is the most overwhelming, not just because of its huge volume but also because of the presence of different types of data and the methodology by which all such data are organized and maintained. In addition, the total amount of data with respect to patients is also rising dynamically, thereby increasing the complexity of healthcare data sectors.

Emerging healthcare industries utilize big data to tackle a broad category of business and medical challenges such as increasing patient safety, avoiding patient readmission, and improving drug selection related to diseases using solutions provided by big data. Data from traditional data banks such as electronically stored patient health records or file-based patient records are imported, gathered, and analyzed for the extraction of information in a reliable and commercial manner using big data. Patient billing history as well as other stakeholder data related to healthcare are also efficiently analyzed using big data. The data in healthcare industries are in both structured and unstructured formats, so the healthcare industries are trying to swift from traditional way of prediction to real time prediction as mentioned by Muñoz-Gea, Aparicio-Pardo, Wehbe, Simon, and Nuaymi (2014).

Researchers analyze and perform actions using big data techniques to change traditional treatments for genetic disorders. Markers for genetic diseases and suitable drugs for those diseases are identified by biologists using the visualization of data from cells. Medicines are produced in a shorter time span by pharmaceutical industries. This in turn benefits patients in disease diagnosis and prognosis in a faster manner using these predictive analytics. Patient lives become better with analytics using big data for the most serious genetic disorders as well as the possibility of curing such disorders in a more favorable manner.

5.2 Elderly health monitoring using big data

The term "population aging" refers to an increase in the median age population in a society. This is due to issues such as a decrease in the fertility rate and a rapid increase in life expectancy. When there is a drastic increase in such a population, that in turn increases challenges such as the health of the elderly, the quality of their life, the medical facilities available, etc. A system that reads the required current measures of the elderly to assess diseases and analyze the data gives a decision as to whether critical care is required. Such medical tracking systems through technologies such as the Internet of Things (IoT) and big data are highly appreciated, as they allow medical technicians and doctors to monitor patients without being in constant direct contact. As per the statistics from (United Nations, 2017), 13% of the world's population was older than 60 as of 2017, and that could reach 25% by 2050.

Because there has been an increase in life expectancy and an increase in the median age population, a system to take care of such geriatric people is mandatory. Handling geriatric persons and their health status with tools and systems involves a large dataset to be processed by professionals to make accurate decisions about their health. An assessment of the day-to-day activities of geriatric persons may include checking the hearing, moving ability, balance, and digestion. A significant measure of geriatric assessment is comprised of a range of well-validated tools for analysis. A complete assessment is needed only when professionals diagnose something that is abnormal. This includes measures of geriatric patients that lie outside the normal range specified by a medical council, as suggested by Elsawy and Higgins (2011).

5.2.1 eHealth

The term eHealth refers to technological support for healthcare, where healthcare services can be provided through technology. The services include monitoring, self-assessment, and the periodic diagnosis of patients. These will be frequently updated to physicians, replacing traditional medical services. eHealth can be referred to as the use of information and communication technology in healthcare, as proposed by Conceição, Wouter, and Alise (2018). Further, patients using eHealth services can decide when to make informed decisions regarding their health measures. This is how data has to be collected from patients. Those data are analyzed and reported in an easily readable form to caretakers, as in Hale et al. (2018).

eHealth should include monitoring patients. As discussed already, the system has to decide when a patient has to be shifted from normal healthcare to intensive healthcare. Medical care is a very sensitive field that requires extremely perfect and appropriate decisions based on the various attributes of the patients. They need to be managed and processed with high concentration to preserve their privacy, easy accessibility, integrity and accuracy as proposed by Ianculescu, Alexandru, and Tudora (2017).

In the technology era, a healthcare system for elderly people is a major requirement. Many have proposed solutions for the home care of such people, including Bajenaru, Ianculescu, and Dobre (2018) and Robert-Andrei, Ciprian, Lidia, and Radu-Ioan (2019). Already existing systems use smartphones to monitor the movement and stability of elderly people through sensors. Also, in the proposed system, all physical activities are considered to develop a monitoring system, including walking, running, sitting, standing, and ascending and descending stairs.

The report concentrates on monitoring the various service components provided by eHealth remotely. The need for frequent monitoring of medical conditions has increased exponentially in recent times. For patients with disorders that need to be monitored continuously and treated for longer periods, this type of monitoring device is highly needed. A study from Eurostat (2020) says that 45% of people in European countries in 2014 were affected with a minimum of one chronic disease. Also, chronic diseases were the cause of 71% of deaths all over the world (World Health Organisation, 2018). There was a 5.5% prevalence of dementia in those 60 and above in the European Union in 2014, as suggested by Health at a Glance Europe 2012 (2012). The advantages of monitoring the health conditions of patients remotely are a reduction in cost, less time, no unnecessary in-person interaction with doctors, etc. To monitor patients remotely, sensors are highly required. They can be standalone or interoperable with smart devices connected through the Internet of Things (IoT) (Năstase, Sandu, & Popescu, 2017).

5.2.2 General health issues in the elderly

Apart from direct medical services, other functionalities can also be added in the service such as nutrition, physiotherapists, counsellors, and other professionals related to a patient's lifestyle. This will help in getting the medical data with still more clear view, which means the medical data along with physical and mental health can also be monitored which has significant effect in human health as mentioned by Elsawy and Higgins (2011) and Glen et al. (2005). Assessing the health conditions of the elderly is a multidiscipline task that requires an evaluation of physical, mental, cognitive, and social health. So, when there is an observable problem, the doctor has to check all interrelated conditions before beginning treatment (Elsawy & Higgins, 2011). The functional capacity of a person refers to the ability of a person to carry out daily activities (ADL, activities of daily living) and handle household machines (IADL, instrumental activities of daily living). The ADL and IADL can be assessed using the Katz ADL scale as proposed by Katz, Downs, Cash, and Grotz (1970) and the Lawton IADL scale as proposed by Lawton and Brody (1969). More knowledge about the patient's work culture and living environment is required when assessing the ADL and IADL. The physical health assessment not only depends on the present health data, but also depends on the previous history of the patient and their family. Also, imbalance and falling are the main reasons for the elderly to be hospitalized, as mentioned by Gillespie et al. (2003).

An important tool to monitor and assess the risk of falling is the Tinetti Balance and Gait Evaluation. The patient will be observed while getting up from chairs without the use of hands, walking for maximum of 10 feet, then turning and returning to their previous position. When patients fail to do this task, then the risk of abnormality is high and they require further diagnosis (Rao, 2005). This risk can be reduced by physical movements and therapies, as suggested by Joanne and Lorraine (2002). Other than physical health, when it comes to cognitive and mental health, there are again many assessment models to read a weakening in the neural system (Bajenaru et al., 2018). Also, the giddiness disorder among the elderly is the fore-symptom for many cognitive disorders. Such depression can lead to many other major disorders such as strokes, Alzheimer's disease, and other neural disorders (Baloh, 1992). To monitor and assess depression, the Geriatric Depression Scale and the Hamilton Depression Scale are widely accepted tools (Baloh, 1992).

Dementia by itself is not a disease but is a condition that may lead to other disorders such as cerebral-vascular disease and Alzheimer's disease (Bajenaru, Tiu, Antochi, & Roceanu, 2012). The risk level of dementia depends on age, physical mobility, diabetes, and similar factors (Glen et al., 2005; Popescu et al., 2009). Also, the relationship between dementia and diabetes is highly complex (Herghelegiu, Nacu, & Prada, 2016). Also, such neural disorders are due to different medical scenarios for which no clear manifest (Dwolatzky, Ruffle, & Zultan, 2018; Marinescu, Bajenaru, & Dobre, 2018). The risk factors are based on the mental state of the patient, and their environment plays an important role in the health of geriatric patients. Further, the medical team handling the geriatric patients should keep track of their data for the future to make decisions on disorders that occur. When the monitoring system starts collecting data from all elderly patients, the data will keep getting bigger and bigger. The main challenge is handling a large volume of data with wide variety. This is the concept of big data to process and extract useful values from the large dataset collected (Gantz & Reinsel, 2011 and Groves, 2016). The parameters considered from the large datasets under consideration are volume, variety, velocity, veracity, and value. With the large volume of data, big data is more accurate when delivering to patients, which in turn increases the safety of patients under monitoring (Ianculescu, Alexandru, & Gheorghe-Moisii, 2017). On an important note, apart from delivering precise decisions, big data also helps identify diseases so that they can be treated early. Collectively, big data provides valuable assistance to the medical community that includes diagnosis, detection, prediction, prevention, and data management.

5.3 **Personalized monitoring and support platform (MONISAN)**

A project developed by the Romanian National Institute called MONISAN supports the transformation of data to a digital platform. The main objective of this project is to collect and monitor patient data and conditions and then apply recent technologies to visualize, analyze, and regulate the stored data. The results of this project are a patient-oriented online platform based on surroundings, endorsements, detection, monitoring, and personal assistance.

The platform is built on a two-level access system. One is for authenticated users while the other is for unauthenticated users. The unauthenticated user can view data related to healthy routine commendations. The authenticated user module has three types of authenticated users—doctors, medical caretakers, and patients. All the users have different access to the system in viewing and analyzing the data collected. Doctors and medical caretakers can view and modify information such as a patient's medical sheet, clinical examination, healthcare indication, doctor recommendation, etc. The patients can access only their own records but they cannot modify the stored data, even if it is their own data.

5.3.1 **Proposed development**

Though the MONISAN project has many advantages, it also has its challenges. As we have already discussed, the parameters for big data are volume, variety, velocity, veracity, and value. Even after required modifications, MONISAN can address variety and veracity but not the other three measures. To make big data features to be addressed by MONISAN, the project is enhanced to handle vast amounts of data along with real-time analysis of the same engendered by the sensors.

Elderly patients were monitored by Kiran, Jibukumar, and Premkumar (2016) and they proposed a system that makes use of big data analytics to predict dementia. There are certain factors identified as conditions for dementia in elderly people, including hypertension, hypercholesterol, and obesity (Exalto et al., 2014; Miia et al., 2006). Also, the main objective is to read the state of the patient via continuous long-term monitoring and to check for the risk of dementia. This is done by comparing the data of the patient under consideration with the data of people who already had dementia.

Research on dementia in integrative approaches uses neural images, preexisting digital health records, cognitive tests, data from sensors, and even gestures (Marcello, Effy, & Alessandro, 2018). Dementia is a complex disorder because the reasons are not identifiable. These reasons include societal, environmental, health factors, and many more. Also there is no proper cure for the disorder so far. One such approach is to combine data from the MONISAN platform, digital records of patients, sensor data, etc. These data are collectively processed using big data analytics to predict the pattern of a neural system with dementia. Moreover, a study says that 44 million people were affected in 2015 (Deetjen, Meyer, &

Schroeder, 2015). The Organization for Economic Cooperation and Development (2015) said dementia is expected to affect 50% of geriatric persons by 2030.

The proposed system enhances the MONISAN platform with big data analytics with the use of the Apache Hadoop framework (Bajenaru & Custura, 2019). Mainly designed to produce more efficient results using Hadoop Distributed File System (HDFS) which is used for storage and MapReduce model has been used for processing. The collected data are separated into blocks that are given to different computers where MapReduce models are executed, as shown in Fig. 5.1.

The system has two different MapReduce modules: one will be running on patient data to check for dementia risk while the other will process all sensor data. Both solutions will collect details from various sources such as sensor readings, clinical examinations, electronic health records, and neuroscience explanatory models provided by medical specialists and give a thorough, analyzed result. The first algorithm identifies the potential dementia of elderly people in the beginning state and also changes the medical treatment and lifestyle to prevent earlier onset. The perspective of how each of the physical and ambient parameters influences the risk of developing dementia is provided by the second algorithm. This is the enhanced version of the MONISAN platform by including big data analytics.

5.4 Patient-centric healthcare provider using big data

Many big data sources such as IoT, healthcare sensors, and smart devices such as smartwatches and smartphones with medical apps are available. The most common problem all these data generators faces is network connectivity. Ensuring the connectivity of the network and providing quality communication are among the biggest

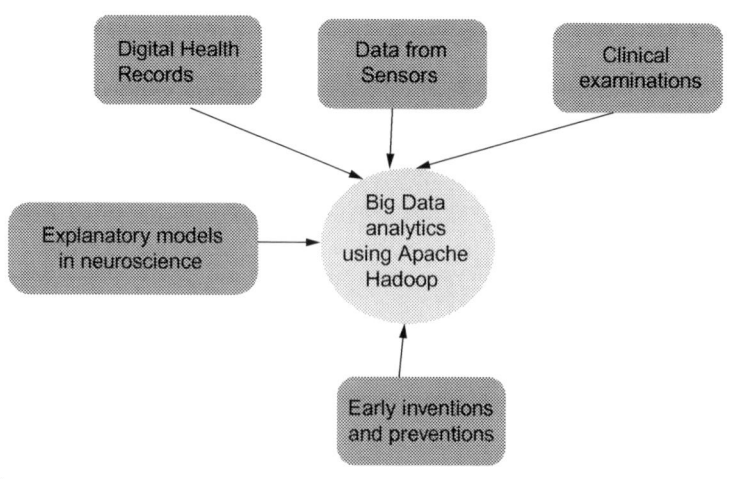

FIG. 5.1

Architecture of the healthcare system with big data analytics.

challenges for researchers and the healthcare sector. Providing an optimal solution for this problem is a challenging research area nowadays. Another challenge is handling the big data produced from the healthcare sector. A network operator should focus on the patients in an emergency and direct resources to the patients. The most important step in personalized healthcare management lies in providing high-quality connectivity between the patient's medical devices and the physicians.

Because many sensor devices are with the patients, ensuring wireless connectivity among these devices is a better option than a wired connection. Mobile networks and Wi-Fi are the two most famous wireless connection technologies. The area of connectivity differs for each wireless technology. So, if Wi-Fi is used for communication, then the patients and the sensor devices should be within Wi-Fi range. Wi-Fi is most suitable for indoor coverage. Whereas if existing mobile networks are utilized, the level of coverage is higher than Wi-Fi due to the availability of a wide range of cell towers. But the major problem faced in mobile networks is path loss and network fading because of blind spots and locations where the connectivity is deeply faded. The connection becomes unreliable and the connectivity could not be established when the signal-to-interference-plus-noise ratio (SINR) level is at a minimum. This in turn causes an additional burden in case of emergency communications about a patient's status to healthcare providers.

Big data is an important technology that could be used to solve problems such as resource sharing, resource allocation and reallocation, and optimized connectivity in a wireless medium (Kiran et al., 2016). User-centric optimization of the mobile networks is an unexplored area. In this chapter, two patient-centric optimizations of the uplink connection of multiple-cell orthogonal frequency division multiple access (OFDMA) networks are proposed. In these approaches, the main methodology lies in improving patient-centric mobile networks, which has high priority, by maximizing the SINR at the base station. This in turn maximizes the overall SINR of the entire network. The mobile network for patients may be dedicated or nondedicated. But the optimization is done with nondedicated mobile networks for certain reasons:

- First, the deployment cost of a nondedicated network is cheaper than dedicated, and the nondedicated networks require only a low commissioning operational time period.
- Second, the service level of other users is not affected while improving the SINR of the patients.
- Third, the proposed approach could be established in recognized networks because this kind of patient-centric beneficiary is an attractive market for networks.
- Finally, if a dedicated network is used, the patient's mobility could fall only within the area where the network coverage is present. Instead, if a nondedicated standardized regulated network is used, then the area of mobility of the patients is nationwide.

Big data technology provides an assignment score in the proposed model where those scores are made in the physical resource block (PRB) of the patients that is proportional to the level of emergency. Equality is maintained to reduce the negative impact among other users for these assignment scores. The main contributions of the chapter lie in: (1) Optimization of the connection of multiple-cell OFDMA networks while the patient's mobile networks are prioritized using big data analytics. (2) Mixed integer linear programming-based optimization to assign scores in a patient's PRB with respect to their health condition. (3) Developing a classification model to decide the likelihood of a heart arrest for a patient using the big data of the patient's dataset.

5.4.1 Resource allocation in mobile networks using big data analytics: A survey

Providing a wireless mobile network seamlessly based on patient priority is taken into account. Some previous work includes use of data like configuration, alarm based priority and log-files are processed using big data for proper identification of the users as well as the network is proposed in the work by Kiran et al. (2016). The major objective lies in minimizing the delay between the resource request and allocation in the radio access network. The resources in heterogeneous networks are managed by collecting and analyzing the user's behavioral and sentimental patterns along with the data communicated in social networks (Zheng et al., 2016). The aim of the proposed approach lies in minimizing the disruption in service while servicing at the right moments.

5.4.2 Healthcare analytics: A survey

Many approaches are proposed to utilize the big data technique in patient-centric treatments. A real-time emergency response model is proposed in the work by Mazhar, Awais, Anand, Jiafu, and Daqiang (2016), where processing the data collected from many wireless sensors is used. The real challenges faced during data processing from various sensors through big data services are discussed in the work by Rudyar, Xavier, Olivier, and Pierre (2015). Real-time data collection and processing are also discussed. The data collected from various sensors and medical devices is analyzed using the Microsoft Kinect Sensor System. Patients could monitor their disease progression and treatment outcomes using this Microsoft system. This methodology is modeled for patients with Parkinson's disease in a work from Ballon (2013). Authors such as Banu and Swamy (2016) proposed a survey for analyzing heart disease in patients at an early level. All healthcare analytics discussed so far utilize machine learning, deep learning, and big data services.

The healthcare analytics models discussed above assume that the networks used have good connectivity. Whereas in real life, networks do not possess good connectivity, and there may be some fading and noise issues in networks. These problems also should be considered while imposing patient-centric treatments. The proposed methodology implies the usage of big data techniques for the optimization of RAN in a long-term evolution-advanced (LTEA) network for a particular group of people such as patients in healthcare analytics. Service is ensured at the correct time for the patients when they are really in urgent need. The patients who are at a greater risk such as a heart attack are obtained by processing the patient's big data. Patients are treated with high priority compared with other users while shifting the network connection. This is one of the major problems among healthcare sectors, as the symptoms for a heart attack begin at home. Many patients would die before treatment starts. This death rate could be minimized in the proposed approach as the resources are allocated in high priority. The proposed method provides an efficient and reliable connection while transferring the data from medical devices to health practitioners. In addition, the symptoms are predicted earlier. The patients are admitted to hospitals earlier and treated with high priority to save lives. In order to satisfy the connection problem and respond according to the patient's needs, mixed integer linear programming is used to provide an optimal solution, whereas heuristics methods are used for a fast response. The dataset for training the model is collected from Kaggle datasets. The proposed dataset contains the data for about 43,400 patients. The dataset contains features such as high blood pressure (BP), smoking rate, BMI, age, gender, glucose level, cholesterol level, hypertension, married, work type, etc. The features are optimized to choose the best attributes for training using the proposed model.

5.5 Patient-centric optimization model

5.5.1 Structure model

A metropolitan city secured by the LTEA network is considered. The base station is located more than 600 m, a number of users spread out at different location and distances. The patients can provide the cellular network using the connection it is a best method. The mobile phone receives a varied level of signal strength based on the patient's random location. The duration of the transmission time interval (TTI) is less than the coherent time. Data from patients with a high possibility of heart arrest must be transmitted at high priority. The cloud-based big data analytics engine is used to analyze the patient data. It operates naïve Bayesian algorithms. The cloud-based engine is used to predict the high possibility of heart arrest for patients. The optimization of the model is to provide the patients who are in emergency, a high score to obtain PRB and increase the signal-to-noise-plus-interference ratio at BS. Needy patients are allocated with maximum resources in a way that the other users are not affected.

5.5.2 Classification using naïve Bayesian

The classification using naïve Bayesian identifies a heart arrest by a set of variables obtained from the patient's records. The classifier can use the patient's medical record to identify a heart arrest and other risk factors. The features used in the classification model are the systolic and diastolic levels, the blood pressure level, the total cholesterol level, and the smoking rate level collected from the human body using the attached IoT sensors. These are input into the big data analytics model for classification using naïve Bayesian. The features taken are independent to one another. The following are the merits of classification using naïve Bayesian compared to other classifiers: (a) low complexity model, (b) one classifier used to analyze all types of disease, (c) the classifier is more accurate, (d) the competitiveness is greater compared to other classifiers such as neural networks and random decision trees, (e) this classifier need minimum datasets for training, and (f) this classifier is a disease risk indicator. Therefore, the classification using naïve Bayesian is most effective and the predictions are most accurate. Table 5.1 highlights some patient data taken for analysis.

The likelihood of occurrence of the target label for the input features is estimated using Eq. (5.1).

$$p(F_i| T)^n = \frac{\sum_{i=1}^{n} T \cap F_i}{\sum_{i=1}^{n} T} \tag{5.1}$$

Where $P(T)$ is the above-mentioned possibility of heart arrest, that is, the total number of days taken for the occurrence of a heart arrest.

Table 5.1 represents a sample of a patient's medical record, and contains five columns of datasets. The patient's conditions should be monitored. The feature variables reading are saved in first 4 columns and the class target variable T of a heart arrest or a critical state mentioned in the 5th column. To prevent heart arrest, the classifier continuously reads the patient's medical records to monitor the patient's actual condition. The total number of days taken into account is 30 days, which implies the total number of columns in Table 5.1. This possibility is converted to a risk factor to compute the assignment score for each patient during PRB assignment using MILP and heuristic techniques. Concerning the security of patient records, the dataset does not hold the data more than 30 days. Framingham techniques as suggested by LaMorte (2017) that are based on data preparation steps have been implemented before classification using naïve Bayesian. Because the amount of data in the Framingham dataset is large, big data analytics are utilized to predict the probability of heart arrest. The reduction of data as well as cleaning and generalization are the few preprocessing steps carried out before applying training. The new data from the patient are taken for testing; this is mentioned in separate rows in Table 5.1. Fig. 5.2 depicts an overview of the calculation of risk factors for prioritizing patients.

Table 5.1 Medical records of patients (sample).

	Blood pressure rate F1	Blood glucose level F2	Body cholesterol level F3	Smoking level F4	Heart arrest (yes/no) T
1	Normal	High	High	Heavy	Yes
2	High	Normal	Prehypertension	Moderate	No
30	Medium	Low	High hypertension	Light	Yes
(Testing)					
	High	High	High hypertension	Light	?

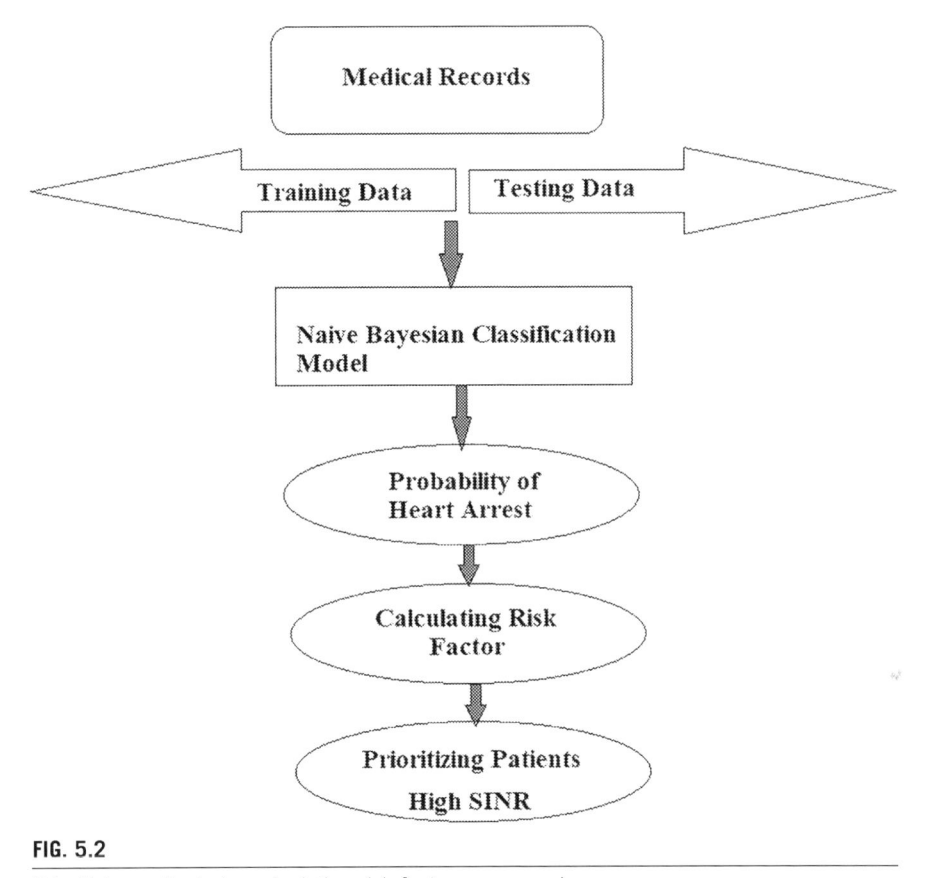

FIG. 5.2

Prioritizing patients by calculating risk factors: an overview.

5.5.3 **Reduction of data**

In the reduction of data process, the distinct features are maintained and others are eliminated. Three main reasons to improve the classifier accuracy are: (1) reducing the number of features impacts the dimensions of the dataset, thereby diminishing the efficiency of the processor and reducing the time. The MILP execution time is also reduced on behalf of data reduction. (2) We are mainly targeting main heart arrest contributors such as hyperlipidemia, BP, and smoking. (3) Only the main contributors are taken as features; other slowly changing features such as weight, age, gender, and BMI are not accountable. The inadequate entries, false entries, and inconsistent data were eliminated. The resultant dataset has no errors and all entries of the value have a complete set. Proper cleaning is done to remove duplicates and make the data ready for processing.

5.5.4 Generalization of data

The discontinuation of the data should be converted into a huge number of continuous feature values and it into lesser value. The discontinuation process can be used in the naïve Bayesian models to get more accuracy. The large data of numeric values in to subset values that are under a same category by using the data reduction mechanism. The American National Institutes of Health and the British Heart arrest Association in *National Institutes of Health. Your Guide to Lowering Your Blood Pressure with DASH* (National Institutes of Health, 2009) define the medical accredited ranges, and these ranges are taken for the dataset. For example, the smoking level is categorized as light, medium, and high as per the National Institutes of Health standards. The medically accredited ranges are shown in Table 5.2.

5.5.5 The naïve Bayesian formulation techniques used to calculate patient priority using MILP

The possibility (PHA_z) of a heart arrest given a current state (CS_i) is calculated using a combination of the naïve Bayesian classifier and the MILP model. The model is converted to user priority (UP_m), as mentioned in Eq. (5.7).

Eq. (5.1) is rewritten as Eq. (5.2) below, and is used to compute the conditional probability:

$$TP_{i,v}^{c,z} = P\left(F_i - V_{F_i}^{j,z} \mid T_i = V_{T_i}^{r,z}\right) = \sum_{d=1}^{|D|} \sum_F \sum_T \frac{S_{F_iT_i}^{j,r,d,z}}{G_{T_i}^{r,d,z}} \tag{5.2}$$

Where the nominator indicates the number of days in total the patient encounters the needed readings ($VF_{ij,z}$) and a heart arrest is indicated as $VC_{11,z}$, where $C1$ indicates the target heart arrest and $r = 1$ represents the occurrence.

The denominator indicates the number of days where heart arrest occurs.

Table 5.2 Medically accredited ranges as per American National Institute of Health Standards (Health at a Glance Europe 2012, 2012) .

Features	Range of values	Standard level
Body cholesterol level (mg/dl)	Less than 200	Medium
	Between 200 and 239	Normal
	Greater than 240	High
Blood pressure rate (mmHg)	Less than 120	Normal
	Between 120 and 139	Prehypertension
	Greater than 140	High hypertension
Smoking level (no. of cigarettes/day)	Within 1 to 10	Light
	Within 11 to 19	Medium
	More than 20	High

Eq. (5.3) represents the method to compute the probability of heart arrest.

$$PS^{z,r} = \left[\sum_{d=1}^{|D|} \frac{G_{T_i}^{r,d,z}}{} \right] \pi_{i=1}^{1} P\left(F_i = V_{F_i}^{CS_{i,z}} | T_i = V_{T_i}^{CS_{i,z}} \right) \tag{5.3}$$

For a current set of variables CS_i, all features are taken into consideration. The LHS indicates the posterior probability that a patient is diagnosed with heart arrest. In RHS, the first part denotes the prior probability of a heart arrest and the second part indicates the combinative probability that the identified patient also has the same feature. The two terms in RHS are multiplied to represent the independent nature of the features taken in the dataset.

Eq. (5.4) is used to calculate the user priority value.

$$UP_m = 1 + \alpha PS_{z,r} \tag{5.4}$$

where $PS_{z,r}$ is the overall obtained probability of a heart arrest. This term is multiplied with a tuning factor to obtain an effective score value.

5.5.6 Formulation of problem

MILP models are developed to enhance the network resource allocation for patients and users. The operation of a patient monitoring system is an LTEA network that contains a base station and channels operating on a bandwidth of 1.4 MHz. The main objective is to optimize the LTEA uplink network so the patients are formulated over other users and they are allocated high-powered PRBs.

The orthogonal frequency division multiplexing network's SINR value at the uplink side is expressed in Eq. (5.5).

$$T_{m,n}^{b} = \frac{Signal}{Interference + Noise} = \frac{Q_{m,n}^{b} X_{m,n}^{b}}{Q_{s,n}^{b} X_{s,n}^{b} + \sigma_{m,n}^{b}} \tag{5.5}$$

The numerator indicates the received signal value at the BS from user m.

The resource allocation problem is solved by two kinds of approaches: (1) increasing the SINR's rate for the needed users, and (2) equality among users by implementing the proportionality fair (PF) function.

5.6 The WSRMAX approach-based MILP formulation
5.6.1 The optimization techniques used before providing priority to patients

This formulation is to maximize the systems over all SINR and individual users' SINR. The patient's risk factors are most preferred. This is used to compute needy patients over others. The formulation of the MILP model is mentioned in Eq. (5.6):

$$\sum_{m \in M} \sum_{n \in N} \sum_{b \in B} T_{m,n}^{b} UP_m \tag{5.6}$$

The above equation represents maximizing the weighted sum of the SINR. These computations are higher for patients compared to healthy users. The patient's proportion is calculated by the risk factor.

For the below-represented values in Eq. (5.7), all users hold equal priority.

$$UP_m = 1$$
$$\forall_{m \in M}$$

(5.7)

The patient's risk factors are converted into weights for patient prioritization over other users during calculating the after patients prioritization.

5.7 MILP formulation-probability fairness approach

5.7.1 The optimization techniques used before providing priority to patients

The objective of these approaches is to increase the arithmetic sum of the user SINR. The overall SINR decreases due to the nature of the natural logarithm (Hadi, Lawey, El-Gorashi, & Elmirghani, 2019). In resource allocation, there is no prioritization as all users are treated equally. The main consideration is each user's SINR and simplifying it as much as possible before the logarithm part is added. The optimization variable to act as the SINR for individual users is represented as S_m in Eq. (5.8).

$$S_m = \sum_{n \in N} \sum_{b \in B} T_{m,n}^b$$

(5.8)

Eq. (5.9) indicates the application of log to the SINR value of each user.

$$L_m = \ln S_m$$
$$\forall_{m \in M}$$

(5.9)

5.7.2 After patients prioritization

Here, the prioritization of patients is considered, and this is represented in Eq. (5.10).

$$L_m = \ln S_m$$
$$\forall_{m \in M} : 1 \le m \le NU$$

(5.10)

Where the log is applied to normal users and weights are indicated for needy patients, as represented in Eq. (5.11).

$$\sum_{m \in M, 1 \le m \le NU} L_m + \sum_{m \in M, m > NU} S_m UP_m$$

(5.11)

The functions in Eq. (5.11) indicate: (i) all users increase the sum of the SINR, (ii) the patient's priority is allocated by PRBs with maximum SINR, and (iii) the fairness of implementation by allocating patients with PRB compared to SINR. These three objectives are implemented by the sum of the healthy user SINR and the weighted sum of the maximum SINR.

5.7.2.1 Receiving power calculation

According to the channel conditions and the distance between the user and the base station, the received signal power is altered. Rayleigh fading and path loss based on distance are considered. The modified received signal energy based on the two considerations is represented in Eq. (5.12).

$$Q_{m,n}^{b} = PH_{m,n}^{b} A_{m}^{b} \qquad (5.12)$$

5.8 Heuristic approach

This is to provide the MILP operation and real-time results as well as to develop the PRB user's priority. In this heuristic approach, the data parameters and variables are initiated and the values of received power are read from another file. The system affects the user's admittance order while checking the priority of the users. If the priority mode is ON, patients in an emergency are served first. Initially, the users are allocated with high values of SINR, as there are a number of free channels and the heuristic is served sequentially. Determining the PRB at which the SINR is high is done by considering a PRB whose interference is a subset of $|B|-1$ interferers with the low interfering power of the user at that PRB. $|B|$ represents the total number of base stations. After allocating the PRB to users, the number of PRBs in total is equal to $(2*Z)$. Free PRB is equal to $[B*N]-[2*Z]$. Finally, a pool is created with a length equal to the free PRB. It comprises the high SINR values to be used on each PRB. The heuristic approach works under the Greedy technique. One SINR is chosen from the best SINR pool. This is done to achieve local equality among the users while assigning SINR to a patient and to maximize the individual's SINR when the overall system SINR is increased. The complete working flow of the heuristic approach is illustrated in Fig. 5.3.

The user and interferer are assigned a SINR and PRBs. The remaining users repeat the same procedure. In a sequential nature, this heuristic iteration was done 1000 times. The system under each iteration randomizes the user's admission order, so the semideterministic nature of the PRB assignment is maintained. The sensitive analysis results in a 95% confidence rate. The heuristic approach is applied over 100 files, with each file representing the different power value collected from the base station. The result can be obtained between the heuristic approach and the MILP model operation. System optimization results in increasing the overall SINR. This will result in an increase of the individual user's SINR value while the priority scheduling of needy patients takes place.

5.9 Results and discussion

The parameters used in the model to gain results are indicated in Table 5.3. A mobile network operating in an urban environment is taken into consideration for the model. The results are obtained in two phases: one method is providing priority to patients

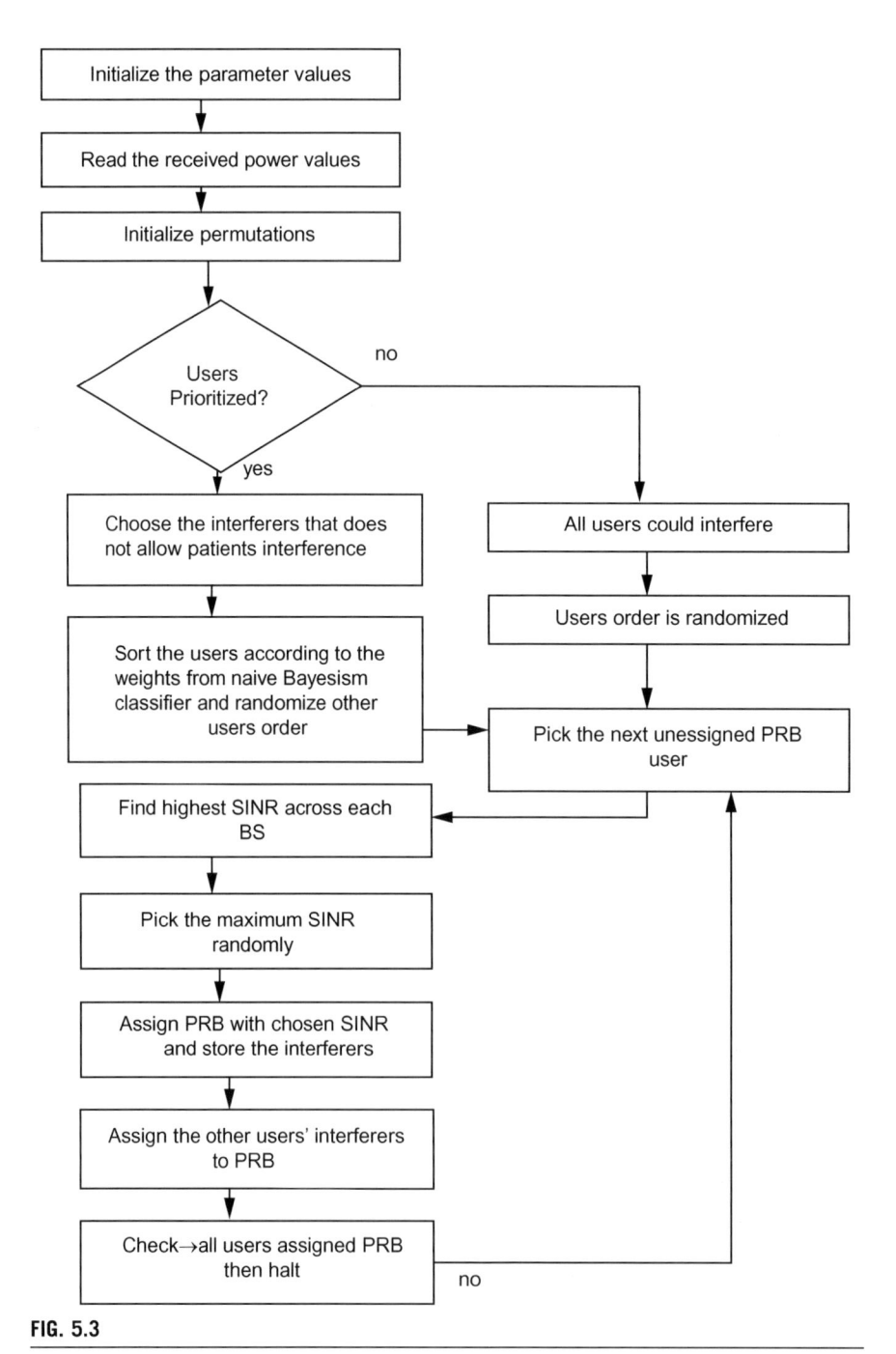

FIG. 5.3

Workflow of the heuristic approach.

Table 5.3 Parameters used in proposed model.

Parameters considered	Description about parameters
Long-term evolution-advanced network range	1.6 MHz
Model of the channel	Rayleigh fading [as suggested by Muñoz-Gea et al., 2014] and distance-based path loss
Base stations	2 in number
Physical resource block allocated per base station	4
Users	7
Health patients	4
Intensive care patients	3
Additive white Gaussian noise in communication	−162 dBM/Hz [as suggested by Muñoz-Gea et al. (2014)]
User and base station distance	200–500 m
Transmission power	23 dBm per connection [as suggested by Muñoz-Gea et al. (2014)]
User sensor transmission energy for a PRB	15 dBm
Weight for normal patients	1
Intensive care patients priority value	Naïve Bayesian result
Observation time period	30 days
Tuning factor values	50, 150, 200, 250, and 500

without utilizing big data and the other method is by utilizing big data to prioritize the patients, depending upon the risk factor and tuning factor calculations. The first method treats all users equally by assigning the same weights. In the proposed model, each patient owns his or her individual cloud for handling and processing their daily data from medical sensors. The dataset is maintained for individual patients for a month with added permissions to include extra observations. The update takes place daily.

The operational dataset and training system are considered and the patient's data are processed using the naïve Bayesian classifier to estimate the exact medical condition. After the dataset is processed, a classifier evaluates the probability of a heart arrest. The patients are continuously monitored and once the probability of a heart arrest is determined, the respective patients are favored. The probabilities of intensive care patients with heart arrest considered in the proposed model are 0.0036, 0.0068, and 0.00214. The training results are obtained by using 10-fold cross-validation in the model. The accuracy, precision, and recall values are calculated for all users. The training accuracies obtained are 65%, 67%, and 97% for users 1, 3, and 4. The accuracy values can be further refined in the future by utilizing other classifiers. The results of Eq. (5.4) to calculate the user priority values are $1.1 \leq UP_m \leq 1.3$, $1.5 \leq UP_m \leq 2.5$ for the tuning factor values 50 and 200, respectively.

5.9.1 The WSRMAX approach-based MILP and heuristic formulation

5.9.1.1 The optimization techniques used before providing priority to patients

Here, users are provided with equal weights according to the base weight (value = 1). The SINR values for all users are nearly equal to the average SINR and the heuristic values maximize the whole SINR. No big data analytics technique is used here. The observation is represented in Fig. 5.4. The average values for both MILP and the heuristic formulations taken are 5.2 and 5.5, respectively. To make the results more prominent, standard deviation (SD) is computed for all patient SINR. The results obtained are 0.3 for MILP and 0.2 for heuristic. The results show that the heuristic mode of SINR calculation is fairer than the MILP formulation. The average SINR for MILP is between 5 and 6 whereas for heuristic it is between 4.5 and 5.5.

5.9.1.2 After patient prioritization

Big data analytics-based patient prioritization using the naïve Bayesian classifier is used in this approach. The results are demonstrated in Fig. 5.5. This implies the priority assigned to the users based on increasing the SINR values of the needed patients. Because the SINR values are increased only for the intensive care patients and the average values of the SINR are reduced, this results in a trade-off in the performance of the system. This trade-off is because of the reduction in overall SINR value. It occurs because the SINR values of individual intensive care patients are increased by the PRB assignment method. From the results, it is clear that the heuristic method outperforms MILP by providing a higher SINR to the patients. The heuristic method usually performs in a sequential manner of serving the most emergency patients first, then prioritizing the rest of the users. Heuristic approaches perform well and operate in a sequential manner by arranging the SINR values that are to

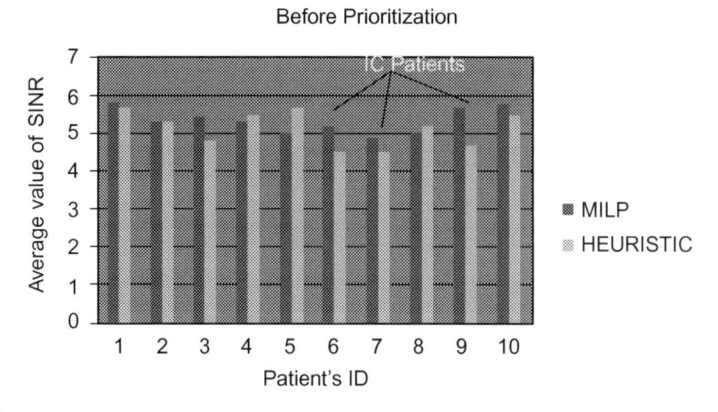

FIG. 5.4

Probability fairness approach before prioritizing patients.

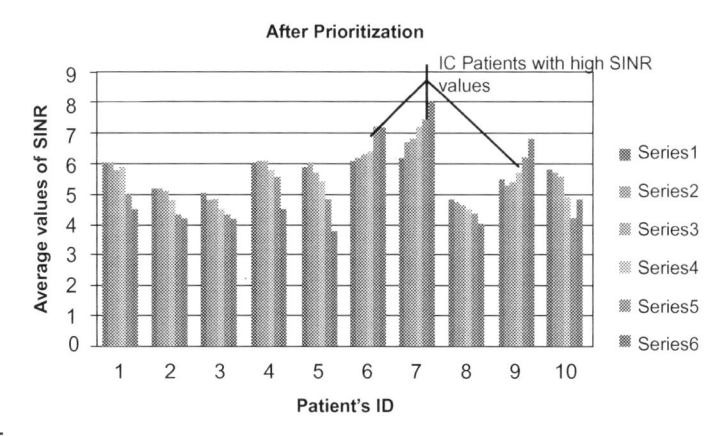

FIG. 5.5

WSRMax approach after prioritization.

be achieved in a list and picking a value from it randomly. The emergency patients with high SINR values are users 6, 7, and 9 in Fig. 5.5.

The results of Fig. 5.4 indicate that the average value of the SINR for MILP ranges from 5.2 to 5.5 based on the tuning factor α value, whereas for heuristic it is about 5. The probability of heart arrest is achieved by the α value to determine the risk factor. The SINR values vary depending upon the tuning factor values in the MILP method. For normal users like 1, 2, and 3, the SINR values are less when the tuning factor α is 500, whereas the SINR values are high when the tuning factor α is 50. So, depending upon the tuning factor, the risk factor is estimated in the MILP approach. For intensive care patients, the average SINR when the tuning factor α is 500 is higher compared to the tuning factor α as 50. When the tuning factor value α is 500, the average SINR values are between 4 and 7 in MILP and between 3.8 and 7.8 in heuristic.

5.9.2 Probability fairness approach
5.9.2.1 The optimization techniques used before providing priority to patients
The objective is to increase the logarithmic sum of the patient's SINR while ensuring user fairness without prioritizing some users. The results are depicted in Fig. 5.6. Applying log in the SINR values of each user helps in the reduction of overall average SINR of around 5% compared to the WSRMax model. The average SINR for MILP after applying fairness is 5.2 and for heuristic it is 5.1.

5.9.2.2 After patient prioritization
Here, big data analytics are applied to determine the likelihood of a heart arrest. Eq. (5.11) is utilized for calculation. The prioritized patients are awarded a higher SINR value than the WSRMax approach, as depicted in Fig. 5.6. The average SINR

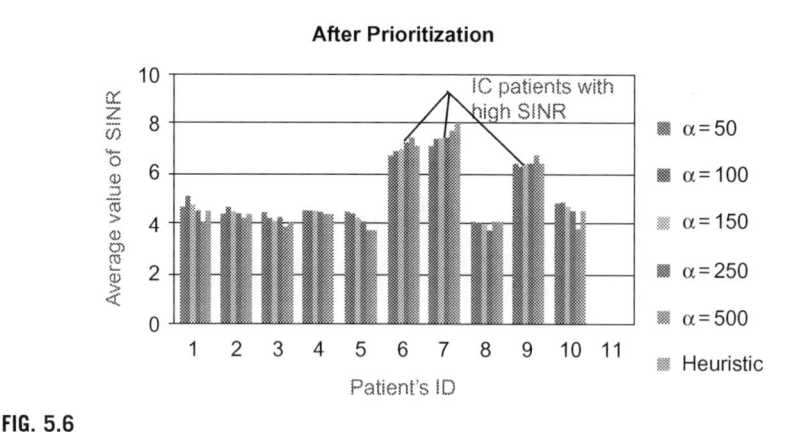

FIG. 5.6

Probability fairness approach after prioritizing patients.

value for the MILP approach based on the tuning factor $\alpha = 50$ is 5.1, whereas when α is 500, the SINR average value is 4.8. For the heuristic model, the average SINR value is 5.1. There is no increase in average SINR because of applying fairness in this approach. The errors obtained are also less when compared to the previous models.

5.10 Future directions

5.10.1 Choice of decision-making platform

Does the computing have to be performed in the cloud or fog? Solving this is an optimization problem that depends upon factors such as the number of patients considered, the number of resources to be shared, and the speed of processing. So the computing platform should be chosen based on the requirements and the platform should be chosen in an optimized manner to increase the accuracy of the results.

5.10.2 Ranking features and selecting the most optimized feature

The features used for determining the risk factor of the probability of heart arrest are treated equally. The features selected are not optimized. So, in the future scope, along with consultation with medical practitioners, the selected features are ranked and only the optimized features are included for training the model.

5.10.3 Integration with 5G

5G networks could be used in the future to minimize traffic and also to reduce the packet loss. The privacy of the data also has to be maintained. Security is achieved by ranking the patients according to the mode of treatment. The patients with high rank are the most intensive care patients and their records are preserved in a safer manner. To transfer the completely private data, a specific route could be utilized that is made possible and effective by 5G networks.

5.10.4 Infrastructure sharing

The patients are broadly located and the transfer of data is made through mobile networks. Mobile operators play a major role in sharing resources. When the amounts of data to be shared are large and when the number of patients is greater in a particular area, then the network infrastructure is shared among the mobile operators to serve the patients at an immediate effect. The mobile operators that share the maximum infrastructure could benefit with huge revenues from the healthcare industries.

5.10.5 Wireless drug injection

During critical scenarios, obtaining sudden treatment from a medical practitioner once the risk factor is determined is practically difficult. To overcome such problems, immediate medication for patients by injecting suitable drugs through body sensors could be integrated in the future. The data loss and delay have to be overcome while incorporating such a methodology.

5.11 Conclusion

Big data helps healthcare sectors change their usual ways of treatment to personalized care, thus achieving far-reaching goals such as improved efficiency, faster diagnoses and treatments, and better outcomes from medication. Healthcare organizations have changed as data-driven sectors due to the rapid alteration from simple reporting systems to deeper insights about new discoveries. The major aim of big data is to highly utilize the available information into actionable perceptions to avoid inefficient treatments and to provide patient happiness. The two major problems are elderly-based monitoring and patient-centric monitoring by the optimization of mobile networks using big data technologies. A platform using big data called MONISAN is utilized to estimate the risks of disorders among the elderly in real time and the impact that affects the mental conditions of the elderly as per the recorded dataset. This is a highly beneficial model for the elderly in both prevention and early treatment. In the latter approach, a system utilizing big data for optimizing the LTEA networks is considered. The patient's records are tracked and analyzed in a big data platform to detect the probability of a heart arrest in real time using medical sensors. The objective is to facilitate needy users with all resources to prevent the high risk of danger to the patients. The prioritized patients are marked with high SINR values, which in turn do not affect the SINR of other users. A real-time heuristic approach is used to verify the operations of MILP. Heuristic scalability is also obtained in this patient-centric model. Thus, big data plays a vital role in both approaches to improve the efficiency of healthcare sectors.

References

Bajenaru, L., Ianculescu, M., & Dobre, C. (2018). A holistic approach for creating a digital ecosystem enabling personalized assistive care for elderly. In *IEEE 16th international conference on embedded and ubiquitous computing (EUC)*.

Bajenaru, O., Tiu, C., Antochi, F., & Roceanu, A. (2012). Neurocognitive disorders in DSM 5 project—Personal comments. *Journal of the Neurological Sciences*, *322*(1–2), 17–19. https://doi.org/10.1016/j.jns.2012.07.067.

Bajenaru, O. L., & Custura, A. M. (2019). Enhanced framework for an elderly-centred platform: Big data in monitoring the health status. In *Proceedings—2019 22nd international conference on control systems and computer science, CSCS 2019* (pp. 643–648). Institute of Electrical and Electronics Engineers Inc. https://doi.org/10.1109/CSCS.2019.00116.

Ballon, M. (2013). *Number crunchers.* https://tfm.usc.edu/numbercrunchers/.

Baloh, R. W. (1992). Dizziness in older people. *Journal of the American Geriatrics Society*, 713–721. https://doi.org/10.1111/j.1532-5415.1992.tb01966.x.

Banu, N., & Swamy, S. (2016). Prediction of heart disease at early stage using data mining and big data analytics: A survey. In *2016 International conference on electrical, electronics, communication, computer and optimization techniques (ICEECCOT)*.

Conceição, G., Wouter, J., & Alise, J. M. (2018). Factors determining the success and failure of ehealth interventions: systematic review of the literature. *Journal of Medical Internet Research*, e10235. https://doi.org/10.2196/10235.

Deetjen, U., Meyer, E. T., & Schroeder, R. (2015). *Big data for advancing dementia research: An evaluation of data sharing practices in research on age-related neurodegenerative diseases.* Paris: OECD Publishing.

Dwolatzky, Y. B., Ruffle, B. J., & Zultan, T. R. (2018). *Time preferences of older people with mild cognitive impairment.* Laurier Centre for Economic Research and Policy Analysis. https://EconPapers.repec.org/RePEc:wlu:lcerpa:0115.

Elsawy, B., & Higgins, K. E. (2011). The geriatric assessment. *American Family Physician*, *83*(1), 48–56. http://www.aafp.org/afp/2011/0101/p48.pdf.

Eurostat. (2020). *Persons reporting a chronic disease, by disease, sex, age and educational attainment level.* http://appsso.eurostat.ec.europa.eu/nui/show.do?dataset=hlth_ehis_cd1e&lang=en.

Exalto, L. G., Quesenberry, C. P., Barnes, D., Kivipelto, M., Biessels, G. J., & Whitmer, R. A. (2014). Midlife risk score for the prediction of dementia four decades later. *Alzheimer's & Dementia*, 562–570. https://doi.org/10.1016/j.jalz.2013.05.1772.

Gantz, J., & Reinsel, D. (2011). *Extracting value from chaos* (pp. 1–12). IDC iView.

Gillespie, L. D., Gillespie, W. J., Robertson, M. C., Lamb, S. E., Cumming, R. G., & Rowe, B. H. (2003). *Interventions for preventing falls in elderly people.* Wiley. https://doi.org/10.1002/14651858.cd000340.

Glen, D., David, Z., Avraham, S., Tzvi, D., Howard, C., Howard, C., & Ely, S. (2005). Towards practical cognitive assessment for detection of early dementia: A 30-minute computerized battery discriminates as well as longer testing. *Current Alzheimer Research*, 117–124. https://doi.org/10.2174/1567205053585792.

Groves, P. (2016). The 'big data' revolution in healthcare: Accelerating value and innovation. *The McKinsey Quarterly, 2*.

Hadi, M. S., Lawey, A. Q., El-Gorashi, T. E. H., & Elmirghani, J. M. H. (2019). Patient-centric cellular networks optimization using big data analytics. *IEEE Access*, 49279–49296. https://doi.org/10.1109/access.2019.2910224.

Hale, T. M., Chou, W.-Y. S., & Cotten, S. R. (2018). *EHealth: Current evidence, promises, perils, and future directions*. Emerald Group Publishing.

Health at a Glance Europe 2012. (2012). *OECD*. https://doi.org/10.1787/23056088.

Herghelegiu, A. M., Nacu, R. M., & Prada, G. I. (2016). Metabolic parameters and cognitive function in a cohort of older diabetic patients. *Aging Clinical and Experimental Research, 28*(6), 1105–1112. https://doi.org/10.1007/s40520-015-0515-0.

Ianculescu, M., Alexandru, A., & Gheorghe-Moisii, M. (2017). Harnessing the potential of big data in Romanian healthcare. In *Vol. 2017. Proceedings—2017 5th international symposium on electrical and electronics engineering, ISEEE 2017* (pp. 1–6). Institute of Electrical and Electronics Engineers Inc. https://doi.org/10.1109/ISEEE.2017.8170630.

Ianculescu, M., Alexandru, A., & Tudora, E. (2017). Opportunities brought by big data in providing silver digital patients with ICT-based services that support independent living and lifelong learning. In *International conference on ubiquitous and future networks, ICUFN* (pp. 404–409). IEEE Computer Society. https://doi.org/10.1109/ICUFN.2017.7993817.

Joanne, K., & Lorraine, T. (2002). A clinical practice guideline approach to treating depression in long-term care. *Journal of the American Medical Directors Association*, 103–110. https://doi.org/10.1016/s1525-8610(04)70422-3.

Katz, S., Downs, T. D., Cash, H. R., & Grotz, R. C. (1970). Progress in development of the index of ADL. *The Gerontologist, 10*(1), 20–30. https://doi.org/10.1093/geront/10.1_Part_1.20.

Kiran, P., Jibukumar, M. G., & Premkumar, C. V. (2016). Resource allocation optimization in LTE-A/5G networks using big data analytics. In *Vol. 2016. International conference on information networking* (pp. 254–259). IEEE Computer Society. https://doi.org/10.1109/ICOIN.2016.7427072.

LaMorte, W. W. (2017). *Using spreadsheets in public health*. School of Public Health, Boston University.

Lawton, M. P., & Brody, E. M. (1969). Assessment of older people: Self-maintaining and instrumental activities of daily living. *Gerontologist, 9*(3), 179–186. https://doi.org/10.1093/geront/9.3_Part_1.179.

Marcello, I., Effy, V., & Alessandro, B. (2018). Big data and dementia: Charting the route ahead for research, ethics, and policy. *Frontiers in Medicine*. https://doi.org/10.3389/fmed.2018.00013.

Marinescu, I. A., Bajenaru, L., & Dobre, C. (2018). Conceptual approaches in quality of life assessment for the elderly. In *Proceedings—16th International conference on embedded and ubiquitous computing, EUC 2018* (pp. 111–116). Institute of Electrical and Electronics Engineers Inc. https://doi.org/10.1109/EUC.2018.00023.

Mazhar, R. M., Awais, A., Anand, P., Jiafu, W., & Daqiang, Z. (2016). Real-time medical emergency response system: Exploiting IoT and big data for public health. *Journal of Medical Systems*. https://doi.org/10.1007/s10916-016-0647-6.

Miia, K., Tiia, N., Tiina, L., Bengt, W., Hilkka, S., & Jaakko, T. (2006). Risk score for the prediction of dementia risk in 20 years among middle aged people: A longitudinal, population-based study. *The Lancet Neurology*, 735–741. https://doi.org/10.1016/s1474-4422(06)70537-3.

Muñoz-Gea, J. P., Aparicio-Pardo, R., Wehbe, H., Simon, G., & Nuaymi, L. (2014). Optimization framework for uplink video transmission in HetNets. In *Proceedings of the 6th ACM mobile video workshop, MoVid 2014*Association for Computing Machinery. https://doi.org/10.1145/2579465.2579467.

Năstase, L., Sandu, I. E., & Popescu, N. (2017). An experimental evaluation of application layer protocols for the internet of things. *Studies in Informatics and Control*, *26*(4), 403–412. https://doi.org/10.24846/v26i4y201704.

National Institutes of Health. (2009). *Your guide to lowering your blood pressure with DASH.* Smashbooks.

Popescu, B. O., Toescu, E. C., Popescu, L. M., Bajenaru, O., Muresanu, D. F., Schultzberg, M., & Bogdanovic, N. (2009). Blood-brain barrier alterations in ageing and dementia. *Journal of the Neurological Sciences*, 99–106. https://doi.org/10.1016/j.jns.2009.02.321.

Rao, S. S. (2005). Prevention of falls in older patients. *American Family Physician*, *72*(1), 81–94.

Robert-Andrei, V., Ciprian, D., Lidia, B., & Radu-Ioan, C. (2019). Human physical activity recognition using smartphone sensors. *Sensors*, *458*. https://doi.org/10.3390/s19030458.

Rudyar, C., Xavier, B., Olivier, M., & Pierre, S. (2015). Stream processing of healthcare sensor data: Studying user traces to identify challenges from a big data perspective. *Procedia Computer Science*, 1004–1009. https://doi.org/10.1016/j.procs.2015.05.093.

United Nations. (2017). *World population prospects: The 2017 revision.* https://www.un.org/development/desa/publications/world-population-prospects-the-2017-revision.html.

World Health Organisation. (2018). *Noncommunicable diseases.* https://www.who.int/news-room/fact-sheets/detail/noncommunicable-diseases.

Zheng, K., Yang, Z., Zhang, K., Chatzimisios, P., Yang, K., & Xiang, W. (2016). Big data-driven optimization for mobile networks toward 5G. *IEEE Network*, *30*(1), 44–51. https://doi.org/10.1109/MNET.2016.7389830.

Emergence of decision support systems in healthcare

A. Reyana[a], Sandeep Kautish[b], and Yogita Gupta[c]

[a]*Department of Computer Science and Engineering, Hindusthan College of Engineering and Technology, Coimbatore, Tamilnadu, India,* [b]*LBEF Campus, Kathmandu, Nepal (In Academic Collaboration with APUTI Malaysia),* [c]*Thapar Institute of Engineering and Technology, Patiala, India*

6.1 Introduction

6.1.1 Overview

Clinical decision support systems (CDSS) impact an individual patient's decision making, with computer systems preventing medical errors. Following growth, these systems are commercialized, reshaping the environment in both negative and positive ways, attracting people to social media and business intelligence. Clinical data are captured and analyzed in the emerging global society for political, data-driven, economic, and social environment wellness. The author (Ando, 2000) has designed maximum consistency gradient operators that can be utilized in medical image processing and analysis exhibiting consistent information. The gradient direction in the pattern is smooth for invariant rotations. Also, the method is easy to implement with less computation time as well as superior accuracy, bandwidth, and isotropy. The decision support analytics by the author Kautish and Thapliyal (2012) and Kautish (2013) will speculate on future health-related societal changes. Due to people's real-time interactions with information technologies, there has been an increase in knowledge and sharing on health data, which medical practitioners capture for detailed analysis. Many clinical organizations show significant changes in the way they capture, store, and retrieve health-related information. Clinical data are captured and analyzed to formulate hypotheses for testing and speculating on long-term consequences. Developing a data-driven clinical DSS necessitates the use of analytics tools, implying a larger technological change.

Most organizations depend on cloud storage for backup. There are greater technology changes with the development of the clinical support system. The cycle of the CDSS consists of four components: (i) capture and storage, (ii) data analysis, (iii) decision support system, and (iv) decision-making system. Each time, there is a technology update in the system (Iyer & Power, 2014).

6.1.2 **Need for CDSS**

Recent research by Sim et al. (2001) has proven that a CDSS optimizes and facilitates disease diagnosis, enabling sustained outcome-based healthcare as shown in Fig. 6.1. The CDSS facilitates evidence-based medicine in improving the quality of healthcare. The author's recommendation was to capture the literature, practice, and evidence-based knowledge interpretation of the machine. Further, the evaluation of the CDSS based on cost and effectiveness was done, establishing public policies in improving healthcare quality. Today, hospital administers and clinicians face a number of challenges for a better healthcare delivery system, as shown in Fig. 6.1. Diagnosis, treatment planning, and follow up are continuous processes.

The ease of clinical decision systems can be leveraged by the use of augmented intelligence enabling standardized patient management systems. These systems help to determine the optimal treatment and care for patients supporting future reviews. As described by Axiomes et équations fondamentales du traitement d'images (Analyse multiéchelle et edp) (1992), the author has transformed the image processing requirements into three categories such as architectural, stability, and morphological requirements. The classification yields new models in projection invariance for shape recognition. The technology supports informed decisions considering the key factors of the patient for guided therapy. Therefore, it provides the benefits and risks for diagnostic testing for recommending treatment options. As specified by Miller (1990), and Miller (1994), the CDSS (as shown in Fig. 6.1) is essential to deliver a value-based system ensuring patient safety and avoiding complications. The use of clinical decision support systems is the over time, according to research findings, whether the diagnostic system is easy or complex. The development of EHR has led to the implementation of large-scale diagnostic systems that have become an essential component in medical technology.

Clinical decision support tools from Osheroff et al. (1991) integrated with EHR improve quality and care. CDSS not only supports doctors and nurses but also patients and caregivers too. Patients get medication instructions, dietary guidelines, and home management tips. Hospitals are quick in applying these imperatives, for example: Alabama hospital has decreased 53% of sepsis mortality through their

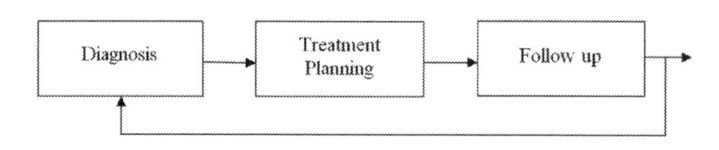

FIG. 6.1

Handling patients using CDSS.

No permission required.

computerized surveillance algorithm. Additional areas where clinical decision support tools could be beneficial include:

- Drug dosage calculation
- Drug formulation
- Severity index of a specific illness
- Collecting samples for a specific illness
- Time-triggered recommendations on medication and dosage
- Preventive care measures
- Educational materials and resources

6.1.3 **Types of CDSS**

CDSS can be potentially supported by a variety of systems. Recent research by the author Kantardzic (2003) proposed data mining approaches supporting the analysis of clinical data as well as for developing guidelines and protocols for decision making. CDSS differ based on the support timing, that is, before, after, or during the medications, decisions are made. It involves how actively or passively it alerts the patient or physician. The next categorization is based on knowledge and nonknowledge systems employing approaches of pattern recognition and machine learning. The approaches for CDSS are discussed below.

Knowledge-based CDSS: A knowledge-based system builds computer programs capable of simulating human thinking. Many developers have adapted the authors' (Forsythe, Buchanan, Osheroff, & Miller, 1992) medical care domain to promote patient care in real life. The intent was to assist medical practitioners in their decision making. It provides information to the user having appropriate expectations and actively interacting with the system. The parts of the knowledge-based system are:

1. Knowledge base.
2. Reasoning engine user communication.

The knowledge base has compiled information in the form of rules designed to prevent duplication testing (Berner, 2007). The reasoning engine has the association to combine rules of patient data. Finally, user communication will give the actual decision from the system. Here, the initial data are entered in electronic form by practitioners, lab technicians, and pharmaceutical systems. These systems assist in diagnosis for users, starting from a patient's symptoms and generating possibilities from the available information, as shown in Fig. 6.2. The other choices from the set of diagnoses are eliminated by the medical practitioner based on the patient's health history.

Input: The user needs to select a standard clinical vocabulary to enter information into clinical systems. The author Kantardzic (2003), in his data mining concepts and model, specified that the goal of developers is to develop a CDSS (shown in Fig. 6.2) operated in natural language providing patients a highly

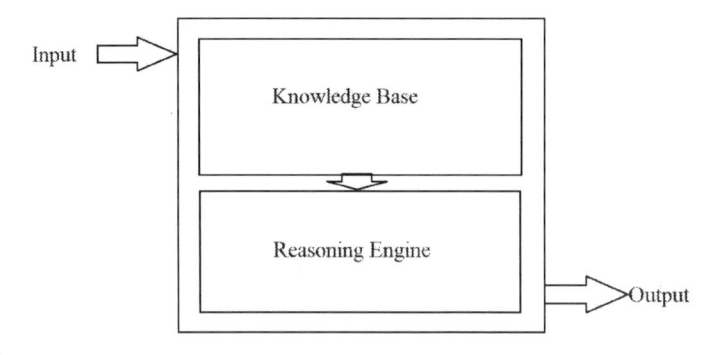

FIG. 6.2

Decision support system.

No permission required.

comfortable and desirable system. Here, the user input is restricted to vocabulary terms that the system accepts. These terms include laboratory examination results, patient history, drug dosage, etc. Certain systems make assumptions on a disease such as acute, subacute, or chronic based on the symptoms and their levels described. Reasoning Engine: This combines the input and related data to obtain conditional probabilities. These are domain-specific. Knowledge Base: Matching the design of the reasoning engine to built medical knowledge. Output: Possibilities ranked in the probability order required to diagnose a disease.

Nonknowledge-based CDSS: These systems use a machine learning approach to learn patterns from clinical data based on past experiences. The two types of algorithms used are artificial intelligence and a genetic algorithm. The artificial algorithm has three layers: data receiving, determining results, and communicating results. The patterns obtained from patient data are associated with the disease symptoms for diagnosis. The systems are limited to a single disease. The process of verifying the corrected output from a large amount of data is network training. Once network trained, the systems can be used on new cases of the same disease type. The system is dynamic, eliminating direct input from experts. However, the training process is time-consuming and less reliable. Despite these drawbacks, they are applied in various applications such as dementia, psychiatric treatments, sexual diseases, etc.

The genetic algorithm is recombination, which allows organisms to evolve faster than their early pioneers. The solutions are to be considered or eliminated depending on the fitness function. Most of the systems are developed using knowledge-based CDSS.

6.1.4 **Effectiveness and applications of CDSS**

The clinical support system reduces caring costs and improves patient outcomes. Because the system alerts physicians to dangerous diagnoses and drug dosages, it minimizes and prevents severity and complications.

The various factors indicating successful implementation are:

- Automatic alerts
- Timed suggestions
- Recommendations for suitable action computerization

However, some implementation challenges exist, including:

- The entry of actual data to the system.
- Adding a user query to the system.
- Manual entry of data.
- Double data entry.
- Integration of CDSS to potentially act on the data.
- Notification on errors.
- Use of universally agreed clinical vocabulary.

Despite these challenges, this has proved to be a significant improvement in quality care. Health administrators and patients use technology to reduce medical errors. Many applications such as drug interaction databases, disease diagnostic systems, etc., use these systems.

6.2 Transformation in healthcare systems

6.2.1 Adoption of CDSS

The issue in the healthcare system is the lack of knowledge in the adoption of a CDSS. Today's environment creates opportunities for new developments in CDSS. Earlier biomedical informatics systems have evolved other methods over the past 50 years related to:

- Information retrieval.
- Conditional logic evaluation.
- Probabilistic classification or prediction.
- Data-driven classification or prediction.
- Artificial intelligence.
- Calculations, algorithms, and multistep processes.
- Visualization and cognitive support.
- Associative relationship-based grouping.

However, many healthcare departments have adopted CDSS for its users based on circumstances such as:

- Alerts/reminders.
- CPOE.
- Medication/dosage support.
- Drug interaction checks.
- Order sets structured data entry forms/templates.

Where this system has shown its efficiency is in quality improvement, cost, and error reduction. In the current decade, CDSS is deployed for EHR adoption, precision medicine, genomics, sensors, technology advances, and knowledge explosion. An aging population and increasing care complexity require care coordination emphasizing wellness, disease prevention, and runaway costs. Pay for what you get in terms of value and performance. These structures have also been drawn in by the empowered user, App culture, and expectation.

6.2.2 Key findings

From various research studies after CDSS implementation, there has been a decrease of 55% in medication errors, a decrease in redundant laboratory tests, appropriate renal dosing, and greater cost-effectiveness.

6.2.3 Enterprise-level adaptation

The major challenge in the enterprise level stated by Bace (Ando, 2000) is to require a major commitment to knowledge management. While there are widely accepted and accessible resources for knowledge management, there is still a need to manage the life cycle from narrative to structured and executable, as well as monitor adaptations and make updates easier. Communal information will continue to require adaptation by and for the local site, including the incorporation of business logic and constraints that represent workflows, integration into proprietary electronic health records, and coordination of internal processes for human-readable knowledge that is compatible with executable versions embedded in vendor systems. CDSS is a service embedded in EHR. The challenges faced in enterprise-level adaptation are listed below and their rule mapping is shown in Fig. 6.3 (CDS by Greenes, 2017b).

- User languages.
- Limitations in sharing.
- Poor design leading to fatigue and useless alerts.
- Adaptation to triggers, workflows, and staffing.
- Time considerations.
- The same rule and single instruction with multiple variations.
- Similar intent with multiple rules.
- Triggering/identification modes.
- User login maintenance.
- Accuracy in laboratory test results.
- An open chart on the latest findings.
- Inclusions/exclusions inpatient list.
- Periodic panel search.
- Patient-specific interaction modes.
- Data availability/sources/entry requirements.
- Changes in user settings.

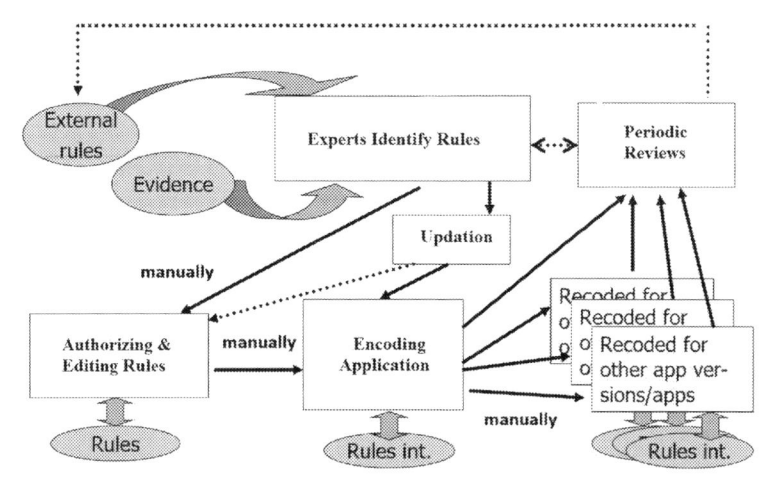

FIG. 6.3

Mapping of rules.

- Constraints and threshold values.
- Actions/notifications.
- Exceptions refusal/losing follow up.

6.2.4 Health IT infrastructure

The health IT infrastructure focuses on wellness, early intervention, and prevention from diseases. Person-centered caring equipment is introduced, including wearable sensors, connected care, genomes, practitioner-patient interaction applications, etc. However, challenges still exist in continuity and coordination, the integration of big data analytics, increased population growth, and the emergence of new diseases. Many new approaches to implementation science are being developed, including knowledge management, e-Health discussions, context awareness paradigms, and rule primitives. CDSS was primarily implemented by specifically emphasizing workflow embedding, using event-driven triggers to invoke it when appropriate conditions occur, customization for SSF adaptation, maintaining a repository of knowledge artefacts indexed by metadata relating to appropriate context, and precoordinating any possible variation of knowledge artefacts for every workflow. The cognitive support uses a logic model to identify other elements that may indirectly impact the focused area to display context-relevant tasks, data needed, possible interventions, and recommendations, as shown in Fig. 6.4 (CDS described in the work of Greenes (2017a)). Klosgen and Zytkow (2002) identified the major problem in the support vector machine that decides the kernel function for a specific task and dataset. Compared to other machine learning algorithms, the SVM focuses on increasing the generalization depending on the risk and machine complexity. The results revealed that SVM using a linear kernel will be the best choice to analyze a diabetes dataset.

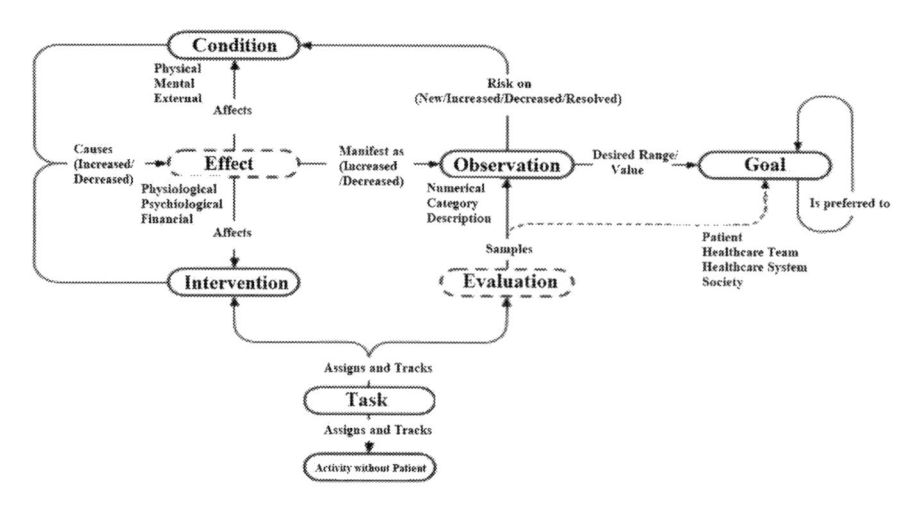

FIG. 6.4

Logical model of a decision support system.

No permission required.

Cai, Wang, Li, and Liu (2019) described the techniques used for decision making in smart healthcare systems. Mobilization in multimodal healthcare data has rapidly increased the existence of clinical data on Internet platforms. These are high-precision, real-time data used for medical services. Data collection can be obtained from a centralized data warehouse such as EHR, wearable devices, installed applications, online forums, etc. Today, IoT-enabled data collection plays a major role in the healthcare industry. Multimodal data-driven applications are becoming an essential technical carrier in terms of consultation, diagnosis, and treatment. When using clinical data, protection and privacy concerns must be addressed.

The healthcare landscape is evolving coordinated care delivery systems where user groups, consortia, professional societies, and enterprise health care systems provide a basis for sharing knowledge as well as implementation, as shown in Fig. 6.5

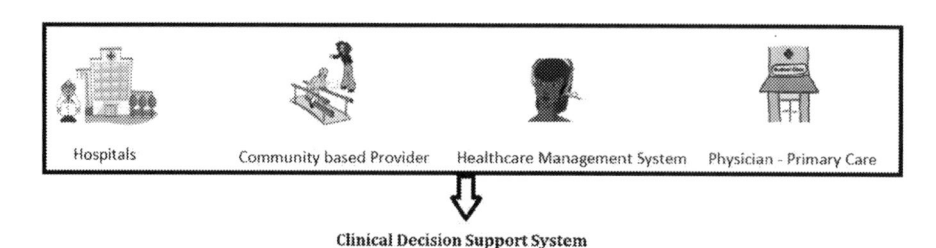

FIG. 6.5

Coordinated care delivery system.

No permission required.

(CDS by Greenes, 2017a). The author has described the support systems that can be incorporated in various domains such as hospitals, community-based providers, healthcare management systems, and physician clinics offering the best-quality healthcare to the public.

6.3 CDS-based technologies

6.3.1 Supervised learning techniques

In supervised learning, the user is aware of the available classes ahead of time, which is referred to as guided data mining. The dataset sample is transferred to the system as training samples consisting of dependent/independent variables. The difference between the desired to actual responses is the error signal. Minimization in error reduces discrepancies, providing accurate pattern recognition. The model predicts the behavior of an entity.

The classification of supervised learning is described below.

6.3.1.1 Decision tree

The decision tree is a graphical representation and a simple way of supervised learning. A simple decision tree adopts a top-down strategy with minimum attributes, as shown in Fig. 6.6. In each node, the predictor attributes are tested and the value of each attribute determines the target variable. The partitioned process is recursively applied to each member of the subgroup.

6.3.1.2 Logistic regression

This models the data into the dependent binary variable that takes the probability value 1 as success and 0 as a failure. The success probability is considered as p and failure probability as 1-p. The binary variable is independent and is in the form of supervised learning. These depict the variable interaction and predict the

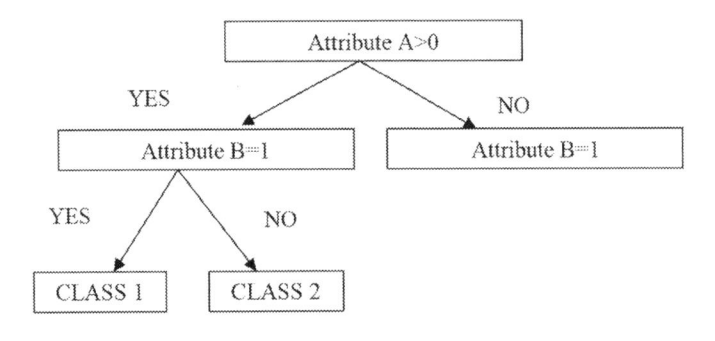

FIG. 6.6

Simple decision tree.

logistic regression estimation for a given value. The probability function Eq. (6.1) is given as:

$$P(y) = 1/(1 + e(-a - bx)) \qquad (6.1)$$

6.3.1.3 Neural networks

This helps to recognize patterns from a large set of neurons having local memory and unidirectional connection. The neuron takes more than one input and operates on connections upon receiving its input. The network generates output from the actual feedback. Adjust contributes to an incorrect parameter since the output varies from the actual value. When the parameters are stabilized, the process terminates. The representation of a training dataset and its size shall refine the training set.

6.3.2 Unsupervised learning techniques

Here, the system is represented with a no information dataset, that is, information not available to group the data into meaningful classes. The system develops classes of defined patterns where there is no target variable and all variables are treated as the same in Fig. 6.7.

During cluster analysis, an unsupervised situation classifies a diverse set of data into various groups. The cluster analysis describes methodologies to measure perceived similarities among the multidimensional data. The identification leads to better understanding and the characteristics provide classifications for further relationships. Clustering enables consistency, speed, and reliability. Despite the availability of vast data collection, the two common approaches followed are hierarchical clustering and nonhierarchical clustering.

For example, on the genome project, about 30,000 gene sequences were analyzed from tissue, blood, and body fluid samples using the gene expression data analysis algorithm. These were classified into various disease sets and the prediction of therapy was done from them. Gene-expressed data discover the unknown categories and the samples are categorized based on similarity. The method grouped into cluster similarity is based on the knowledge of any predefined classification.

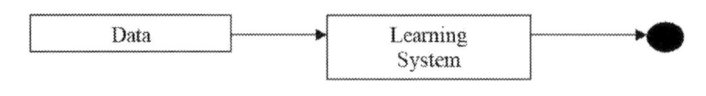

FIG. 6.7

Unsupervised learning system.

6.3.3 **Disease diagnosis techniques**

6.3.3.1 Domain selection

In order to facilitate disease diagnosis, a knowledge-base framework must be established in an academic setting. Due to the inadequate availability of domain expertise and funding, many projects do not survive. Considering real-world demands, the clinical expertise is adequate and most of them earn their wages through patient care. Commercial vendors hire qualified staff or physicians to maintain their knowledge base. The income generated through the number of users shall update and scale up the maintenance department. Different problems narrow the audience. The problem with successful experimentation is that it requires specialized systems and people to develop the system. The choir includes glaucoma specialists and ophthalmologists who diagnose glaucoma on a regular basis. Programs including CASNET will find use in clinical practice and are extremely robust.

6.3.3.2 Knowledge base-construction

The knowledge database is hard to maintain when the first reports of new diseases arrive, and these require a confirmation from other groups before they can be added to the knowledge base. The current understandings of disease change from patient to patient, resulting in an unrecognized form of illness. Hence, the knowledge base must be kept up to date. If the knowledge base construction depends on an individual or institution, then the system's survival is difficult.

6.3.3.3 Algorithms and user interface

Complex algorithms involve many trade-offs for a real-world diagnostic system and involve the practical ability for the system to create user responses. These complexities make the system reliable. However, building such types of systems for patient care is not practical. Hence, most clinicians purchase machines to run these applications and use them successfully.

6.3.4 **CDS-related issues**

Even knowing the signs and symptoms, an accurate diagnosis is difficult. This requires experience and great knowledge. Patient-nurse/physician experiences can only be trusted if professional practices are followed. The decision support system follows ethics to perform decision making. From history, it is known that health computation raises ethical issues, and these threaten physicians. The standard view speaks on technological changes, the risks, and standards in terms of ethics, computing efforts, and medicine. Patients are seen as helpless and should be respected. Each physician/nurse should do his or her best in providing treatments irrespective of economic concerns. For this, ensuring computing tools meet such requirements is a must. Failing to meet the responsibility increases error risks and questions on ethical standards. New evidence causes changes leading to risk in invoking ethical and legal issues. We sometimes deviate from norms for good reason, such as to demonstrate clinical improvement that justifies norm modifications. Only with an accurate decision can optimal treatments be given. Some patients are treated for an incorrect diagnosis, increasing patient risk. Using a tool without adequate training causes shortcomings.

The authors Fried and Zuckerman (2000) have stated that the use of tools for other than their intended purpose, a lack of training, and careless use are called usage in an unintended context. A tool designed for one purpose has fewer chances to work on another purpose, and this is also unethical. Ethically, the use of CDSS assures its success in many models, including genetics, organ transplantation, and other health science diagnoses. It is important that only qualified professionals are allowed to use new tools. The challenge still exists in identifying qualified healthcare professionals to use DSS in healthcare. The novice fails to understand the system error or flawed output during its operation. The other ethical focus is using support software properly. This shall be considered as an area with a better scope for future research. Medical decisions are also of ethical significance in terms of quality assessment, practice evaluation, and professional service.

Today, as stated by Goldschmidt (2006), patients and physicians are considered partners in assessing the computer system. The two areas of concern are the conceptual and interpersonal distances between the patients and the physicians. Communication uncertainty falls on the diagnosis of whether the result is valid or reliable. Physicians/nurses can eliminate some of these tensions by not disclosing the use of decision support software to patients. But is this ethical? As a result, the emphasis is solely on the complexity and communication techniques employed. Providing inappropriate data to the patient has to be restricted. Hence, healthcare professionals must provide quality care and compassionate communication on illness diagnosis.

6.4 Clinical data-driven society
6.4.1 Information extraction

A unique challenge in CDSS is dealing with unstructured data. The decision-making process requires information extraction from these unstructured/semistructured data. The knowledge based-information extraction methodologies are described here to support efficient decision making. In text mining, components such as preprocessing, pattern discovery, and visualization are detailed to extract text from unstructured language. Here, the objective is to transform stored data into an intermediate form where analytical techniques applied. This helps in the identification of classes/events/relationships relevant to the event. The various phases of information extraction are: Named entity recognition, coreference task, template element production, template relation production, and template extraction.

The named entity shall extract the noun phrases of the domain. The CO task identifies the occurrence of the same entities. The third shall associate the attributes to the entities extracted. Finally, the context-dependent relations between entities are identified to express domain and task-specific entities/relationships. The information extraction is processed following: (1) knowledge-based and (2) machine learning/statistical approaches. Ontologies play a vital role in information extraction acting as a:

- Guide
- Repository
- Representation scheme
- Basis for reasoning

Ontology detects entities and attributes for creating noun resolutions. This is otherwise called semantic annotation. Ontology limits the span of predefined texts and semantics. The two types of ontology presented in the literature are: (i) lexicon-based and (ii) thesauri-based. Both provide aliases for processing domain-specific resources. They contain a sequence of lemmas and token strings matched with the enlisted ontology. The thesauri-based lists the possible aliases for implicit semantic information.

6.4.2 CDS today and tomorrow

General drug information to assist patients along with symptom management and limited diagnosis are provided by today's decision support system. The health risks are collected in terms of questionnaires of patient behaviors and family records. This enables one to prioritize health goals. These are provided as a web service by many organizations. The patient undergoes an online assessment in the various health portals. The decision support system pursues a diagnosis and provides advice to the patient on when to seek treatment. Nowadays, scheduling appointments and checking the availability of practitioners worldwide are being done by these systems. Today there are a number of health portals to calculate body mass, pressure, and pulse rate as well as sites to monitor pregnancies that are easily amenable to patients. The usual approach of knowing symptoms paired with a list of possible causes is now guided to various links for further reading. Diagnosis, treatment options, drug descriptions, and pros and cons are the available options. There are also online video options to educate patients on their necessity. This approach doesn't diagnose or provide recommendations on their treatment. Rather, it prepares them on the consequences and concerns that have to be addressed during their next visit to the clinician. The future is focused on treatment decisions and health outcomes after undergoing treatment. Further, this shall be extended by showing a video of their therapy or treatment to understand their health outcomes and make them participate in shared decision making. The organization FIMDM has provided tools for coronary artery disease, breast cancer, and knee osteoarthritis as well as for reproductive conditions. Treatment decisions may often be severe for sensitive patients, so they must be considered an integral part of such processes. The DSS package consists of a broad range of problem flowcharts that are in alphabetic order for body function diagnosis. The home medical analyzer provides a program for analyzing single and multiple symptoms using rule-based algorithms. The analysis is categorized as disease match and diagnosis. The drug interactions provide user feedback on possible interactions and patient medical packages with complex diagnostic features. These are designed by expert clinicians for the consumer to give answers. Once inputs are given, the system ranks the disease disorders and symptoms based on likelihood. The software also suggests whether there is a need for physical examination/tests. The diagnosis is based on the historical contributions of the patient. The criteria guidelines relevant to information accuracy and relevancy support useful accessibility on the quality of the health application system. Most web-oriented materials are tailored by individuals to replicate the information found in textbooks promoting healthcare-based services. Sutton et al. (2020) augmented the healthcare providers on (i) patient safety in reducing medication errors and adverse effects, (ii) clinical management as well as guidelines on review follow-ups and treatment procedures, (iii) cost containment to reduce test duplication and workload reduction, and (iv) diagnosis support in providing automated test results. Other factors include documentation, improvement in workflow, etc. These considerations shall reduce the pitfalls on resource wastage, quality care, etc.

6.5 Future of decision support system

The advancements in information technology have changed the way of practicing medicine. The impact of health information and medical care interactions empowers a patient's active role in decision making. To achieve significant improvements, researchers have gained knowledge based on the previous behavior of the system. This has been commercialized rapidly, providing real benefits showing healthcare effectiveness. However, the deployment of these systems requires a critical check on consumer health IT. A careful assessment of development as well as usability testing provide the best assistance to patients. Khodaei, Candelino, Mehrvarz, and Jalili (2020) described growing fields such as the artificial pancreas and automated anesthesia. The concept of feedback control in regulating the physiological variables such as blood glucose, blood pressure, and anesthesia was considered to assess the treatment. Incorporating closed-loop medical devices and control systems will be the future practice. However, all these systems will maintain patient data in the electronic health network. The remote treatment and data consistency will open new medical treatment challenges.

6.6 Example: Decision support system

Diagnosis is placing an interpretation over a set of primitive observations. Skilled practitioners will have an understanding of how the illness began and what has affected the patient. The practitioner should also know how the patient responds to the illness. Therefore, the diagnosis is a set of tasks: evolving illness history, physical findings, and integrating information to evaluate hypotheses concerning the patient's therapy and illness. These start from questionnaires and end in diagnosis. The diagnosis may be evaluated differently on the same patients based on the experts.

It is possible that these are the results of asking different questions and taking the same actions. Some of the described components involve both established and unknown elements in problem-solving formations. Experts have related the difficulty of gaining medical knowledge in machines for certain specific cases. Clinicians using decision support systems are slow in recognizing the diagnostic system functionality. Inexperienced users may have unrealistic expectations and frustrations. Expecting a DDSS with minimal input to solve a vague problem is unrealistic. Every illness has one or more related disorders that enable physicians to frame the training set and guess the mismatch between the user and the system functionalities. Hence, it is the mind of the clinician that solves the problem. If the diagnostic problem is globally connected, then information from all patient findings shall lead to a new diagnosis. Before the entire diagnosis, clinicians cover them, evaluate them portion by portion, and overcome the difficulty through assistance. In the current decision support systems, the user interaction is iterative and the user needs to understand the system assumptions on the entire patient information. Treatments are initiated before the completion of the entire diagnosis. The real risk exists in the elicitation of patient data ethically and economically. Maximum diagnostic samples are stored and maintained in the medical database for future comparisons. Careful diagnoses involve gathering every piece of information for each individual and justifying appropriate therapies after evaluation. This knowledge and ability to work upon findings and generate disease hypotheses are otherwise called disease diagnostic reasoning.

The other type is human diagnostic reasoning involving cognitive activities such as:

- Information gathering
- Pattern recognition.
- Problem solving
- Decision making
- Judgment and uncertainty

Human diagnostic reasoning is:

- Probabilistic
- Casual
- Deterministic

However, as described by Herrmann (model 2006), diagnosticians look for the existence of probabilistic concepts during reasoning. In deterministic models, appropriate actions are considered basic building blocks to solve human problems. Other compiled knowledge is obtained from if-then rules, logical flow charts, and algorithms. The greatest challenge is that production rules are not effective to deal with uncertainty. Focused decisions are usually deviated from by a patient's ill-structured problems and vague and undifferentiated complaints. Therefore, one has to use common sense and a few computer-based systems to make meaningful progress.

This has to be taken into account when designing a DSS by Herrmann (2006) for it is useful both for novices and experts. Due to anticipation from experts, they will be able to manually modify schedules to prevent them from further problems. Therefore, the possibility of manually modifying the schedules must be present in the system. The unstructured cognitive domain can function only with a huge amount of relative knowledge. The human psychological knowledge and problem-solving behavior can be included. Rather than novices, the medical literature has published on identifying trends. Researchers use behavioral methods on complex tasks to perform cognitive diagnosis. Most of the models include working on hypotheses and generating process information upon adding or rejecting the hypothesis. An early hypothesis is accomplished by compiling expert knowledge and pattern recognition.

6.6.1 CDSS for liver disorder identification

The next section presents the decision support system by Baitharu and Pani (2016) for data mining the dataset of liver disease. To learn from a dataset, accuracy is a major concern. Today, many humans suffer from liver problems and many lose their lives. Healthcare industries fail to collect proper data for optimum use. Most of these data remain unused due to the nonexploitation of relationships in hidden patterns. Therefore, the problem has to be addressed in developing a clinical decision support system that supports physicians. Baitharu and Pani (2016) considered the problem understudy and has used various algorithms to compare their effective rate thereby performing required corrections helping the patient to cure at the beginning stage of the illness. However, data classification accuracy outperformed human community support in the diagnosis and rehabilitation of liver disease.

The dataset obtained by any clinical servicing company is its most valuable asset, and its success is determined by how well it is used by business start-ups. Hence, for strategic decision making, datasets play a vital role. Also, storing a customer database enables predicting the preference of any customer to offer better services.

Here, Baitharu and Pani (2016) identified a disease prediction system using various classification algorithms that are described in detail in the forthcoming sections. The study's main goal was to identify liver diseases using a liver function test from the Kaggle dataset of patients with liver disease caused by excessive alcohol intake and inhalation of harmful gases, and categorize them into the following types:

- Cirrhosis
- Bile duct
- Chronic hepatitis
- Liver cancer
- Acute hepatitis

In the human body, life plays a key role in metabolism, immunity development, storage of nutrients, and other essential functions. Hence, dealing with such disorders is a constant need in today's society. Knowledge discovery requires analyzing the data from different perspectives to formulate valuable insights. The system performance is improved by the use of various techniques and algorithms. Learning and classification are the two steps involved to analyze or classify data. Some of the classification algorithms that can be used for the problem under study by Baitharu and Pani (2016) are:

- Decision tree J48
- Naïve Bayes
- Multilayer perceptron
- ZeroR

Decision trees J48: A predictive machine learning algorithm makes decisions using training data. The terminal node generates the final value of the classification.

Naïve Bayes: This is used for huge datasets. It computes the probability $P(c\,|\,x)$ for a given class (c), and predictor (x). Jiang, Qiu, Xu, and Li (2017) proposed a three-layer knowledge base model to analyze patients based on multisymptoms. The naïve Bayes algorithm was used to estimate the degree of symptom contribution. The algorithm has reduced attribute dependencies and the clinical predictions provide effective recommendations to doctors.

Multilayer perceptron: This works on an artificial neural network model to set charts of outputs for the input given. It consists of three layers: input, hidden, and output. Here, each node in the neuron uses an activation function. This belongs to a category of supervised learning techniques.

It all consists of activation function mapping the input to the output of each neuron. The activation function sigmoids are given in Eq. (6.2).

$$y(vi) = \tanh(vi) \tag{6.2}$$

ZeroR: The easiest classification method predicts the majority category (class) and performs classification by selecting the most frequent value.

Using the above algorithms, an experimental study was done by Baitharu and Pani (2016) using the WEKA tool, an open-source data mining tool implemented in the java language. The tool is used for most machine learning algorithms that rely on real-time data mining. The tool is implemented with various algorithms that are directly applied to the dataset given in Table 6.1. The tool implements data preprocessing, classification, and association rules. The classifier's qualitative performance is analyzed based on the margin curve, which describes the probability between the predicted and the actual class. This is shown in Fig. 6.8, where X: Margin in Number and. Y: Cumulative in Number.

Table 6.1 Performance results of data mining using CDSS (Baitharu & Pani, 2016).

Data mining algorithm	Mean absolute error	Root mean squared error	Relative absolute error	Accuracy
Decision tree J48	0.37	0.5	75	69
Naive Bayes	0.5	0.5	103	55
Multilayer perceptron	0.35	0.45	73	72
ZeroR	0.5	0.5	100	58

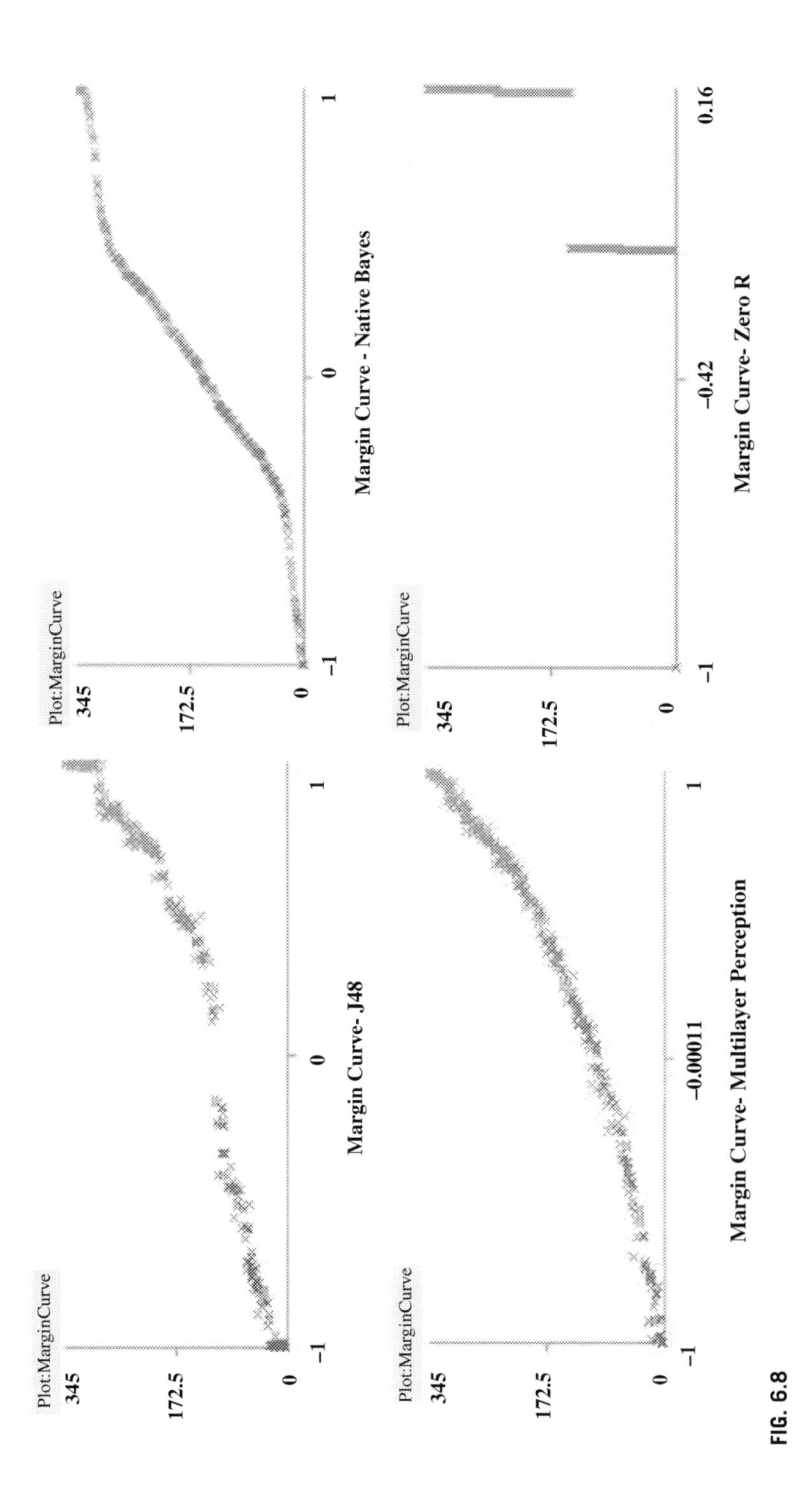

FIG. 6.8

Margin curve validation of the classifier algorithms.

Based on the curve in Fig. 6.8, the performance analyses of classifying datasets of liver disease.

From the study by Baitharu and Pani (2016), it is understood that decision support systems can be used in various classification algorithms. The four popularly used classification algorithms are considered for analysis and the computer simulation results are shown, proving that compared to all classification algorithms, the Naïve Bayes classification is not significant in classifying liver disorder datasets. At the same time, Multilayer perceptron use will provide better accuracy for liver disease prediction.

6.7 Conclusion

The use of a specialized decision support system shall be applied to a wide variety of applications. Nowadays, most automated medical devices use DSSS software that helps physicians interpret a patient outcome. The future development is promising and the field is growing rapidly. This attracts the keen interest of researchers to develop more CDSS, expanding its commercialization in healthcare IT and supporting the human community at a global level. From the rapid developments, this has proven to provide cost-effective and better prediction results.

References

Ando, S. (2000). Consistent gradient operators. *IEEE Transactions on Pattern Analysis and Machine Intelligence*, *22*(3), 252–265. https://ieeexplore.ieee.org/abstract/document/841757.

Axiomes et équations fondamentales du traitement d'images (Analyse multiéchelle et edp). (1992). Comptes Rendus de l'Académie Des Sciences. Série 1. *Mathématiques*, *315*(2), 135–138. https://pascal-francis.inist.fr/vibad/index.php?action=getRecordDetail&idt=4740065.

Baitharu, T. R., & Pani, S. K. (2016). Analysis of data mining techniques for healthcare decision support system using liver disorder dataset. *Procedia Computer Science*, *85*, 862–870. https://www.sciencedirect.com/science/article/pii/S1877050916306263.

Berner, E. S. (2007). *Clinical decision support systems: Vol. 233*. New York: Springer.

Cai, Q., Wang, H., Li, Z., & Liu, X. (2019). A survey on multimodal data-driven smart healthcare systems: Approaches and applications. *IEEE Access*, *8*, 23965–24005.

Forsythe, D. E., Buchanan, B. G., Osheroff, J. A., & Miller, R. A. (1992). Expanding the concept of medical information: An observational study of physicians' information needs. *Computers and Biomedical Research*, *25*, 181–200.

Fried, B. M., & Zuckerman, J. M. (2000). FDA regulation of medical software. *Journal of Health Law*, *33*, 129–140.

Goldschmidt, P. S. (2006). *Compliance monitoring for anomaly detection. Patent no. US 6983266 B1*. http://www.freepatentsonline.com/6983266.html.

Greenes, R. A. (2017a). Clinical decision support and knowledge management. In *Key advances in clinical informatics* (pp. 161–182). Academic Press.

Greenes, R. A. (2017b). Webinar on clinical decision support system for a transforming health/healthcare system. *Patient Centred Outcomes Research Clinical Decision Support Learning Network*, 1–18.

Hermann, M. L. (2006). Technology and reflective practice: the use of online discussion to enhance post conference clinical learning. *Nurse Educator*, *31*(5), 190–191.

Iyer, L. S., & Power, D. J. (Eds.). (2014). *Vol. 18. Reshaping society through analytics, collaboration, and decision support: Role of business intelligence and social media* Springer.

Jiang, Y., Qiu, B., Xu, C., & Li, C. (2017). The research of clinical decision support system based on three-layer knowledge base model. *Journal of Healthcare Engineering*, *2017*. https://doi.org/10.1155/2017/6535286, 6535286.

Kantardzic, M. (2003). *Data mining: Concepts, models, methods, and algorithms*. New Jersey: John Wiley.

Kautish, S. (2013). Knowledge sharing a contemporary review of literature in context of information systems designing. *International Multidisciplinary Research Journal*, *3*(1), 101–114.

Kautish, S., & Thapliyal, M. P. (Eds.). (2012). Concept of decision support systems in relation with knowledge management—Fundamentals, theories, frameworks and practices. *International Journal of Application or Innovation in Engineering & Management*, *1*, 1–9.

Khodaei, M. J., Candelino, N., Mehrvarz, A., & Jalili, N. (2020). Physiological closed-loop control (PCLC) systems: Review of a modern frontier in automation. *IEEE Access*, *7*, 133583–133599.

Klosgen, W., & Zytkow, J. M. (2002). *Handbook of data mining and knowledge discovery*. Oxford: OUP.

Miller, R. A. (1990). Why the standard view is standard: People, not machines, understand patients' problems. *Journal of Medicine and Philosophy*, *15*, 581–591.

Miller, R. A. (1994). Medical diagnostic decision support systems—Past, present, and future: A threaded bibliography and commentary. *Journal of the American Medical Informatics Association*, *1*, 8–27.

Osheroff, J. A., Forsythe, D. E., Buchanan, B. G., Bankowitz, R. A., Blumenfeld, B. H., & Miller, R. A. (1991). Physicians' information needs: An analysis of questions posed during clinical teaching in internal medicine. *Annals of Internal Medicine*, *114*, 576–581.

Sim, I., Gorman, P., Greenes, R. A., Haynes, R. B., Kaplan, B., Lehmann, H., & Tang, P. C. (2001). Clinical decision support for the practice of evidence-based medicine. *Journal of the American Medical Informatics Association*, *8*, 527–534.

Sutton, R. T., Pincock, D., Baumgart, D. C., Sadowski, D. C., Fedorak, R. N., & Kroeker, K. I. (2020). An overview of clinical decision support systems: Benefits, risks, and strategies for success. *NPJ Digital Medicine*, *3*(1), 1–10.

Machine learning and deep learning for healthcare

A comprehensive review on deep learning techniques for a BCI-based communication system

7

M. Bhuvaneshwari, E. Grace Mary Kanaga, J. Anitha, Kumudha Raimond, and S. Thomas George

Karunya Institute of Technology and Sciences, Coimbatore, Tamilnadu, India

7.1 Introduction

7.1.1 Brain signals

The human brain has four lobes that are responsible for certain activities of the brain. The frontal region of the brain is responsible for walking, conscious thought, and talking. The temporal lobe assists in speech hearing and perception. The parietal lobe controls sensation and coordination. The occipital lobe controls the vision process. Each lobe exerts rhythmic waves at different frequencies in various circumstances, which are listed in Table 7.1.

These waves can be captured using noninvasive and invasive methods as well as sensor-based methods. The waves acquired from the brain are decoded outside the brain and fed into the target application. The brain-computer interface (BCI) system based on electroencephalography (EEG) is used to design wheelchairs for paralytic people (Al-qaysi, Zaidan, Zaidan, & Suzani, 2018), emotion recognition (Liang, Oba, & Ishii, 2019), a wide variety of prosthetic devices (Alazrai, Alwanni, & Daoud, 2019), and cognitive biometrics (Bansal & Mahajan, 2019). In order to design this type of BCI system, brain waves have to be recorded from various neuroparadigms.

7.1.1.1 Brain computer interface

This is the system that interprets the electric potential generated by the brain into directives to perform a specific task. Nijboer et al. (2009) proved that BCIs can also be used to make changes in brain activities through a method called neurofeedback. The electric potential termed here is the brain wave, which can be recorded using noninvasive or invasive techniques. The invasive technique implants electrodes into the brain through surgery and store the electropotential activity of the brain with the help of a chip. Because electrodes are implanted directly into the brain, it has the

Demystifying Big Data, Machine Learning, and Deep Learning for Healthcare Analytics.
https://doi.org/10.1016/B978-0-12-821633-0.00013-1

Table 7.1 Frequency ranges of brain waves.

Brain waves	Frequency range	State
Infralow	($<.5$ Hz)	–
Delta band	(.5–3 Hz)	Deep sleep
Theta band	(3–8 Hz)	Sleep and deep meditation
Alpha band	(8–12 Hz)	Mood swings
Beta band	(12–38 Hz)	Problem-solving
Beta1 band	(12–15 Hz)	Fast and idle
Beta2 band	(15–22 Hz)	Figuring out something
Beta3 band	(22–38 Hz)	Anxious
Gamma	(38–42 Hz)	Spiritual emergence

potential to record every single neuron of the brain. Despite its accurate results, it is not being used because of its complication.

Noninvasive techniques measure electrical activity by placing the electrode on the scalp. Near-infrared spectroscopy (NIRS), one of the noninvasive technique, acts as a delegate for neural activity measures the oxyhemoglobin (HbO_2) and deoxyhemoglobin (Hb) composition in the brain which was experimented by Birbaumer and Rana (2019). Despite the low-bandwidth and SNR ratio results of the noninvasive techniques, Bansal and Mahajan (2019) overcame this by adapting hierarchical BCI (HBCI). EEG is another technique to noninvasively measure the electrical activity of the brain. Although it has a lower spatial resolution, it is still widely used in BCI systems because of its simple acquisition and harmless nature. This paper surveys the challenges and techniques to develop a communication device for paralytic people using BCI.

7.1.1.2 Electric potential source for BCI

Based on the work of Kaongoen and Jo (2017) on mental tasks and the use of the neuroparadigm, BCIs are categorized as active, passive, and reactive. Active BCIs record and translate responses generated by the voluntary control of the brain into directives for the device to perform a specific task such as motor imagery or mental calculation (Nijboer et al., 2009). Passive BCI works on the nonvoluntary control of the brain such as motion detection. Reactive BCI records the electric potential in response to stimuli to accomplish a certain task. The classifications of an electric potential source are depicted in Fig. 7.1.

Many neuroparadigms exist to capture brain activity as depicted in Fig. 7.1. It is categorized as evoked potential (EP) and event-related potential (ERP). Both systems generate electric potential based on the type of stimuli. EP is categorized as somatosensory evoked potential (SEP), visual evoked potential (VEP), and auditory evoked potential (AEP). AEP originates from the brain stem concerning sound stimuli but the latter originates from the peripheral nerve. Another brain wave component is P300, a positive deflection recorded at the 300ms later a stimulus that has a direct

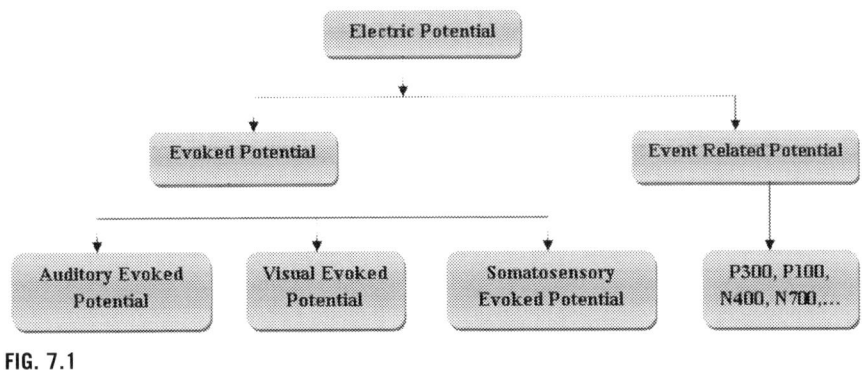

FIG. 7.1

Neuroparadigm.

No permission required.

relation with cognitive functions hence employed in EP based studies by Güven and Batbat (2019). But the information in a high-frequency band may be lost at times. This is one of the promising waves with high accuracy and ITR for auditory-based BCIs that was discussed in Kaongoen and Jo (2017). According to Güven and Batbat (2019), the subject experiences discomfort while using lengthy stimuli, hence it is better to use short stimuli.

7.1.1.3 Evoked potential

In Güven and Batbat (2019), VEP is generated by flashing or flickering the image at a certain frequency whose potential is higher than the potential in the normal state. VEP amplitude steady-state visually evoked potential (SSVEP) is a steady-state response to a visual stimuli range between 3.5–75 Hz. When the retina is fired up with the visual stimulus, an electric potential of the same or multiple frequencies is generated in the brain and serves as the BCI input (Cao et al., 2019). SSVEP is widely used in BCI systems for its high information transfer rate (ITR), which was elaborated upon by Lin, Gao, and Gao (2019) and Safi, Pooyan, and Motie Nasrabadi (2018). But as per Güven and Batbat (2019), its computational complexity is high and not flexible at all times. Recent studies by Lin et al. (2019) and Qin and Li (2018) proved that upon increasing the ITR, the classification accuracy also increased. SEP is also the stimulation of peripheral nerves. Imagined or motor imagery are often used paradigm, according to (García-Salinas, Villaseñor-Pineda, Reyes-García, & Torres-García, 2019) only a few imagined movements are available. Another widely used paradigm is recording the electrical activity based on the internal stimulus. This paradigm is found more suitable for real-time BCI on excluding frequency-related features. The EP and ERP are low-amplitude signals hence McFarland and Wolpaw (2017) assumed an average response from the target. The EP of visual stimulus is shown in Fig. 7.2.

7.1.1.4 Event-related potential (ERP)

The ERP is the direct response generated by the brain for a spontaneous or triggered event identified as a positive or negative amplitude shift in the EEG signal. There are many components such as P300 (Nakao, Barsky, Nishikitani, Yano, & Murata, 2007), P100 (Kunita & Fujiwara, 2005), N400 (Weimer, Clark, & Freitas, 2019), and N700 (Althen, Banaschewski, Brandeis, & Bender, 2020) associated with ERP based on the response time of the recorded signal. P300, a form of VEP, can be divided into endogenous (early 150 ms) and exogenous (latter 150 ms) components based on the latency duration. Though it elicits at 300 ms, the latency ranges between 250 to 750 ms (Fajardo, García-Galvan, Barranco, Galvan, & Batlle, 2016). These P300 waves are prominent in the frontal lobe of the brain. Once acquired and identified from the EEG signal, these components are processed using various soft computing techniques and are used in various BCI-based studies such as schizophrenia (Ahuja et al., 2020), Alzheimer's, alcoholism detection (Rodrigues, Filho, Peixoto, Arun Kumar, & de Albuquerque, 2019), and speller-based applications (Arican & Polat, 2019).

From Figs. 7.2 and 7.3, both ERP and EP look similar and even theoretically both seem to be similar. However, in practice, both are different in the context of arose. ERP is the potential generated as a result of certain sensory or motor events whereas EP is the brain response triggered by a certain stimulus.

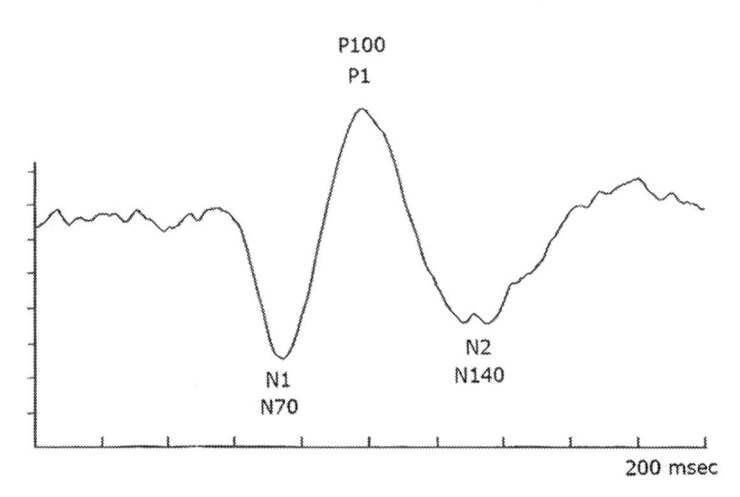

FIG. 7.2

EP.

Source: Retrieved March 2020, 1AD, from https://www.researchgate.net/post/visual_evoked_potential_P100_N70_Top-Down_or_Down-Top.

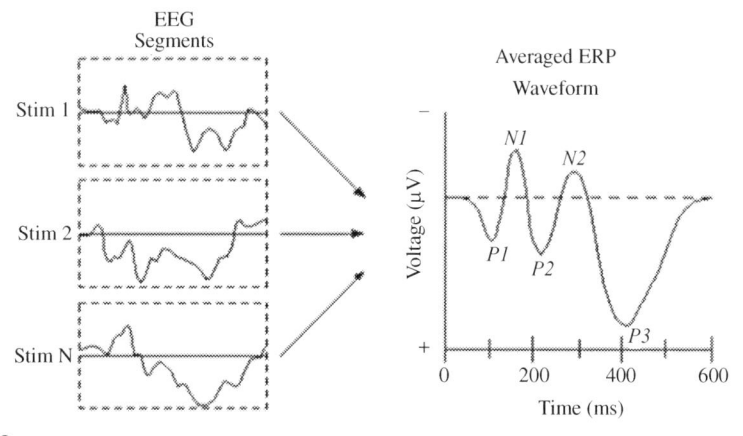

FIG. 7.3

Event related potential.

From https:/www.researchgate.net/publication/260595680_Precision_is_in_the_Eye_of_the_Beholder_
Application_of_Eye_Fixation-Related_Potentials_to_Information_Systems_Research/figures?lo=1.

7.2 Communication system for paralytic people

Many researchers designed assistive tools for patients suffering from neurodegenerative diseases. They have adopted various methods and technologies to develop these systems. The following sections describe various technologies used for developing assistive tools for disabled people.

7.2.1 Oculography-based control systems

Lacourse and Hludik (1990) designed a system to help disabled people to use the potential of the corneal retina (CRP) measured by biopotential skin electrodes placed around the eyes. These electrodes are formed by silver chloride mounted on a pair of sports glasses and designed in such a way that they do not distract vision. The displacement of the eyeball from the target to the random positions generates the steady-state CRP that was measured using a Princeton Applied Research (PAR) amplifier. The CRP relies on the distance of the eye from the target and the angle of rotation of the eyeball. The peak voltage and response time tell the caregiver that the subject needs attention. Bhat, Abhilash, Shivakumar, Vinay, and Santosh (2017) proposed an idea for the wheelchair movement of paralyzed people through their eye gaze. This system used a video processor system instead of electrodes. Desai (2013) made clear that the electrooculography-based method experiences baseline drifts due to the friction between the skin and the ions of the electrode.

7.2.2 Morse code-based assistive tool

Morse code is a method of encoding text characters based on signal duration. Alphabets and numbers are represented as dots and dashes. There is a standard code pattern for alphabets and numbers, for example, A is represented as .-, B as -..., 1 as .- -. Similarly, the code is available for all alphabets and numbers. Yang, Lin, Lin, and Lee (2013) suggested that Morse code thus generated requires a stable writing speed, which is difficult for the disabled. Yang, Chuang, Yang, and Luo (2003) solved this problem by using an adaptive signal processing algorithm called least mean square (LMS) that adapts to unstable writing speed.

7.2.3 Sensor-based systems

Sensors are electrical devices that take input from the desired environment and transforms it into a machine-accessible form of output. Many sensors are available to sense pressure, temperature, moisture content, nutrition content, etc. The modern medical world uses many healthcare-related sensors for an effective and reliable treatment process. The sensors used in healthcare monitoring are classified as wearable or implanted. The former monitors the patient externally while the latter is implanted into the human body through surgery. Shi (2015) discussed various assistive technologies for humans using sensors. Machangpa and Chingtham (2018) designed a robotic wheelchair using MPU 6050, which detects movement of the head based on which control is given to the raspberry pi processor for the movement of the wheelchair. Using this technique, direction can be shaped out, but determining the speed of the chair is complex work. Although it is low cost, it is not popular because of its lesser reliability.

7.2.4 EEG-based systems

In recent days, many research works such as Al-qaysi et al. (2018), Alazrai et al. (2019), and Nijboer et al. (2009) proposed assistive systems for the disabled using EEG-based BCI systems. Most of the systems were prosthetic devices and few were designed to control wheelchair directions. There are few speller-based systems available to assist the disabled in communication.

Table 7.2 summarizes various assistive tools available for disabled people using different technologies. Though there are many communication tools available, some drawbacks exist in the available technologies such as carrying electrodes in the face in oculography-based systems. Morse code-based systems require fast typing speed whereas sensor-based systems experience battery and power-related issues. EEG systems were found to be a better platform for developing communication-based systems.

Table 7.2 Summary of various assistive tools.

System	Application	Author
Oculography	Communication tool	Yang et al. (2013)
	Communication tool	Wu, Chuang, Hsieh, and Chang (2013)
Morse code	Communication tool	Yang et al. (2003)
	Communication tool	Das and Shivakumar (2019)
	Communication tool	Shantha Selva Kumari (2018)
Sensor	Health monitoring	Boyanapalli and Patil (2019)
	Wheelchair control	Machangpa and Chingtham (2018)
	Wheelchair control	Pu et al. (2018)
EEG	Wheelchair control	Chaudhary, Taran, Bajaj, and Siuly (2020)
	Wheelchair control	Al-qaysi et al. (2018)
	Lower limb	Tariq, Trivailo, and Simic (2018)
	Communication tool	Spüler, López-Larraz, and Ramos-Murguialday (2018)
	Speller	Saravanakumar and Ramasubba Reddy (2020)
	Speller	Chang, Lee, Heo, and Park (2016)
	SSVEP-based speller	Cecotti (2010)
	Webcam-based eye tracking	Lim, Lee, Hwang, Kim, and Im (2015)
	Mobile keyboard	Hwang, Hwan Kim, Han, and Im (2013)
	P300-based speller	Fernández-Rodríguez, Velasco-Álvarez, Medina-Juliá, and Ron-Angevin (2019)
	P300-based speller	Velasco-Álvarez et al. (2019)
	SSVEP based speller	Sadeghi and Maleki (2020)

7.3 **Acquisition system**

Data can be collected through an invasive or a noninvasive type of signal acquisition method such as EEG. The electrical activity of the brain is regarded as the potential difference here. This can be measured by attaching the electrode to the scalp. This has been designed in the form of a cap-like structure in recent

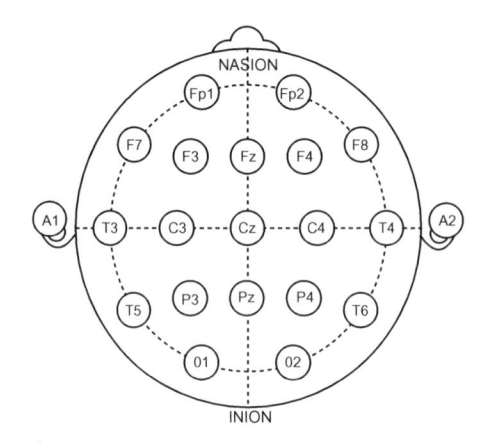

FIG. 7.4

International standard 10-20% electrode placement.

Danny Oude Bos (2006). EEG-based emotion recognition. Research Gate.

instruments. The electrodes record the electrical action and send it to the computer. The subjects wear this electrode cap like a normal cap. The position of the electrodes can be adjusted. It is placed based on the 10/20 system, as depicted in Fig. 7.4. This system is the international standard for placing electrodes, which states that the distance between the neighbor electrodes must be 10% or 20% of the total skull area from the nasion to the inion.

The electrodes are designed using steel coated with silver chloride. Electrodes are named with alphabets and numbers; the alphabet refers to the lobes (frontal, temporal, parietal, occipital) of the brain and the number refers to the side (right or left). EEG provides better time resolution to detect cortical activities, even at subsecond timescales. A gel acts as a transducer between the scalp and the electrodes. Based on the conductive connection, electrodes are commonly classified as dry electrodes, gel-based electrodes, and water-based electrodes (Pedroni, Bahreini, & Langer, 2019), as depicted in Table 7.3.

7.3.1 Benchmark datasets

Because medical research is an interdisciplinary field of study, it is beneficial to share datasets among peer researchers in a common repository. The next generation of researchers can explore their ideologies using those databases. Table 7.4 refers to some of the publicly available EEG datasets. Fig. 7.5 represents the basic architecture diagram of the BCI-DL system in which the signals acquired from the human brain undergo preprocessing before employing the deep learning architectures.

Table 7.3 EEG acquisition system.

Name of the amplifier	Electrode type	Authors
Acticap system	Gel based	Spüler et al. (2018)
Compumedics neuroscan	Gel-based	Weimer et al. (2019)
g.GAMMA sys (g. tec)	Hydrogel-based	Pinegger, Wriessnegger, Faller, and Müller-Putz (2016)
g.sahara (g.tec)	Hydrogel-based	Pinegger et al. (2016)
Porti 7	Water-based	Volosyak, Valbuena, Malechka, Peuscher, and Gräser (2010)
OPA378	Gel based	Guermandi, Bigucci, Scarselli, and Guerrieri (2015)
Compact PCI	Gel-based	Chen, Li, Mi, and Pan (2014)

Table 7.4 EEG public datasets.

Name of the dataset	Description
DEAP	Multimodal dataset for human emotion analysis using EEG
PhysioNet	Medical databases
BCI-competition IV 2a	MotorImagery database
BCI-competition III	P300 speller
MIMIC III	Critical care database
ABIDE I	EEG database for autism
Sleep, MI, Epilepsy, Psychophysics	EEG database of various disorders
SEED	EEG dataset for various disorders
BNCI Horizon 2020 consortium	BCI dataset
Pattern visual evoked potentials	EEG datasets for Evoked Potential
EEGLearn	Neuroscan dataset
VEP data	Evoked potential data
BCI data	EEG dataset for BCI
P300 dataset	P300 dataset
MI	MotorImagery dataset

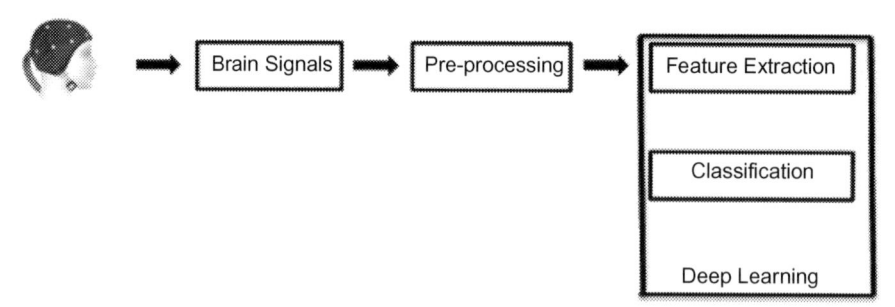

FIG. 7.5

Architecture diagram of the BCI-DL system.

No permission required.

7.4 Machine learning techniques in EEG signal processing

Machine learning (ML) is a method of automating the analytical model using intelligent algorithms. ML classifiers applied in various research works are shown in Table 7.5.

7.4.1 Support vector machine

The support vector machine (SVM) is the supervision-based ML technique in which the data points are separated by a hyperplane. The data items are plotted in the p-dimensional region where p represents the features. Kernel functions are used to plot the low-dimensional input data into a higher-dimension vector space. The kernel function may be sigmoid, linear, nonlinear, polynomial, or radial basis function (RBF). SVM is a widely applied classifier for signal processing simplicity and reliability, even for smaller datasets.

Paul et al. (2019) used the nonlinear polynomial 3 kernel function and achieved an accuracy of 96.15%. According to Yeo, Li, Shen, and Wilder-Smith (2009), SVM reduces the empirical classification and increases the geometric margin of the hyperplane between the classes, which eventually better classifies the data. SVM works better for multiclass classification.

7.4.2 k-NN

This is a parameterless classification algorithm executed with low complexity. It is also called a lazy algorithm, as it achieves generalization at the latter stage of classification. It identifies the k-element of the training set first and forms a class with its neighboring elements. The most frequently used distance measurement method between the k-element and the unknown element is the Manhattan distance and Euclidean distance.

Rahman, Gosh, Shuvo, and Rahman (2020) used the Euclidean distance formula and achieved an accuracy of 81.98%. Awan, Rajput, Syed, Iqbal, and Sabat (2016)

Table 7.5 Machine learning classifiers in various research works.

Classifier	Author	Application	Performance
SVM	Paul et al. (2019)	Characterization of fibromyalgia using sleep EEG signals	Accuracy 96.15%
	Yeo et al. (2009)	Drowsiness detection during travel	Accuracy 99.3%
	Bablani, Edla, and Dodia (2018)	Facial expressions	Accuracy 90.698%
Naïve Bayes	Rodrigues et al. (2019)	Detecting alcoholism	Accuracy 99.78%
k-NN	Li, Hu, Sun, and Cai (2016)	Mild depressive detection	Accuracy 92%
	Bablani et al. (2018)	Concealed Information Test	Accuracy 96.7%
	Awan et al. (2016)	Facial expression classification	Accuracy 96.1%
	Rahman et al. (2020)	Mental stress recognition	Accuracy 81.98%
Logistic regression	Asif et al. (2019)	Human stress classification	Accuracy 98.5%
	Bachmann et al. (2018)	Classifying depression	Accuracy 90%
	Ding et al. (2019)	Classifying major depression patients and healthy controls	F1 scores 80.70%
	Huang et al. (2019)	Identity authentication system	Accuracy 81.59%
Random forest classifier	Edla, Mangalorekar, Dhavalikar, and Dodia (2018)	Human mental states classification	Accuracy 75%
	Yang, Duan, and Zhang (2016)	Automated seizure detection	Accuracy 97.352%
Ensemble learning	Abdulhay, Elamaran, Chandrasekar, Balaji, and Narasimhan (2017)	Automated diagnosis of epilepsy	Accuracy 98.5%

calculated the distance and compared it with the test signal. Based on the voting of KNN for the least distance, the neighbors are selected.

7.4.3 Logistic regression

LR is a statistical model that defines the probability an event occurs by

$$P(P_i) = y + \sum X_i W_i \tag{7.1}$$

where y is the intercept, X_i is the input, and W_i is the weight. Logistic regression removes the outliers and produces the outcome with fewer or no errors.

Asif, Majid, and Anwar (2019) used this model for classifying human stress types.

7.4.4 Naïve Bayes

This is a probabilistic classifier that preassumes certain features that are not related to the occurrence of any other feature. The feature-pair should be distinct classes. Each feature shares an independent and equal contribution to the outcome. Naive Bayes (NB) with class c and predictor x is given by

$$P(a/b) = \frac{P(b/a)P(a)}{P(b)} \qquad (7.2)$$

where $P(b/a)$ is the prior probability and $P(a/b)$ is the posterior probability. The computational cost of NB is high because it has to determine the joint probabilities for all the values. Machado and Balbinot (2014) proved that NB gives poor classification results for a dataset with dependent features.

7.5 Deep learning techniques in EEG signal processing

There are ample ML and DL algorithms available to distinguish the brain potentials of varying frequencies. In ML algorithms, feature extraction and classification rely upon different algorithms, which increases the complexity of classification. DL is also an ML technique that trains the system to detect the features and classify automatically by processing the inputs over multiple layers. In DL, the layers are connected as neurons that represent the mathematical computation. Here, learning can be supervised, semisupervised or unsupervised, which is elaborated upon in the following section.

7.5.1 Deep learning models

DL concerns algorithms inspired by the functioning of the human brain that abstract the features from the raw data by incorporating multiple layers. The DL methods are divided into supervised, semisupervised, and unsupervised methods, as listed in Fig. 7.6.

7.5.1.1 Supervised deep learning

Supervised DL methods translate the raw data into intermediate representation, thus removing redundancy. It is quite popular as it merges both feature extraction and classification into one model.

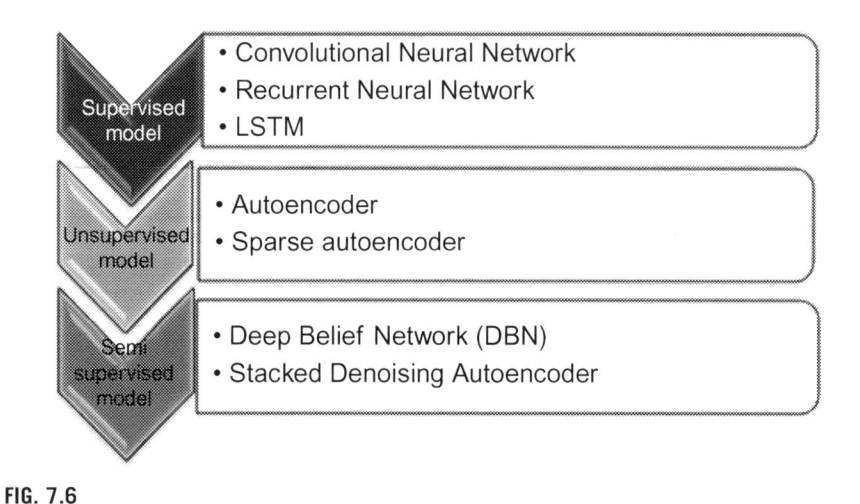

FIG. 7.6

Deep learning models.

7.5.1.2 Convolutional neural network (CNN)

A CNN is a frequently used DL architecture, which extracts the features or classify the data. There are generous numbers of researches on EEG classification using machine learning methods. Several studies (Chen, Song, & Li, 2019; Dose, Møller, Iversen, & Puthusserypady, 2018; Liu et al., 2018) worked on DL-based BCI systems. The most prominent algorithm for EEG classification is the common spatial pattern (CSP), which focuses more on frequency. Recent studies have proven that time-frequency domain analysis gives more accurate results than frequency analysis. Recent studies by Dose et al. (2018) employed DL to classify the EEG signal. CNN and convolutional fully connected (FC) layers perform dimensionality reduction and classification, respectively. This model used data from PhysioNet, which again classifies two, three, and four classes based on MI. The model achieves mean accuracies of about 80.38%, 69.82%, and 58.58% for each class, respectively.

In order to train the neural network, a deep feed-forward neural network is a popular model. In this model, the layers and neuron numbers are directly proportional, which increases the size of the training network. This is not advisable for high-dimensional data such as EEG. Dose et al. (2018) solved this issue by employing CNN with a convolutional window to determine a set of weights and a bias for all neurons in the hidden layer. The training process can be geared up by giving fewer gradients. This can be made possible by applying optimization techniques. The more complex features are learned by stacking CNN layers. The dimensions of the EEG data can be reduced by applying a pooling operation, which aggregates the adjacent values in the feature map. The activation function is used to gear up the training. Ma et al. (2017) combined multilevel compressed sensing and a genetic algorithm to enhance the efficiency of feature extraction. Ma et al. (2017) proposed a feature extraction framework called multimodal feature extraction, in which the first step senses the available feature and the second step senses the hidden information. This type of feature improves the accuracy of the classifier to 87.5% (Dose et al., 2018).

7.5.1.3 RNN

CNN is the most commonly used architecture for DL. Apart from CNN, there are other DL architectures such as autoencoder, deep generative models, and the recurrent neural network (RNN). CNN and RNN experience the vanishing gradient problem, which gets resolved by a specially designed RNN model called a long short-term memory network. Toraman, Tuncer, and Balgetir (2019) used various CNN architectures such as VGG16, VGG19 (Liu et al., 2018), ResNet (Liu et al., 2018), DenseNet, MobileNet, NasNetMobile, and NasNetlarge to extract features from different spectrograms. Dose et al. (2018) used shallow CNN to classify the EEG signal. In order to enhance the quality and quantity of feature vector information, Ma et al. (2017) used the deep belief network (DBN) to map the feature. Table 7.6 summarizes the architectures used in recent studies.

7.5.1.4 Unsupervised deep learning

The modern era of research is coupled with ample unstructured data, which can be resolved by these types of models. This type of model aims to build a generic classification model that could be trained with little data. The unsupervised learning method hierarchically discards irrelevant features to select informative features. The following are some of the unsupervised DL models.

7.5.1.5 Autoencoder

Autoencoder is a neural network that uses a backpropagation algorithm for feature learning. It works in two phases: encoding and decoding. In the encoding phase, the input data are mapped to a low-dimensional representation space to obtain the most appropriate feature, which again maps to the input space in the decoding phase. Fig. 7.7 shows a pictorial representation of the autoencoder. Boloukian and Safi-Esfahani (2020) applied the autoencoder to recognize the unspoken words of

Table 7.6 Deep learning architectures in various research works.

Architecture	Authors
Convolutional neural network	Dose et al. (2018), Liu et al. (2018), Liu, Zhou, Zhang, and Xiong (2020) and Chen et al. (2019)
Deep belief network	Ma et al. (2017) and Längkvist, Karlsson, and Loutfi (2012)
Recurrent neural network	Reddy and Delen (2018),Tsiouris et al. (2018), Michielli, Acharya, and Molinari (2019), and Abbasvandi and Nasrabadi (2019)
LSTM	Hosseini, Tran, Pompili, Elisevich, and Soltanian-Zadeh (2020) and Zheng et al. (2019)
Autoencoder	Boloukian and Safi-Esfahani (2020)
Sparse autoencoder	Kundu and Ari (2020)

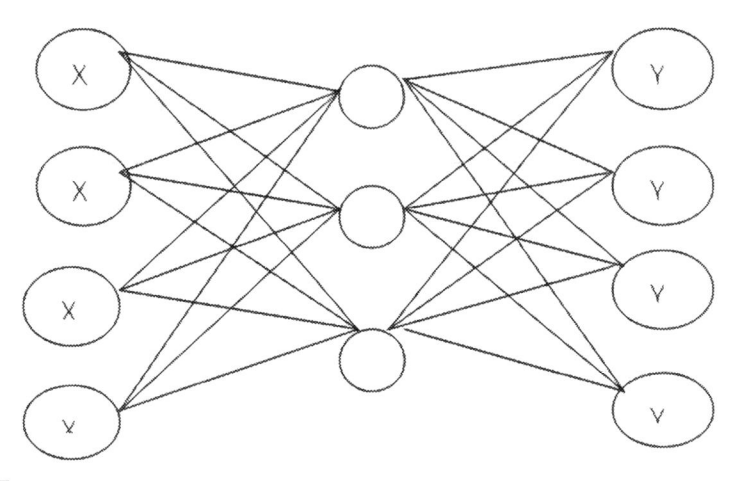

FIG. 7.7

Autoencoder.

No permission required.

speech-impaired people. There are different autoencoders, namely the sparse autoencoder and the denoising autoencoder (Yang et al., 2019).

7.5.1.6 Semisupervised deep learning

These types of algorithms are used widely due to their generalization for both labeled and unlabeled types of data. This method applies the supervised classification strategy for pretraining and the unsupervised strategy to fine-tune the parameters.

Jamshid and Hasan (2019) discussed various techniques of semisupervised models such as iterative modeling as well as graph-based and generative-based methods. Some common semisupervised architectures are the deep belief network and stack denoising. Autoencoder is applied in the EEG-based study from Xu and Plataniotis (2017).

7.5.1.7 Deep belief network

The deep belief network (DBN) is a set of interconnected layers that sequentially pretrains and fine-tunes the model training. Hassan et al. (2019) used DBN for human recognition. As it supervises the fine-tuning and unsupervised preretraining, the feature extraction and classification results are appropriate. Yin and Zhang (2017) utilized DBN for cross-subject recognition.

7.5.2 Deep learning in feature extraction

DL algorithms are accustomed to feature extraction from EEG signals. This is a benefit over ML algorithms, as DL algorithms extract the features appreciably. The extracted features are classified using classifiers. Classifiers such as SVM with various kernel functions such as linear, polynomial, and RBF are used to classify the

Table 7.7 Classifiers in deep learning-based BCI.

Classifiers	Authors
SVM	Acı, Kaya, and Mishchenko (2019), Gao, Guan, Gao, and Zhou (2015), Ma et al. (2017), and Li, Fan, Wang, and Wang (2019)
kNN	Ibrahim, Djemal, Alsuwailem, and Gannouni (2017)
Adaptive neurofuzzy system (ANFIS)	Deivasigamani, Senthilpari, and Yong (2016)
Principal component-based covariate shift	Jirayucharoensak, Pan-Ngum, and Israsena (2014)
Softmax classifier	Jirayucharoensak et al. (2014)

extracted features. Other classifiers such as Naïve Bayes and k-NN are also applied. Table 7.7 gives various classifiers employed in recent works. Owing to their appreciable performance, DL algorithms are used only for extracting features from the EEG signal.

Dose et al. (2018) used the backpropagation algorithm to compute the cost function. For noisy and sparse gradients, the combined Adam and stochastic gradient descent (SGD) can be adapted. Zheng et al. (2020) combined the Swish function-based LSTM with an ensemble model to extract the visual features from the EEG data. Recent studies apply DL algorithms in healthcare applications to improve their accuracy and efficiency. Wei, Chen, Song, Lou, and Li (2020) proposed combined recurrent and ensemble methods to learn the time and frequency-based features of MI EEG data.

7.5.3 Deep learning for classification

EEG data are sparse and dense, thereby paving the way for countless ML and DL algorithms. Though there are quantifiable DL algorithms for classification, many new algorithms are emerging for application and to enhance the adaptability and accuracy. The DL architecture itself will learn the features and classify the data based on the similarity observed in the learning phase. Many states of the algorithm are also used along with DL methods to classify the data. Liu et al. (2020) proposed a semisupervised Cartesian K-means algorithm for small datasets. The ensemble methods in various combinations have been used in recent studies and have shown remarkable results. In a study from Kundu and Ari (2020), other classifiers such as SVM, ensemble SVM, and Fisher linear discriminant showed better results in DL-based BCI systems. DL algorithms employed in recent studies are shown in Table 7.8.

Table 7.8 Recent research on deep learning algorithms with EEG signals.

Application	Algorithm	Performance analysis	Author/year
P300 signal detection	CNN	BN^3	Liu et al. (2018)
BCI-cinematics	H_2O DL algorithm	AC-85%	Sundhara, Krishna, Bhalaji, and Chithra (2019)
MI-EEG signal classification	CNN	AC-80.38%, 69.82%, 58.58% for different classes	Dose et al. (2018)
Identifying ADHD in children	CNN	AC-94.67%	Chen et al. (2019)
Identifying cerebral dominance	Various CNN Models	AUC-0.6127–0.7603	Toraman et al. (2019)
Ocular artifact removal	Deep learning network (DLN)	Varies with subject	Yang, Duan, Fan, Hu, and Wang (2018)
Focal epileptiform discharge detection	Combination of CNN and RNN	AUC-0.94	Tjepkema-Cloostermans, de Carvalho, and van Putten (2018)
Early Alzheimer's disease diagnosis	Contractive Slab and spike convolutional deep Boltzmann machine (CssCDBM)	AC-95.04%	Bi and Wang (2019)
Emotion analysis	Deep learning network (DLN)	AC-49.52%	Jirayucharoensak et al. (2014)
P300 signal classification	Artificial neural network (ANN)	AC-79.4%	Gao et al. (2015)

AC, *accuracy*; AUC, *area under the curve*.

7.6 Performance metrics

Evaluating the DL algorithm is an indispensable part of research. The known metric to measure the efficiency of the model is the classification accuracy. Because deployment is in a real-world environment, the efficiency of the EEG system has to be evaluated. There are a variety of metrics available to evaluate the performance of the system depending on the system where it is applied. Classification accuracy, one of the common metrics used to analyze the classification efficiency, is the number of correct predictions over the entire results. The other metrics used for the evaluation are shown in Table 7.9.

Table 7.9 Performance metrics.

Metrics	Method/formula
Classification accuracy	Cross-validation-evaluating the algorithm on the complementary set of input data $$\text{Accuracy} = \frac{\text{Correct prediction}}{\text{Total prediction}}$$
Confusion matrix	$$\text{Accuracy} = \frac{TP + FN}{\text{Total samples}}$$
Area under curve	The region under the ROC curve
F1 score	The harmonic mean between recall and precision
Mean absolute error	Average difference between the actual and predicted value
Mean squared error	Average square difference between the actual and predicted value
Receiver operating characteristics	Graphical depiction of sensitivity
Sensitivity	$$\text{Sensitivity} = \frac{TP}{TP + FN} \times 100$$
Specificity	$$\text{Specificity} = \frac{TN}{TN + FP} \times 100$$
Recall	$$\text{Recall} = \frac{TP}{TP + TN}$$
Precision	$$\text{Precision} = \frac{TP}{TP + FP}$$
ITR	$$\text{ITR} = \frac{\left(\log_2 N + P \log_2 P + (1 - P) \log_2 \left[\frac{1 - P}{N - 1} \right] \right)}{T}$$

FN, *false negative;* FP, *false positive;* TN, *true negative;* TP, *true positive.*

7.7 Inferences

This chapter aimed to review the feasibility of developing an EEG-based communication system using DL methods. The study also briefs assistive tools developed using various technologies for disabled people with their shortcomings. Because other methods of assistive technologies are unreliable and EEG-related studies have proliferated recently, analyzing the feasibility of EEG-based assistive tools would be appropriate now. This study reveals that the communication system using BCI is possible. Also, to obtain more accurate results, DL methods would be helpful. DL methods can also be combined with an ensemble approach to produce more accurate results. Compared to the ML techniques, the processing is simpler in DL methods. Both feature extraction and classification can be done with one algorithm, which reduces the computational cost and complexity.

Based on the review and the mathematical expressions of various features of EEG signals by Boonyakitanont, Lek-uthai, Chomtho, and Songsiri (2020), the computational complexity of the statistical and entropy-related features takes O (N) and O (N log N), respectively. Using DL, this complexity can be reduced as the DL methods will extract the features based on the learning index.

7.8 Research challenges and opportunities

Based on a detailed literature survey, the following challenges were observed.

7.8.1 Using multivariate system

The problem concerning multivariate data such as EEG is that although it increases the precision of the system, the computational complexity is high for this type of data. The acquisition process is also difficult. Because the acquisition system captures human brain signals, its fabrication cost will be naturally higher and the maintenance of the system is also difficult.

7.8.2 The dimensionality of the data

The major challenge while using the EEG signal is its dimensionality. Signals from various spots of the brain are acquired in the form of channel data. For a 16-channel system, the brain signals are acquired from 16 regions of the brain, which are commonly referred to as dimensions. As the recent study uses multivariate systems for better precision, the dimensionality of the data increases abruptly, making the classification process more complex.

7.8.3 Artifacts

Another challenge is the ocular and muscular artifacts acquired along with the EEG signal. The brain signals acquired through noninvasive techniques are captured from the surface of the scalp. The acquired signals hold body movement data, muscular artifacts, and eye blink data; eliminating these artifacts is complicated. An efficient artifact removal method has to be identified or designed to remove unwanted data.

7.8.4 Unexplored areas

The EEG signals are explored in all healthcare-related applications using ML methods. Most recent research combined ensemble models with DL methods, which showed notable results. The review reveals that semisupervised models are less explored with EEG-based study. There are only a few DL models coupled with ERP- and EP-based applications.

7.9 Future scope

Event-related and evoked potential data are alternative clinical methods of diagnosing diseases. In certain cases, the actual condition of the diseases is difficult to diagnose using normal clinical methods. The future scope of this review aims at exploring ERP and EP data for healthcare-related applications such as rehabilitation, designing assistive tools for disabled people, disease diagnoses, etc., by applying DL techniques. Because DL-based methods are less explored in this area, the future scope of this study moves toward applying DL-based models for ERP/EP-based healthcare applications.

7.10 Conclusion

The chapter explored the various architectures of DL and ML techniques that would be used for EEG-based communication systems. From the study, it is observed that a few kinds of research used DL techniques for feature extraction. In some studies, they have been also applied as classifiers. It has been observed that ML techniques can be applied for smaller datasets. This chapter also lists the open research challenges and future directions for researchers who work in the same domain. As the DL -based EEG classification is in its emerging stage, there exists a high scope for researchers to dig further in these directions.

Acknowledgments

This study is financially supported by the Department of Science and Technology (DST)—(ICPS) (Grant Number).

References

Abbasvandi, Z., & Nasrabadi, A. M. (2019). A self-organized recurrent neural network for estimating the effective connectivity and its application to EEG data. *Computers in Biology and Medicine, 110*(May), 93–107. https://doi.org/10.1016/j.compbiomed.2019.05.012.

Abdulhay, E., Elamaran, V., Chandrasekar, M., Balaji, V. S., & Narasimhan, K. (2017). Automated diagnosis of epilepsy from EEG signals using ensemble learning approach. *Pattern Recognition Letters*. https://doi.org/10.1016/j.patrec.2017.05.021.

Acı, Ç.İ., Kaya, M., & Mishchenko, Y. (2019). Distinguishing mental attention states of humans via an EEG-based passive BCI using machine learning methods. *Expert Systems with Applications, 134*, 153–166. https://doi.org/10.1016/j.eswa.2019.05.057.

Ahuja, S., Gupta, R. K., Damodharan, D., Philip, M., Venkatasubramanian, G., & Hegde, S. (2020). Effect of music listening on P300 event-related potential in patients with schizophrenia: A pilot study. *Schizophrenia Research, 216*, 85–96. https://doi.org/10.1016/j.schres.2019.12.026.

Al-qaysi, Z. T., Zaidan, B. B., Zaidan, A. A., & Suzani, M. S. (2018). A review of disability EEG based wheelchair control system: Coherent taxonomy, open challenges and recommendations. *Computer Methods and Programs in Biomedicine, 164*, 221–237. https://doi.org/10.1016/j.cmpb.2018.06.012.

Alazrai, R., Alwanni, H., & Daoud, M. I. (2019). EEG-based BCI system for decoding finger movements within the same hand. *Neuroscience Letters, 698*, 113–120. https://doi.org/10.1016/j.neulet.2018.12.045.

Althen, H., Banaschewski, T., Brandeis, D., & Bender, S. (2020). Stimulus probability affects the visual N700 component of the event-related potential. *Clinical Neurophysiology, 131* (3), 655–664. https://doi.org/10.1016/j.clinph.2019.11.059.

Arican, M., & Polat, K. (2019). Pairwise and variance based signal compression algorithm (PVBSC) in the P300 based speller systems using EEG signals. *Computer Methods and Programs in Biomedicine, 176,* 149–157. https://doi.org/10.1016/j.cmpb.2019.05.011.

Asif, A., Majid, M., & Anwar, S. M. (2019). Human stress classification using EEG signals in response to music tracks. *Computers in Biology and Medicine, 107,* 182–196. https://doi.org/10.1016/j.compbiomed.2019.02.015.

Awan, U. I., Rajput, U. H., Syed, G., Iqbal, R., & Sabat, I. (2016). *Effective Classification of EEG Signals using K-Nearest Neighbor Algorithm.* 120–124. In *doi:10.1109/FIT. 2016.28.*

Bablani, A., Edla, D. R., & Dodia, S. (2018). Classification of EEG data using k-nearest neighbor approach for concealed information test. *Procedia Computer Science, 143,* 242–249. https://doi.org/10.1016/j.procs.2018.10.392.

Bachmann, M., Päeske, L., Kalev, K., Aarma, K., Lehtmets, A., Ööpik, P., … Hinrikus, H. (2018). Methods for classifying depression in single channel EEG using linear and nonlinear signal analysis. *Computer Methods and Programs in Biomedicine, 155,* 11–17. https://doi.org/10.1016/j.cmpb.2017.11.023.

Bansal, D., & Mahajan, R. (2019). EEG-based brain-computer interfacing (BCI). In *EEG-Based Brain-Computer Interfaces.* https://doi.org/10.1016/b978-0-12-814687-3.00002-8.

Bhat, G. M., Abhilash, A., Shivakumar, M., Vinay, B., & Santosh, D. (2017). *Eye Gaze Recognition System, 1*(4), 39–41.

Bi, X., & Wang, H. (2019). Early Alzheimer's disease diagnosis based on EEG spectral images using deep learning. *Neural Networks, 114,* 119–135. https://doi.org/10.1016/j.neunet.2019.02.005.

Birbaumer, N., & Rana, A. (2019). Brain–computer interfaces for communication in paralysis. In *Casting light on the dark side of brain imaging.* https://doi.org/10.1016/b978-0-12-816179-1.00003-7.

Boloukian, B., & Safi-Esfahani, F. (2020). Recognition of words from brain-generated signals of speech-impaired people: Application of autoencoders as a neural Turing machine controller in deep neural networks. *Neural Networks, 121,* 186–207. https://doi.org/10.1016/j.neunet.2019.07.012.

Boonyakitanont, P., Lek-uthai, A., Chomtho, K., & Songsiri, J. (2020). A review of feature extraction and performance evaluation in epileptic seizure detection using EEG. *Biomedical Signal Processing and Control, 57.* https://doi.org/10.1016/j.bspc.2019.101702.

Boyanapalli, A., & Patil, R. (2019). Assistive technology using IoT for physically disabled people. *International Journal of Innovative Technology and Exploring Engineering, 8* (7), 903–907.

Cao, Z., Ding, W., Wang, Y.-K., Hussain, F. K., Al-Jumaily, A., & Lin, C.-T. (2019). Effects of repetitive SSVEPs on EEG complexity using multiscale inherent fuzzy entropy. In *Neurocomputing.*

Cecotti, H. (2010). A self-paced and calibration-less SSVEP-based brain-computer interface speller. *IEEE Transactions on Neural Systems and Rehabilitation Engineering, 18*(2), 127–133. https://doi.org/10.1109/TNSRE.2009.2039594.

Chang, M. H., Lee, J. S., Heo, J., & Park, K. S. (2016). Eliciting dual-frequency SSVEP using a hybrid SSVEP-P300 BCI. *Journal of Neuroscience Methods, 258,* 104–113. https://doi.org/10.1016/j.jneumeth.2015.11.001.

Chaudhary, S., Taran, S., Bajaj, V., & Siuly, S. (2020). A flexible analytic wavelet transform based approach for motor-imagery tasks classification in BCI applications. *Computer Methods and Programs in Biomedicine*, 105325. https://doi.org/10.1016/j.cmpb.2020.105325.

Chen, J., Li, X., Mi, X., & Pan, S. (2014). A high precision EEG acquisition system based on the compact PCI platform. In *Proceedings—2014 7th international conference on Bio-Medical Engineering and Informatics, BMEI 2014* (pp. 511–516). Bmei. https://doi.org/10.1109/BMEI.2014.7002828.

Chen, H., Song, Y., & Li, X. (2019). A deep learning framework for identifying children with ADHD using an EEG-based brain network. *Neurocomputing, 356*, 83–96. https://doi.org/10.1016/j.neucom.2019.04.058.

Das, R., & Shivakumar, K. B. (2019). Headspeak: Morse code based head gesture to speech conversion using intel realsense[TM] technology. *International Journal of Recent Technology and Engineering, 8*(2), 2866–2874. https://doi.org/10.35940/ijrteB2140.078219.

Deivasigamani, S., Senthilpari, C., & Yong, W. H. (2016). Classification of focal and nonfocal EEG signals using ANFIS classifier for epilepsy detection. *International Journal of Imaging Systems and Technology, 26*(4), 277–283. https://doi.org/10.1002/ima.22199.

Desai, Y. S. (2013). Natural Eye Movement and its application for paralyzed patients. *International Journal of Engineering Trends and Technology (IJETT), 4*(4), 679–686.

Ding, X., Yue, X., Zheng, R., Bi, C., Li, D., & Yao, G. (2019). Classifying major depression patients and healthy controls using EEG, eye tracking and galvanic skin response data. *Journal of Affective Disorders, 251*, 156–161. https://doi.org/10.1016/j.jad.2019.03.058.

Dose, H., Møller, J. S., Iversen, H. K., & Puthusserypady, S. (2018). An end-to-end deep learning approach to MI-EEG signal classification for BCIs. *Expert Systems with Applications, 114*, 532–542. https://doi.org/10.1016/j.eswa.2018.08.031.

Edla, D. R., Mangalorekar, K., Dhavalikar, G., & Dodia, S. (2018). Classification of EEG data for human mental state analysis using Random Forest Classifier. *Procedia Computer Science, 132*(Iccids), 1523–1532. https://doi.org/10.1016/j.procs.2018.05.116.

Fajardo, S., García-Galvan, F. R., Barranco, V., Galvan, J. C., & Batlle, S. F. (2016). New materials for thin film solar cells. *Intech, 13*. https://doi.org/10.5772/57353.

Fernández-Rodríguez, Á., Velasco-Álvarez, F., Medina-Juliá, M. T., & Ron-Angevin, R. (2019). Evaluation of flashing stimuli shape and colour heterogeneity using a P300 brain-computer interface speller. *Neuroscience Letters, 709*(June), 134385. https://doi.org/10.1016/j.neulet.2019.134385.

Gao, W., Guan, J. A., Gao, J., & Zhou, D. (2015). Multi-ganglion ANN based feature learning with application to P300-BCI signal classification. *Biomedical Signal Processing and Control, 18*, 127–137. https://doi.org/10.1016/j.bspc.2014.12.007.

García-Salinas, J. S., Villaseñor-Pineda, L., Reyes-García, C. A., & Torres-García, A. A. (2019). Transfer learning in imagined speech EEG-based BCIs. *Biomedical Signal Processing and Control, 50*, 151–157. https://doi.org/10.1016/j.bspc.2019.01.006.

Guermandi, M., Bigucci, A., Scarselli, E. F., & Guerrieri, R. (2015). EEG acquisition system based on active electrodes with common-mode interference suppression by Driving Right Leg circuit. In *Proceedings of the annual international conference of the IEEE Engineering in Medicine and Biology Society, EMBS, 2015-Novem(c)* (pp. 3169–3172). https://doi.org/10.1109/EMBC.2015.7319065.

Güven, A., & Batbat, T. (2019). Evaluation of filters over different stimulation models in evoked potentials. *Biocybernetics and Biomedical Engineering, 39*(2), 339–349. https://doi.org/10.1016/j.bbe.2018.08.007.

Hassan, M. M., Alam, M. G. R., Uddin, M. Z., Huda, S., Almogren, A., & Fortino, G. (2019). Human emotion recognition using deep belief network architecture. *Information Fusion*, *51*, 10–18. https://doi.org/10.1016/j.inffus.2018.10.009.

Hosseini, M.-P., Tran, T. X., Pompili, D., Elisevich, K., & Soltanian-Zadeh, H. (2020). Multimodal data analysis of epileptic EEG and rs-fMRI via deep learning and edge computing. *Artificial Intelligence in Medicine*, *104*(December 2019), 101813. https://doi.org/10.1016/j.artmed.2020.101813.

Huang, H., Hu, L., Xiao, F., Du, A., Ye, N., & He, F. (2019). An EEG-based identity authentication system with audiovisual paradigm in IoT. *Sensors (Switzerland)*, *19*(7). https://doi.org/10.3390/s19071664.

Hwang, H. J., Hwan Kim, D., Han, C. H., & Im, C. H. (2013). A new dual-frequency stimulation method to increase the number of visual stimuli for multi-class SSVEP-based brain-computer interface (BCI). *Brain Research*, *1515*, 66–77. https://doi.org/10.1016/j.brainres.2013.03.050.

Ibrahim, S., Djemal, R., Alsuwailem, A., & Gannouni, S. (2017). Electroencephalography (EEG)-based epileptic seizure prediction using entropy and K-nearest neighbor (KNN). *Communications in Science and Technology*, *2*(1), 6–10. https://doi.org/10.21924/cst.2.1.2017.44.

Jamshid, B., & Hasan, A. (2019). A review of various semi-supervised learning models with a deep learning and memory approach. *Iran Journal of Computer Science*, 65–80. https://doi.org/10.1007/s42044-018-00027-6.

Jirayucharoensak, S., Pan-Ngum, S., & Israsena, P. (2014). EEG-based emotion recognition using deep learning network with principal component based covariate shift adaptation. *Scientific World Journal*, 2014. https://doi.org/10.1155/2014/627892.

Kaongoen, N., & Jo, S. (2017). A novel hybrid auditory BCI paradigm combining ASSR and P300. *Journal of Neuroscience Methods*, *279*, 44–51. https://doi.org/10.1016/j.jneumeth.2017.01.011.

Kundu, S., & Ari, S. (2020). A Deep Learning Architecture for P300 Detection with Brain-Computer Interface Application. *IRBM*, *41*(1), 31–38. https://doi.org/10.1016/j.irbm.2019.08.001.

Kunita, K., & Fujiwara, K. (2005). P100 latency of the visual evoked potential by hemifield pattern reversal stimulation during isometric contraction of the unilateral shoulder girdle elevator. *International Congress Series*, *1278*, 65–68. https://doi.org/10.1016/j.ics.2004.11.067.

Lacourse, J. R., & Hludik, F. C. (1990). An Eye Movement Communication-Control System for the Disabled. *IEEE Transactions on Biomedical Engineering*, *37*(12), 1215–1220. https://doi.org/10.1109/10.64465.

Längkvist, M., Karlsson, L., & Loutfi, A. (2012). Sleep stage classification using unsupervised feature learning. *Advances in Artificial Neural Systems*, *2012*, 1–9. https://doi.org/10.1155/2012/107046.

Li, X., Fan, H., Wang, H., & Wang, L. (2019). Common spatial patterns combined with phase synchronization information for classification of EEG signals. *Biomedical Signal Processing and Control*, *52*, 248–256. https://doi.org/10.1016/j.bspc.2019.04.034.

Li, X., Hu, B., Sun, S., & Cai, H. (2016). EEG-based mild depressive detection using feature selection methods and classifiers. *Computer Methods and Programs in Biomedicine*, *136*, 151–161. https://doi.org/10.1016/j.cmpb.2016.08.010.

Liang, Z., Oba, S., & Ishii, S. (2019). An unsupervised EEG decoding system for human emotion recognition. *Neural Networks*, *116*, 257–268. https://doi.org/10.1016/j.neunet.2019.04.003.

Lim, J. H., Lee, J. H., Hwang, H. J., Kim, D. H., & Im, C. H. (2015). Development of a hybrid mental spelling system combining SSVEP-based brain-computer interface and webcam-based eye tracking. *Biomedical Signal Processing and Control, 21*, 99–104. https://doi.org/10.1016/j.bspc.2015.05.012.

Lin, K., Gao, S., & Gao, X. (2019). Boosting the information transfer rate of an SSVEP-BCI system using maximal-phase-locking value and minimal-distance spatial filter banks. *Tsinghua Science and Technology, 24*(3), 262–270. https://doi.org/10.26599/TST.2018.9010010.

Liu, M., Zhou, M., Zhang, T., & Xiong, N. (2020). Semi-supervised learning quantization algorithm with deep features for motor imagery EEG recognition in smart healthcare application. *Applied Soft Computing Journal, 89*, 106071. https://doi.org/10.1016/j.asoc.2020.106071.

Liu, M., Zhou, M., Zhang, T., & Xiong, N. (2020). Semi-supervised learning quantization algorithm with deep features for motor imagery EEG recognition in smart healthcare application. *Applied Soft Computing Journal, 89*, 106071. https://doi.org/10.1016/j.asoc.2020.106071.

Liu, M., Wu, W., Gu, Z., Yu, Z., Qi, F. F., & Li, Y. (2018). Deep learning based on Batch Normalization for P300 signal detection. *Neurocomputing, 275*, 288–297. https://doi.org/10.1016/j.neucom.2017.08.039.

Ma, T., Li, H., Yang, H., Lv, X., Li, P., Liu, T., … Xu, P. (2017). The extraction of motion-onset VEP BCI features based on deep learning and compressed sensing. *Journal of Neuroscience Methods, 275*, 80–92. https://doi.org/10.1016/j.jneumeth.2016.11.002.

Machado, J., & Balbinot, A. (2014). Executed Movement Using EEG Signals through a Naive Bayes Classifier. *Micromachines, 5*, 1082–1105. https://doi.org/10.3390/mi5041082.

Machangpa, J. W., & Chingtham, T. S. (2018). Head Gesture Controlled Wheelchair for Quadriplegic Patients. In *Procedia Computer Science* (Vol. 132, pp. 342–351). *Elsevier B., V.* https://doi.org/10.1016/j.procs.2018.05.189.

McFarland, D. J., & Wolpaw, J. R. (2017). EEG-based brain–computer interfaces. *Current Opinion in Biomedical Engineering, 4*, 194–200. https://doi.org/10.1016/j.cobme.2017.11.004.

Michielli, N., Acharya, U. R., & Molinari, F. (2019). Cascaded LSTM recurrent neural network for automated sleep stage classification using single-channel EEG signals. *Computers in Biology and Medicine, 106*(December 2018), 71–81. https://doi.org/10.1016/j.compbiomed.2019.01.013.

Nakao, M., Barsky, A. J., Nishikitani, M., Yano, E., & Murata, K. (2007). Somatosensory amplification and its relationship to somatosensory, auditory, and visual evoked and event-related potentials (P300). *Neuroscience Letters, 415*(2), 185–189. https://doi.org/10.1016/j.neulet.2007.01.021.

Nijboer, F., Carmien, S. P., Leon, E., Morin, F. O., Koene, R. A., & Hoffmann, U. (2009). (2009). Affective brain-computer interfaces: Psychophysiological markers of emotion in healthy persons and in persons with amyotrophic lateral sclerosis. In *Proceedings - 2009 3rd International Conference on Affective Computing and Intelligent Interaction and Workshops, ACII.* https://doi.org/10.1109/ACII.2009.5349479.

Paul, J. K., Iype, T., Dileep, R., Hagiwara, Y., Koh, J. E. W., & Acharya, U. R. (2019). Characterization of fibromyalgia using sleep EEG signals with nonlinear dynamical features. *Computers in Biology and Medicine, 4*, 111. https://doi.org/10.1016/j.compbiomed.2019.103331.

Pedroni, A., Bahreini, A., & Langer, N. (2019). Automagic: Standardized preprocessing of big EEG data. *NeuroImage*, *200*, 460–473. https://doi.org/10.1016/j.neuroimage.2019.06.046.

Pinegger, A., Wriessnegger, S. C., Faller, J., & Müller-Putz, G. R. (2016). Evaluation of different EEG acquisition systems concerning their suitability for building a brain. *Computer Interface : Case Studies*, *10*(September), 1–11. https://doi.org/10.3389/fnins.2016.00441.

Pu, J., Jiang, Y., Xie, X., Chen, X., Liu, M., & Xu, S. (2018). Low cost sensor network for obstacle avoidance in share-controlled smart wheelchairs under daily scenarios. *Microelectronics Reliability*, *83*(March), 180–186. https://doi.org/10.1016/j.microrel.2018.03.003.

Qin, Z., & Li, Q. (2018). High rate BCI with portable devices based on EEG. *Smart Health*, *9–10*, 115–128. https://doi.org/10.1016/j.smhl.2018.07.006.

Rahman, T., Gosh, A. K., Shuvo, M. H., & Rahman, M. (2020). Mental Stress Recognition using K-Nearest Neighbor (KNN) Classifier on EEG Signals. In *3*.

Reddy, B. K., & Delen, D. (2018). Predicting hospital readmission for lupus patients: An RNN-LSTM-based deep-learning methodology. *Computers in Biology and Medicine*, *101*(May), 199–209. https://doi.org/10.1016/j.compbiomed.2018.08.029.

Rodrigues, J.d. C., Filho, P. P. R., Peixoto, E., Arun Kumar, N., & de Albuquerque, V. H. C. (2019). Classification of EEG signals to detect alcoholism using machine learning techniques. *Pattern Recognition Letters*, *125*, 140–149. https://doi.org/10.1016/j.patrec.2019.04.019.

Sadeghi, S., & Maleki, A. (2020). Character encoding based on occurrence probability enhances the performance of SSVEP-based BCI spellers. *Biomedical Signal Processing and Control*, *58*, 101888. https://doi.org/10.1016/j.bspc.2020.101888.

Safi, S. M. M., Pooyan, M., & Motie Nasrabadi, A. (2018). Improving the performance of the SSVEP-based BCI system using optimized singular spectrum analysis (OSSA). *Biomedical Signal Processing and Control*, *46*, 46–58. https://doi.org/10.1016/j.bspc.2018.06.010.

Saravanakumar, D., & Ramasubba Reddy, M. (2020). A virtual speller system using SSVEP and electrooculogram. *Advanced Engineering Informatics*, *44*(February), 101059. https://doi.org/10.1016/j.aei.2020.101059.

Shantha Selva Kumari, R. (2018). FPGA based communication for a disabled person using morse. *Code*, *119*(7), 1173–1178.

Shi, W. (2015). Recent Advances of Sensors for Assistive Technologies. *Journal of Computer and Communications*, *3*(5), 80–87. https://doi.org/10.4236/jcc.2015.35010.

Spüler, M., López-Larraz, E., & Ramos-Murguialday, A. (2018). On the design of EEG-based movement decoders for completely paralyzed stroke patients. *Journal of NeuroEngineering and Rehabilitation*, *15*(1), 1–12. https://doi.org/10.1186/s12984-018-0438-z.

Sundhara, S. K., Krishna, G., Bhalaji, N., & Chithra, S. (2019). BCI cinematics—A pre-release analyser for movies using H_2O deep learning platform. *Computers and Electrical Engineering*, *74*, 547–556. https://doi.org/10.1016/j.compeleceng.2018.03.015.

Tariq, M., Trivailo, P. M., & Simic, M. (2018). EEG-based BCI control schemes for lower-limb assistive-robots. *Frontiers in Human Neuroscience*, *12*(August). https://doi.org/10.3389/fnhum.2018.00312.

Tjepkema-Cloostermans, M. C., de Carvalho, R. C. V., & van Putten, M. J. A. M. (2018). Deep learning for detection of focal epileptiform discharges from scalp EEG recordings. *Clinical Neurophysiology*, *129*(10), 2191–2196. https://doi.org/10.1016/j.clinph.2018.06.024.

Toraman, S., Tuncer, S. A., & Balgetir, F. (2019). Is it possible to detect cerebral dominance via EEG signals by using deep learning? *Medical Hypotheses*, 131. https://doi.org/10.1016/j.mehy.2019.109315.

Tsiouris, K., Pezoulas, V. C., Zervakis, M., Konitsiotis, S., Koutsouris, D. D., & Fotiadis, D. I. (2018). A long short-term memory deep learning network for the prediction of epileptic seizures using EEG signals. *Computers in Biology and Medicine*, *99*(May), 24–37. https://doi.org/10.1016/j.compbiomed.2018.05.019.

Velasco-Álvarez, F., Sancha-Ros, S., García-Garaluz, E., Fernández-Rodríguez, Á., Medina-Juliá, M. T., & Ron-Angevin, R. (2019). UMA-BCI Speller: An easily configurable P300 speller tool for end users. *Computer Methods and Programs in Biomedicine*, *172*, 127–138. https://doi.org/10.1016/j.cmpb.2019.02.015.

Volosyak, I., Valbuena, D., Malechka, T., Peuscher, J., & Gräser, A. (2010). Brain-computer interface using water-based electrodes. *Journal of Neural Engineering*, *7*(6). https://doi.org/10.1088/1741-2560/7/6/066007.

Wei, C., Chen, L., Song, Z., Lou, X., & Li, D. (2020). Biomedical Signal Processing and Control EEG-based emotion recognition using simple recurrent units network and ensemble learning. *Biomedical Signal Processing and Control*, *58*. https://doi.org/10.1016/j.bspc.2019.101756.

Weimer, N. R., Clark, S. L., & Freitas, A. L. (2019). Distinct neural responses to social and semantic violations: An N400 study. *International Journal of Psychophysiology*, *137*, 72–81. https://doi.org/10.1016/j.ijpsycho.2018.12.006.

Wu, C. M., Chuang, C. Y., Hsieh, M. C., & Chang, S. H. (2013). An eye input device for persons with the motor neuron diseases. *Biomedical Engineering - Applications, Basis and Communications*, *25*(1), 11–16. https://doi.org/10.4015/S1016237213500063.

Xu, H., & Plataniotis, K. N. (2017). Affective states classification using EEG and semi-supervised deep learning approaches. *IEEE 18th International Workshop on Multimedia Signal Processing*. In *doi:10.1109/MMSP.2016.7813351*.

Yang, B., Duan, K., & Zhang, T. (2016). Removal of EOG artifacts from EEG using a cascade of sparse autoencoder and recursive least squares adaptive filter. *Neurocomputing*, *214*, 1053–1060. https://doi.org/10.1016/j.neucom.2016.06.067.

Yang, B., Duan, K., Fan, C., Hu, C., & Wang, J. (2018). Automatic ocular artifacts removal in EEG using deep learning. *Biomedical Signal Processing and Control*, *43*, 148–158. https://doi.org/10.1016/j.bspc.2018.02.021.

Yang, C. H., Chuang, L. Y., Yang, C. H., & Luo, C. H. (2003). Morse Code Application for Wireless Environmental Control Systems for Severely Disabled Individuals. *IEEE Transactions on Neural Systems and Rehabilitation Engineering*, *11*(4), 463–469. https://doi.org/10.1109/TNSRE.2003.819905.

Yang, S. W., Lin, C. S., Lin, S. K., & Lee, C. H. (2013). Design of virtual keyboard using blink control method for the severely disabled. *Computer Methods and Programs in Biomedicine*, *111*(2), 410–418. https://doi.org/10.1016/j.cmpb.2013.04.012.

Yang, S., Yin, Z., Wang, Y., Zhang, W., Wang, Y., & Zhang, J. (2019). Assessing cognitive mental workload via EEG signals and an ensemble deep learning classifier based on denoising autoencoders. *Computers in Biology and Medicine*, *109*, 159–170. https://doi.org/10.1016/j.compbiomed.2019.04.034.

Yeo, M. V. M., Li, X., Shen, K., & Wilder-Smith, E. P. V. (2009). Can SVM be used for automatic EEG detection of drowsiness during car driving? *Safety Science*, *47*(1), 115–124. https://doi.org/10.1016/j.ssci.2008.01.007.

Yin, Z., & Zhang, J. (2017). Cross-subject recognition of operator functional states via EEG and switching deep belief networks with adaptive weights. *Neurocomputing*, *260*, 349–366. https://doi.org/10.1016/j.neucom.2017.05.002.

Zheng, X., Chen, W., You, Y., Jiang, Y., Li, M., & Zhang, T. (2019). Ensemble deep learning for automated visual classification using EEG signals. *Pattern Recognition*, *102*, 107147. https://doi.org/10.1016/j.patcog.2019.107147.

Zheng, X., Chen, W., You, Y., Jiang, Y., Li, M., & Zhang, T. (2020). Ensemble deep learning for automated visual classification using EEG signals. *Pattern Recognition*, 102. https://doi.org/10.1016/j.patcog.2019.107147.

Clinical diagnostic systems based on machine learning and deep learning

Sanjeevakumar M. Hatture and Nagaveni Kadakol

Basaveshwar Engineering College (Autonomous), Bagalkot, Karnataka, India

8.1 Introduction

Revolutions in technology, including Internet applications, embedded systems, the Internet of Things (IoT), and others have made medical services easier with handheld devices. Many mobile healthcare applications are becoming the essential services of mankind. A healthcare system offers wellness programs in order to provide a healthy life to all people. A healthcare system is the organization of people, institutions, and resources for improving or maintaining health by preventing, diagnosing, and treating as well as helping in recovery from disease, illness, and physical and mental impairments. Advancement of health wards that is well-being for the people, openness to the hopes of the residents, reasonable means of finance procedures are depends on four important functions such as giving wellness programs, generation of resources, invest in, and stewardship. Healthcare is envisaged for institutions and individuals. An individual means health specialists and associated health works, working as an employee in hospital, clinic, or other healthcare institutes, or they can be freelancing. Nowadays, there have been huge technical improvements in computational power, fast data storage, and parallelization. These improvements allow machine learning (ML) and deep learning (DL) technologies to be used effectively and efficiently in healthcare systems. Since 2016, a significant amount of investment in artificial intelligence (AI) research has been done in healthcare applications as compared with other sectors. ML and DL help clinicians in early detection, identification, and treatment of diseases as well as the prediction and prognosis of a particular disease. In healthcare systems, ML and DL techniques are used significantly in medical imaging, electronic health records (EHR), genomics, treatment design, consultation using digital media, simulated nurses, management of medication, drug creation, monitoring health, analyses of healthcare systems, sensing, and the online communication of health. ML and DL techniques offer efficient and effective systems for medical diagnosis. Clinical diagnosis is the method of determining the disease or the state that defines symptoms and signs. Various diagnostic procedures are employed in diagnosis, including analysis at the cellular and chemical levels,

Demystifying Big Data, Machine Learning, and Deep Learning for Healthcare Analytics.
https://doi.org/10.1016/B978-0-12-821633-0.00011-8

imaging, testing at the genetic level, measurement, and physical and visual examinations. Diagnostic or medical imaging is a process that creates pictorial demonstrations of the interior body and some organ or tissue functions while also exposing inner structures concealed by the skin and bones. This helps clinicians in analysis, medical intervention, and identifying and treating disease. Medical imaging has greater involvement in improving the health of all population groups; this has frequently been proved in the follow-up of identified and/or treated diseases. Different medical image modalities that help in the early diagnosis of disease are MRI, PET, CT, X-ray, and ultrasound. The medical imaging technique requires a group of clinicians who are experts in radiology as well as radiographers (X-ray technologists), a doctor specialized in the use of ultrasound images in identifying diseases, medical physicists, nurses, engineers in biomedical specialization, and other support staff who work together to optimize the well-being of patients. For decision-making and to reduce unnecessary medical procedures, safe, effective, and high-quality imaging procedures are used. Among different medical imaging techniques, ultrasound imaging is painless, safe, allows real-time imaging, and does not expose the patient to ionizing radiation. Because of these features, ultrasound imaging is the preferred medical imaging technique in clinical practice for diagnosis purposes. Ultrasound imaging is widely accessible and less expensive than other methods. On the other hand, understanding and interpreting ultrasound images requires well-trained radiologists. It also takes more time for diagnosis. It is difficult to provide the well-trained team required for imaging and scanning centers in villages and small towns, as this requires a huge investment of money and manpower. There is also a scarcity of well-trained clinicians, many of whom are unwilling to work in villages and small towns. Hence, in developing countries with digitization such as India, an advisory system is needed that helps in identifying organs and associated abnormalities in a short duration. In the era of modern technology, people become lazy due to unhealthy practices and unhealthy diets. Also, the use of pesticides and chemical composed water for the agriculture and environmental pollution most of people face a problem with intraabdominal organs. Some people won't visit doctors until a serious condition arises. Some people are unaware of serious diseases such as cancer. Others may have economic problems that prevent visiting higher healthcare centers for early screening. ML and DL techniques play vital roles in analyzing different medical image modalities. Hence, there is a scope to develop an advisory system that identifies intraabdominal organs and abnormalities that can be used for early screening in villages and small towns. Intraabdominal organs and their associated diseases that can be viewed significantly by using ultrasound imaging are shown in Table 8.1.

The proposed automated healthcare system using ultrasound imaging facilitates an advisory system for the diagnosis of abnormalities in the organs. People in villages and small towns can benefit from automated healthcare centers for early screening through ultrasound imaging. If any abnormalities are found in screening, the patient can be referred to higher healthcare systems for treatment. The proposed automated healthcare system also helps technicians or new radiologists diagnose the intraabdominal organ diseases. It also helps medical professionals understand the

Table 8.1 Intraabdominal organs and their associated diseases.

Organ	Diseases	Disease identified with ultrasound imaging
Liver	Hepatitis A	Hepatitis A
	Hepatitis B	Hepatitis B
	Hepatitis C	Hepatitis C
	Fatty liver	Fatty liver
	Cirrhosis	Cirrhosis
	Liver cancer	Liver cancer
	Hemochromatosis	–
	Wilson disease	–
Kidneys	Chronic kidney disease	Chronic kidney disease
	Kidney stones	Kidney stones
	Polycystic kidney disease	Polycystic kidney disease
	Urinary tract infections	Urinary tract infections
	Glomerulonephritis	–
Gallbladder	Gallstones	Gallstones
	Cholecystitis	Cholecystitis
	Choledocholithiasis	Choledocholithiasis
	Acalculous gallbladder disease	Acalculous gallbladder disease
	Gallbladder cancer	Gallbladder cancer
	Gallbladder polyps	Gallbladder polyps
	Gangrene of the gallbladder	Gangrene of the gallbladder
	Abscess of the gallbladder	Abscess of the gallbladder
	Biliary dyskinesia	–
	Sclerosing cholangitis	–
Spleen	Cancers (lymphoma and leukemia)	Cancers (lymphoma and leukemia)
	Liver diseases (cirrhosis due to excessive alcohol use)	Liver diseases (cirrhosis)
	Inflammatory diseases (rheumatoid arthritis, sarcoidosis)	Inflammatory diseases
	Trauma	Trauma
	Infections (mononucleosis, toxoplasmosis, endocarditis)	–
Pancreas	Pancreatitis	Pancreatitis
	Pseudocysts	Pseudocysts
	Cysts	Cysts
	Neoplasms	Neoplasms
	Diabetes mellitus	–
	Exocrine pancreatic insufficiency	–
	Cystic fibrosis	–
	Hemosuccus pancreaticus	–

Continued

Table 8.1 Intraabdominal organs and their associated diseases—cont'd

Organ	Diseases	Disease identified with ultrasound imaging
Appendix	Appendicitis	Appendicitis
	Fecolith	Fecolith
	Hamburger sign	–
	Intussusception	–
Small intestine	Infections	Infections
	Intestinal cancer	Intestinal cancer
	Intestinal obstruction	Intestinal obstruction
	Bleeding	–
	Celiac disease	–
	Crohn's disease	–
	Irritable bowel syndrome	–
	Ulcers such as peptic ulcer	–
Large intestine	Colorectal cancer	Colorectal cancer
	Crohn's disease	Crohn's disease
	Intestinal obstruction	Intestinal obstruction
	Ulcerative colitis	Ulcerative colitis
	Colonic polyps	–
	Colonoscopy	–
	Diverticulosis and diverticulitis	–
	Irritable bowel syndrome	–
Stomach	Stomach cancer	Stomach cancer
	Gastritis	–
	Gastroenteritis	–
	Gastroparesis	–
	Nonulcer dyspepsia	–
	Peptic ulcers	–

practical functioning of intraabdominal organs and encourages research/practice. The proposed automated healthcare system can also be utilized by inexperienced radiologists to make comparative studies on images for better decision making. In order to develop an automated healthcare system, exhaustive normal and abnormal intraabdominal ultrasound images containing the organs were collected by scanning centers with the consent of patients. The collected ultrasound images were preprocessed and the region of interest was segmented. A DL algorithm was employed for intraabdominal organ identification. Further, the identified organ was tested for abnormalities. Finally if an intraabdominal abnormality exists, then the type of abnormality is identified and suitable advice is provided through the proposed automated healthcare system.

8.2 Literature review and discussion

An automated healthcare system with ultrasound imaging involves several phases of processing to identify abnormalities in intraabdominal organs and provide advice to the victim. An exhaustive survey was carried out for identifying issues, challenges, and the scope for the development of an automated healthcare system. This is illustrated in the following. Zhang, Zhu, and Yang (2019) proposed a model that uses a convolutional neural network (CNN) to classify fatty liver into four types: normal, low-grade, moderate, and severe fatty liver. This model uses the texture features and gray features of the ultrasound image. The texture features of liver ultrasound images are used in learning the CNN for the classification process. The texture of a fatty liver ultrasound image is nonuniform due to the presence of fat particles; this concept is used for training the network. The authors claim that around 90% accuracy is achieved for fatty liver identification. DL gives its best in making automated solutions, while solving medical difficulties where it takes lot of long-standing accumulation of proficiency. Chen, Hong, Wu, and Mupparapu (2019) discussed the applications of a deep CNN in biomedical image techniques in medicine and dentistry. Zhou et al. (2019) employed AI techniques to detect liver diseases in medical imaging using the CNN model. The authors developed a system for identifying and assessing focal liver lesions; it also eases the treatment and calculates a response for liver treatment.

Kumar and Bindu (2019) reviewed image analysis using DL, which involves identification, classification, and measuring patterns in medical images. Various microservices such as localization in the image, segmentation, detection, and classification are discussed. The authors claim that DL is the strongest concept for the analysis of images. A CNN will find more recognizable features from filtered images that are passed from the pooling process. Akkus et al. (2019) have given various DL applications in ultrasound images. Arora and Mittal (2019) proposed enhancement methods for intraabdominal ultrasound images. The authors used middle, unsharp, and wiener filters for enhancing ultrasound images. They claim that unsharp filtering shows a better entropy value and visibility qualities. Further, Zheng, Furth, Tasian, and Fan (2019) used deep transfer learning techniques in classifying genetic abnormalities of the kidney and urinary tract. The transfer learning approach and imaging features yield the best classification system for differentiating children with genetic abnormalities of the kidney and urinary tract from ultrasound images of kidneys. The method is extended by a model built by using DL algorithms that uses texture features and geometric features of the images.

Hesamian, Jia, He, and Kennedy (2019) proposed a critical appraisal of popular methods, a DL technique for medical image segmentation. The selection of the network structure and training techniques used for deep neural network models is discussed in the paper. The authors explored various challenges and effective solutions correlated with medical image segmentation using DL techniques such as overfitting, training time, gradient vanishing, and organ appearance. Leong et al. (2019) presented a model which uses text on filter banks and features of local configuration patterns for the identification of breast lesions in ultrasound images without the segmentation concept. The exemplar is a best tool for the identification of breast nodules during whole breast scanning, which enables the sonographer to concentrate on a lesion present in images.

Van Sloun, Cohen, and Eldar (2020) discussed the use of DL in medical ultrasound imaging techniques. Liu et al. (2019) described applications of DL architectures in the analysis of US images such as classification, identification, and segmentation. Hussein, Kandel, Bolan, Wallace, and Bagci (2019) explored the tumor characterization as a part of precision medicine. This uses noninvasive cancer staging and prediction. It also gives adoptive personalized treatment planning. The authors propose two types of ML techniques for improving tumor categorization: supervised and unsupervised strategies. Further, the lesser availability of labeled training data is handled by unsupervised learning algorithms. Huang, Zhang, and Li (2018) presented research on the traditional and the DL ultrasound computer-aided diagnostic system. The feature and the classifier are used in the traditional US computer-aided diagnostic system; the DL ultrasound computer-aided diagnostic systems are discussed. Brattain, Telfer, Dhyani, Grajo, and Samir (2018) reviewed research applications in ultrasound images that use ML approaches. The authors also present their forthcoming openings for ML techniques to advance medical workflow and the analysis of ultrasound images. Because of the restricted computational power and memory resources, it becomes very difficult to implement multiple networks to overcome such limits. Xu et al. (2018) developed a framework on multitask learning to handle all tasks by a single network for classification and landmark recognition simultaneously. A network is built by using global convolutional kernels and coordinate constraints; also, a network has a conditional adversarial module to influence the performances. Kaur, Singh, and Kaur (2018) have made an attempt in radiology-based medical application; authors developed a method to denoise the image using suitable ML approach. This review focuses on six radiology applications: fetus development; computer-based diagnosis and the identification of lesions in breasts and the skin, MRI to diagnose brain tumors, analysis of chest X-rays, and MRI imaging in finding breast cancer. Wavelet and curvelet optimization techniques are used to denoise the image. Precious and Selvan (2018) employed the texture and shape features of ultrasound images in the identification of abnormal masses. Speckle noise, which reduces the quality of an ultrasound image, is removed by the SRAD filter and the level set method is used for segmenting the abnormal region. Then, from the segmented region, texture features are extracted using gray level run length matrix (GRLM) and the gray level cooccurrence matrix (GLCM); features of the shape are extracted. After the extraction of features, statistical analysis is carried out to eliminate the overlying features. These features are used for benign and malignant classifications. Jabarulla and Lee (2018) proposed the elimination of multiplicative speckle noise in liver US images. This is based on a dictionary learning approach where a new signal is reconstructed, known as sparse representation. Using a homomorphic filter, additive noise is generated using the nonuniform multiplicative signal. This is followed by pixel-based total variation regularization and patch-based sparse representation over a dictionary, and is trained using K-singular value decomposition. In conclusion, to solve the optimization problem, the split Bregman algorithm is used and the despeckled image is estimated. Gupta and Garg (2017) suggested a method to despeckle the noise and preserve the edges by the combination of two filters with the help of a homogeneity map. One filter is an edge filter known as a detail-preserving anisotropic diffusion and another is a smoothing filter known as an optimized Bayesian nonlocal mean filter.

Raghesh Krishnan and Radhakrishnan (2017) presented a model for the classification of 10 different types of liver ailments, such as focal and diffused livers, using US images. The active contour segmentation approach is used for isolating the diseased portion from the ultrasound image. Biorthogonal wavelet transform is used for decomposing the segmented region into horizontal, vertical, and diagonal components. GRLM features are extracted from the wavelet-filtered component images, and by applying 10-fold cross-validation strategy, classification is done using random forests. Zhang et al. (2017) employed a technique for removing the liver region from US images using uneven illumination variation in US image; this makes the illumination of the liver region in images stable. This technique uses the fuzzy C mean (FCM) algorithm to obtain the shape information for segmentation. Nikolić, Tuba, and Tuba (2016) developed a modified canny edge detection algorithm that softens edges using Gaussian smoothing and determines the edges of organs. Further speckle noise is removed by a modified median filter with minute deprivation of the edges. A weak weighted smoothing filter is used to remove other noise, with much less destruction to the edges.

Garg and Khandelwal (2016) showed a new technique for denoising the image using a bilateral filter and detail-preserving anisotropic diffusion. This combination works on the binary classifier map (BCM). The performance is evaluated by computing peak signal-to-noise ratio (PSNR), signal-to-noise ratio (SNR), and edge keeping index (EKI). Bhavana, Varthamanan, and Sonti (2016) employed a method known as the wavelet-based image fusion technique for improving and denoising bowel gas detection in US scanning. It provides better visualization and also reduces the speckle noise in US images, which eases the identification of good or odd tissues existing over the bowel gas. Zhang et al. (2016) proposed liver segmentation from an ultrasound image using shape information. Marsousi, Plataniotis, and Stergiopoulos (2014) proposed a method in the three-dimensional (3D) abdominal Morison's pouch US image for the identification and segmentation of kidneys based on shape. Manually diagnosing using 3D ultrasound images needs an extremely well-trained doctor specialized in ultrasonography. Therefore, this computer-aided system will automatically diagnose the kidney.

Li, Ross, and Kruusmaa (2013) suggested Bhattacharyya distance with Rayleigh distribution for the segmentation of ultrasound images. The energy function of the algorithm uses the Chan-Vese energy function and the Bhattacharyya distance. This method minimizes the differences between each part and maximizes the distance of density function between each part. It is also able to deal with blurry boundaries in an ultrasound image and provides good results on both phantom and patient US images.

Sridevi and Sundaresan (2013) discussed different ultrasound image segmentation algorithms and also explored a few application-specific segmentation algorithms. Suganya and Rajaram (2012) proposed a model for the classification of diseases in liver US images. A hybrid self-organizing map and a speckle reducing method known as Laplacian pyramid nonlinear diffusion (LPND) are used. Liver disease such as cysts, hepatoma, and hemangioma are dangerous diseases in the liver. Speckle noise is eliminated from US images by the LPND filter; this also conserves the projecting edge data in the classification process. A variation of the median absolute deviation estimator is used for determining the gradient threshold. In each pyramid layer, the gradient threshold is used for nonlinear diffusion. Further, from the normal and abnormal US images, the average gray level, entropy, local variance, cooccurrence matrix, first-order statistics, gradient features, and fractal dimensions are collected. An unsupervised self-organizing map is developed for the classification of normal and abnormal liver diseases from US. Angadi and Hatture (2019) and Shanmukhappa and Sanjeevakumar (2018) employed soft computing techniques and symbolic data analysis for person identification.

8.2.1 Major findings in problem domain

The literature review explores several issues and challenges dealing with ultrasound imaging. Some of the issues are characteristic artifacts such as SNR and the speckle noise having nonzero correlation over relatively large distances. These issues complicate the segmentation task. The identification of abdominal organs from ultrasound image using neural network requires a large training set and opacity may cause improper performance of the neural network. Due to the incomplete volume effects, the gray-level relationships of neighboring organs, the disparity of the media affect, and the quite high dissimilarities of organ positions and shapes, it is a continuously challenging task for the automatic identification of abdominal organs. Using filters in denoising the ultrasound images smooths the edges of organs, which in turn makes it difficult to identify the shape of the organ. In the proposed healthcare advisory system, these issues are alleviated by employing efficient algorithms in different processing stages of abnormality identification in intraabdominal organs.

8.3 Applications of machine learning and deep learning in healthcare systems

With the rapid growth in population, recording and analyzing huge amounts of patient information are challenging tasks. ML and DL provide algorithms that help in finding patient data and processing data automatically, which helps in making healthcare systems more dynamic and robust.

8.3.1 Heart disease diagnosis

The heart is the important muscular organ of the body. Cardiovascular disease is the most common cause of death in people. Medical imaging is used for the diagnosis of cardiovascular disease, but it needs the qualitative assessment of images and computable procedures of the heart structure and heart function. Hence, there is a need for advanced imaging techniques that identify complex and hidden image patterns from the image. Presently, ML plays an important role in identifying cardiovascular diseases such as coronary artery disease (CAD) and coronary heart disease (CHD).

8.3.2 Predicting diabetes

The unbalanced secretion of insulin and impaired biological effects cause blood glucose levels to increase, which leads to diabetes in people. Diabetes causes major health issues such as damaging the large blood vessels of the heart, brain, and legs while damage to small blood vessels causes difficulties in the eyes, kidneys, feet, and nervous system. Diabetes is commonly seen in people. Presently, numerous ML algorithms are implemented in the early identification of diabetes so that it can be controlled as early as possible to avoid future severe consequences.

8.3.3 Prediction of liver disease

The liver is a major internal organ of the human body. It has an important role in metabolism while also regulating the chemicals in the body. Cirrhosis, chronic hepatitis, and liver cancer are several liver diseases. A diagnostic test is used to identify liver disease, and this is a serious job that requires large amounts of data. Hence, ML and DL are used as classifiers for extracting data from the test report for the analysis and detection of disease.

8.3.4 Robotic surgery

The application of robotics in medical surgery has steadily grown. Suturing is one of the categories in robotic surgery. Suturing is the process of sewing up an open wound, and it can take a long time. ML and DL are used in the automation of suturing, which may reduce the surgical procedure length and surgeon fatigue. Surgical trauma is also reduced. ML and DL techniques are also used for the early detection of complications that may occur after surgery.

8.3.5 Cancer detection and prediction

Cancer is a condition where abnormal cells grow uncontrollably, spreading across the body and damaging tissue. Medical imaging and biological tests are used for detecting cancer, and these require time and experts for diagnosis. Presently, ML and DL are used in medical imaging and biological testing for the early detection of cancer. This helps clinicians with proper treatment and increases the survival rate of patients.

8.3.6 **Personalized treatment**

Recently, medical imaging and molecular medicine have moved toward healthcare services. This provides diagnostic tools and information to individuals for personalized management and suggests personalized medicine. In personalized medicine, the daily activity of a person including diet, environmental conditions, gene variability, age, gender, and location need to be considered. Hence, the major problem in personalized medicine is translating this large variable dataset into information that helps in making proper decisions. ML and DL techniques are used for translating this large variable dataset into information suitable for personalized medicine.

8.3.7 **Drug discovery**

Drug discovery is a standard application of ML and DL in medicine. Manual drug discovery takes more time and is very expensive. ML and DL techniques are used to overcome this drawback. They are used in designing the chemical structure of a drug and in the analysis of drug effects in both preclinical research and clinical trials.

8.3.8 **Smart EHR**

ML opportunities such as document classification of and recognition of optical characters can be used in developing an intelligent electronic health record system. This application can be employed to sort patient requests via electronic mail or transform a physical record system into an automated system to construct a harmless and effortlessly manageable system. The fast growth of EHR has enhanced the amount of patient medical information, which can be used for refining healthcare.

8.4 **Proposed methodology**

In this work, a new model for intraabdominal organ recognition and abnormality identification using ultrasound imaging is proposed. The steps involved in the proposed methodology include ultrasound image acquisition, preprocessing (image enhancement by employing an NLM filter and a bilateral filter), region-of-interest (RoI) image segmentation, construction of a deep neural network for intraabdominal organ recognition, feature extraction and identification, and the categorization of abnormalities in the organ. The overall methodology is depicted in Fig. 8.1.

In the literature, various researchers have employed ML and DL algorithms for identifying diseases of human organs such as the breast, kidney, liver, stomach, pancreas, etc. The proposed methodology uses a DL algorithm to identify intraabdominal organs. Once the organ is identified, the further RoI is segmented by employing an adaptive method to ultrasound imaging. Finally, the abnormalities in intraabdominal organs are identified and a suitable advisory is provided to the patient. Initially, sufficient datasets of intraabdominal ultrasound images of the kidney, liver, pancreas, spleen, and bladder are collected. These images initially do not contain any

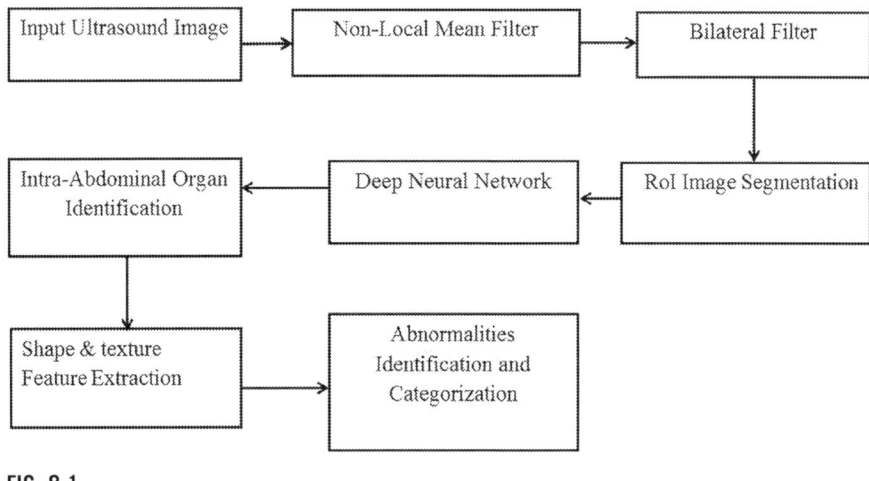

FIG. 8.1

Block diagram of proposed methodology.

No permission required.

information related to the patient. During the acquisition of ultrasound images, the images were degraded with Gaussian noise and speckle noise. Hence, to improve the ultrasound images, denoising of the images is performed. To preserve edges, the NLM filter and bilateral filter are employed. The enhanced ultrasound images are further processed for RoI image segmentation. Around 70% of the RoI images of intraabdominal ultrasound images are used as a training set for DL algorithms. Using a trained DL system, intraabdominal organs are identified from ultrasound images. Further, the abnormalities in the organs are identified by the extracted features from the segmented image. A detailed description of each phase of the proposed system is presented in the following subsections.

8.4.1 Intraabdominal ultrasound image acquisition

The proposed intraabdominal organ recognition and abnormality identification system uses a dataset of ultrasound images. The dataset comprises ultrasound images of six intraabdominal organs: the kidney, liver, gallbladder, pancreas, spleen, and bladder. The database is collected at an ultrasound center to carry out experimentation with the consent of patients for 1 year. The images do not contain any information that exposes patient details. The SMH-NK-DB image database is used for experimentation and database is constructed by collecting ultrasound images of six organs namely kidney, liver, gallbladder, pancreas, spleen, and urinary bladder. The dataset contains ultrasound images of the kidney (3356), liver (1312), gallbladder (517), pancreas (1307), spleen (1186), and urinary bladder (1119). The sample ultrasound images of SMH-NK-DB are shown in Fig. 8.2.

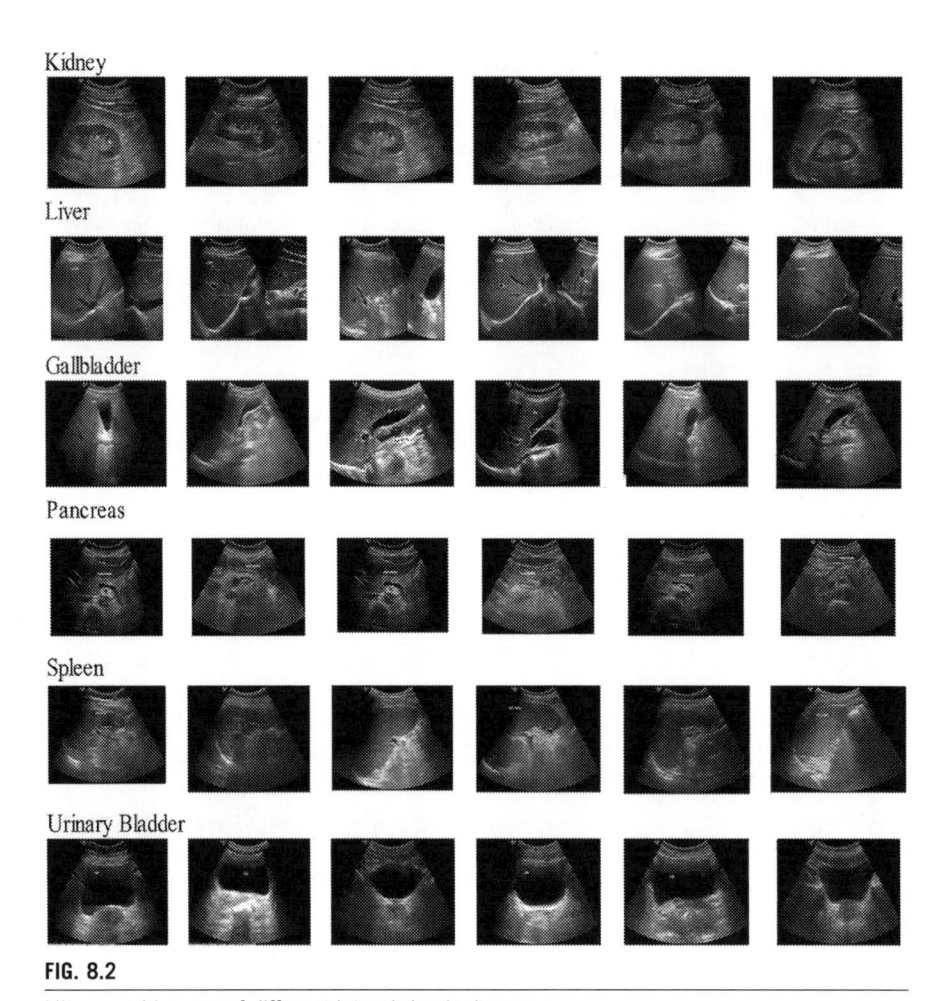

FIG. 8.2

Ultrasound images of different intraabdominal organs.

Courtesy Irappa Madabhavi, Dr., Ultrasound Images, Kerudi Cancer and Heart Hospital, July 16, 2020, Bagalkot, Karnataka.

In order to perform an open-set identification of the intraabdominal organ, a watchlist dataset is also collected separately. Further, the description of the ultrasound image enhancement of SMH-NK-DB is provided in the following section.

8.4.2 Ultrasound image enhancement

In order to segment the RoI of the intraabdominal organ ultrasound images, they are preprocessed. Ultrasound images are preprocessed to explore the intraabdominal organ information for the identification of abnormalities by employing an NLM filter and a bilateral filter. The NLM filtering computes the mean of all pixels in the image,

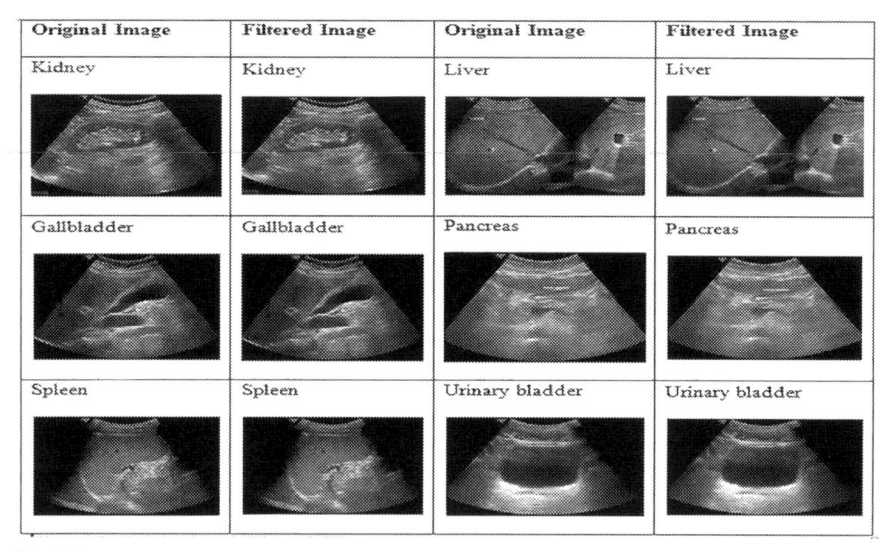

FIG. 8.3

Original and filtered ultrasound images.

Courtesy Irappa Madabhavi, Dr., Ultrasound Image, Kerudi Cancer and Heart Hospital, July 16, 2020,
Bagalkot, Karnataka.

weighted by how analogous these pixels are to the objective pixel. This results in far greater postfiltering clearness and less loss of feature in the image compared with local mean algorithms. An additional level of filtering is performed to preserve the edges of the organs by employing a bilateral filter. A bilateral filter alters the intensity of each pixel with a weighted mean of intensity value from neighboring pixels. The resulting filtered sample ultrasound images after enhancement are shown in Fig. 8.3.

Further, the enhanced ultrasound image is processed to segment the RoI containing the organ. The quality of the ultrasound image is enhanced by retaining the detail information in the image and preserving the edges. The segmentation of the RoI of the organ in the ultrasound image is described in the next subsection.

8.4.3 Segmentation of the RoI image

In order to identify the organ and its abnormality, the RoI of an organ is segmented. In the literature, several image segmentation techniques are presented by many authors, including the thresholding method, the watershed segmentation method, etc. In the proposed segmentation method, several morphological processing steps are introduced. The segmented RoI image of an organ is collected and further used for training the deep neural network. The following algorithm performs the segmentation of a kidney RoI of US. Segmented images are used as training sets for a deep neural network to identify the organ.

Algorithm for segmentation of RoI from ultrasound image

Let I be an input preprocessed color image of size 700×950 and I_s the segmented output image. The following steps give the segmentation of the organ from the ultrasound image.

Step 1: Read the preprocessed ultrasound image I.

Step 2: Convert the color image I to gray image I_g.

Step 3: Perform a TOP HAT morphological operation on image I_g with the structuring element "square" of size 100.

```
Ig= TOPHAT (Ig);
```

Step 4: Perform an open morphological operation on the previous step image I_g.

```
Igo=OPEN(Ig);
```

Step 5: Reconstruct the image I_c from I_{go} and I_g.

```
Ic=RECONSTRUCT (Igo,Ig);
```

Step 6: Perform a BOTTOM HAT morphological operation on image I_c with the structuring element "square" of size 100.

```
Ic= BOTTOMHAT (Ic);
```

Step 7: Perform an open morphological operation on the previous step image I_c.

```
Ico=OPEN(Ic);
```

Step 8: Reconstruct the image I_c from I_{co} and I_c.

```
Ic=RECONSTRUCT (Ico, Ic);
```

Step 9: Convert the previous step image I_c to the binary image I_b.

Step 10: Perform morphological open, close, and dilation operations to get the segmented kidney image I_s from the ultrasound image.

The proposed algorithm is applied for segmenting the RoI of the organ from ultrasound imaging. Fig. 8.4 shows a pictorial representation of the segmentation method.

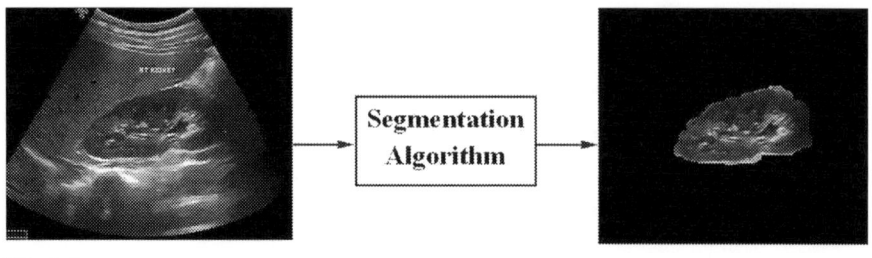

FIG. 8.4

Pictorial representation of the segmentation method.

Courtesy Irappa Madabhavi, Dr., Ultrasound Image, Kerudi Cancer and Heart Hospital, July 16, 2020, Bagalkot, Karnataka.

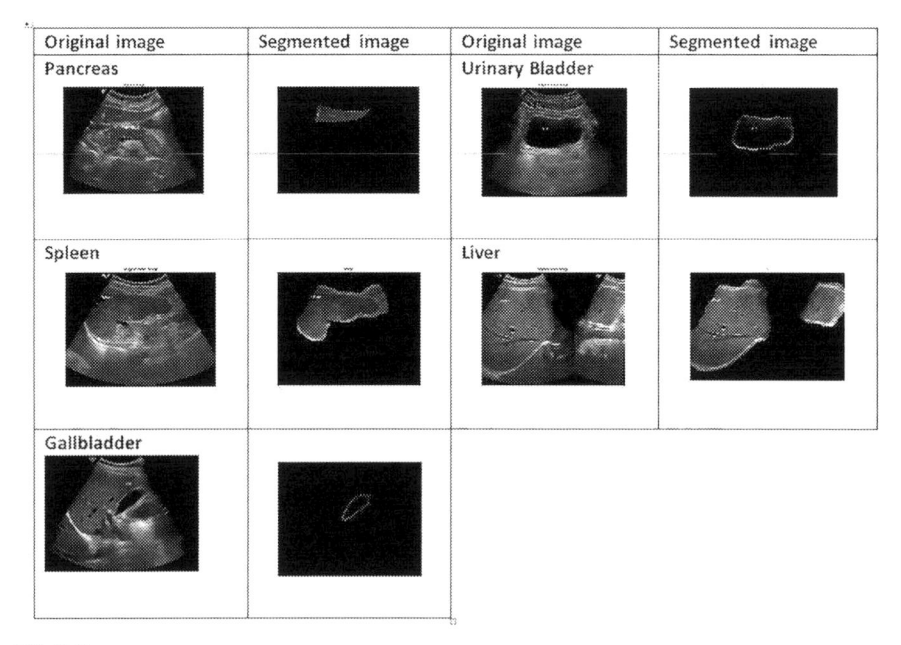

FIG. 8.5

Original and segmented ultrasound images.
Courtesy Irappa Madabhavi, Dr., Ultrasound Image, Kerudi Cancer and Heart Hospital, July 16, 2020, Bagalkot, Karnataka.

A similar segmentation algorithm is applied on the pancreas, liver, gallbladder, spleen, and bladder; this is shown in Fig. 8.5.

To identify the intraabdominal organ from the segmented ultrasound image, a deep neural network is used. By employing the DL technique, a deep neural network is designed and trained. In the following section, intraabdominal organ identification using a deep neural network is explored.

8.4.4 Intraabdominal organ identification using deep neural network

To identify the organ from a segmented ultrasound image, GoogLeNet is used. GoogLeNet is a pretrained CNN with 22 layers. The network is trained to classify the images into 1000 object categories. The network has learned different feature representations for a large variety of images. The input size of the image to the network is 224-by-224. This network is retrained by using intraabdominal ultrasound images to classify the image among six categories: kidney, liver, gallbladder, pancreas, spleen, and bladder. Once the network is retrained, preprocessed segmented ultrasound images are input into the network. The network classifies the input image as one of the classifiers. GoogLeNet is retrained by the transfer learning approach. Transfer learning is most generally used in DL applications.

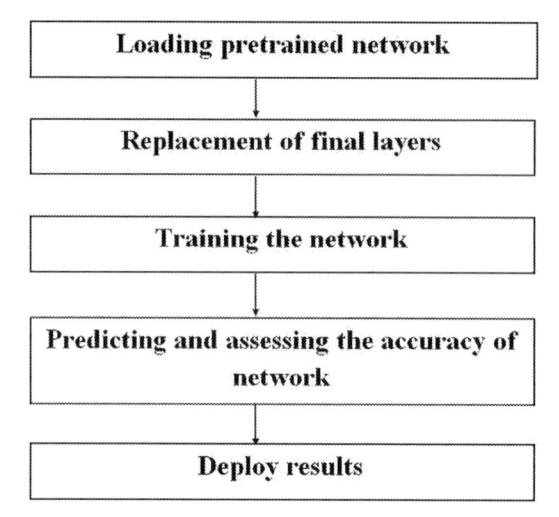

FIG. 8.6

Retraining the GoogLeNet convolutional neural network.

No permission required.

A pretrained network is used as an opening point to study a new task. This fine-tuning of a network with transfer learning is typically far quicker and easier than a network that is trained from scratch with arbitrarily initialized weights.

Retraining the network is depicted in Fig. 8.6.

An analysis of the network gives a pictorial representation of the network architecture and information on the network layers, as shown in Fig. 8.7.

In Fig. 8.7, the first component of the network is the image input layer. This layer needs input images of size 224-by-224-by-3.

The last learnable layer and the final classification layer use the convolutional layers of the network for extracting image features for classification of the input image. These two layers in GoogLeNet are "loss3-Classifier" and "output," as shown in Fig. 8.8. These two layers contain data on how to combine the features that the network extracts into class chances, a loss value, and expected labels. In the proposed methodology, the pretrained network is retrained to classify new images as well as change these two layers with new layers adjusted to the new dataset, as shown in Fig. 8.9. Once the network is retrained, 70% of the images are used for training the network and 30% of the images are used for validation purposes. Fig. 8.10 shows the training progress of the retrained network.

8.4.5 Feature extraction

Once the organ is identified, the image shape and texture features are extracted for identifying abnormalities. Based on the available ultrasound images, the following abnormalities are detected. For identifying kidney stones, the MSER, SURF, and

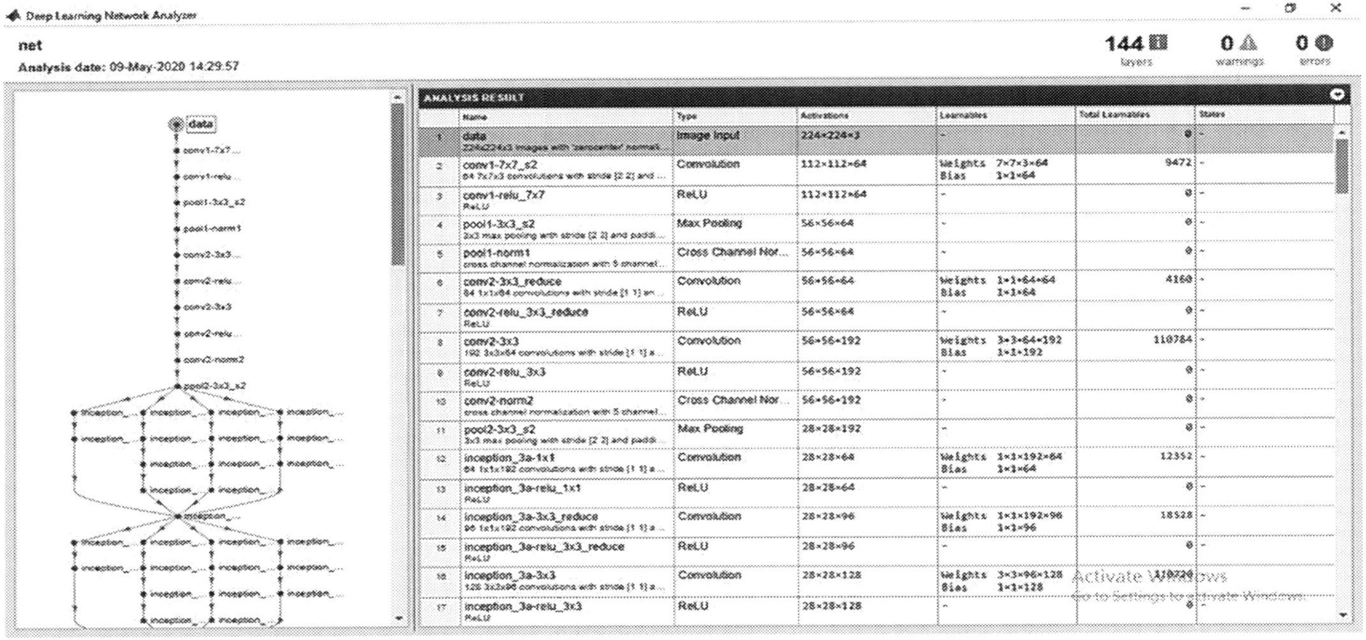

FIG. 8.7

Analysis of pretrained GoogLeNet.

No permission required.

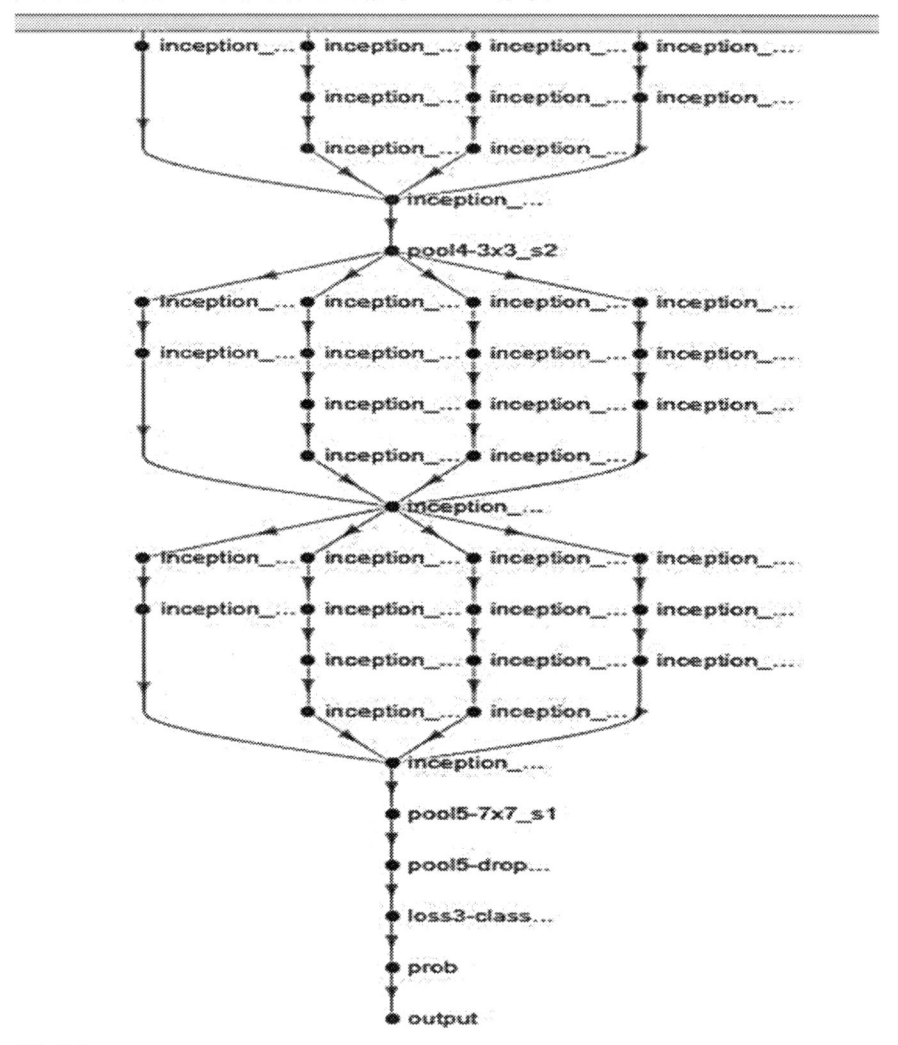

FIG. 8.8

Final layer of the pretrained GoogLeNet.

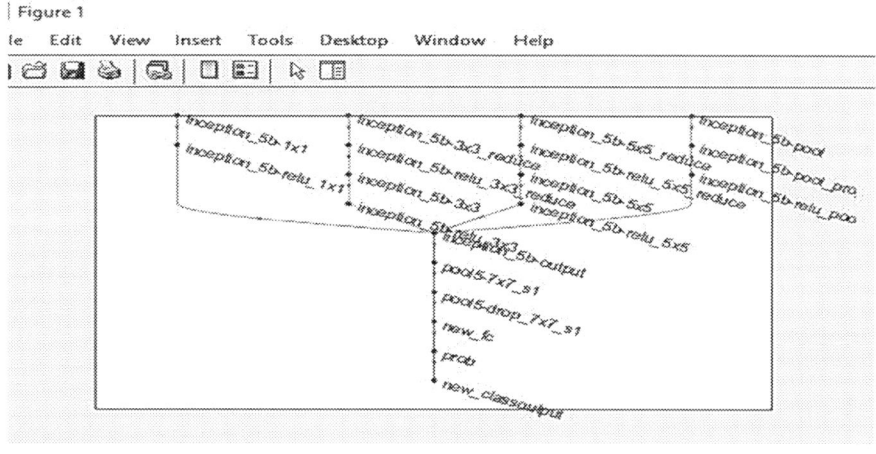

FIG. 8.9

Final layer of the retrained GoogLeNet.

No permission required.

KAZE features are used. For identifying fatty liver, statistical features such as central moment, dispersion, variance, standard deviation, and contrast are considered. For identifying stones in the gallbladder, the shape features circularity and solidity are used.

8.4.6 Abnormality identification and categorization

Based on available ultrasound images, abnormalities such as kidney stones, fatty liver, and gallbladder stones are identified.

A detailed representation of the proposed methodology is shown in Fig. 8.11.

8.5 Results and discussion

For retraining the GoogleNet, ultrasound images of the kidney (3356), liver (1312), gallbladder (517), pancreas (1307), spleen (1186), and bladder (1119) are used; a validation accuracy of 98.30% is achieved. Table 8.2 illustrates the total number of images and the number of images correctly identified by the network.

The performance of the proposed method for the intraabdominal organ for liver identification is better. The system will work efficiently for all six intraabdominal organs. Once the organ is identified, its features are extracted for identifying abnormalities, if any. Based on the data available, kidney stones, gallbladder stones, and fatty liver are identified.

Fig. 8.12 shows a sample of kidney stones, gallbladder stones, and fatty liver.

For identifying kidney abnormalities, it was tested on 64 kidney stone images, where it correctly identified 44 images as abnormal. Similarly, it was tested on 190 normal kidney images, where it correctly identified 84 images as normal. For

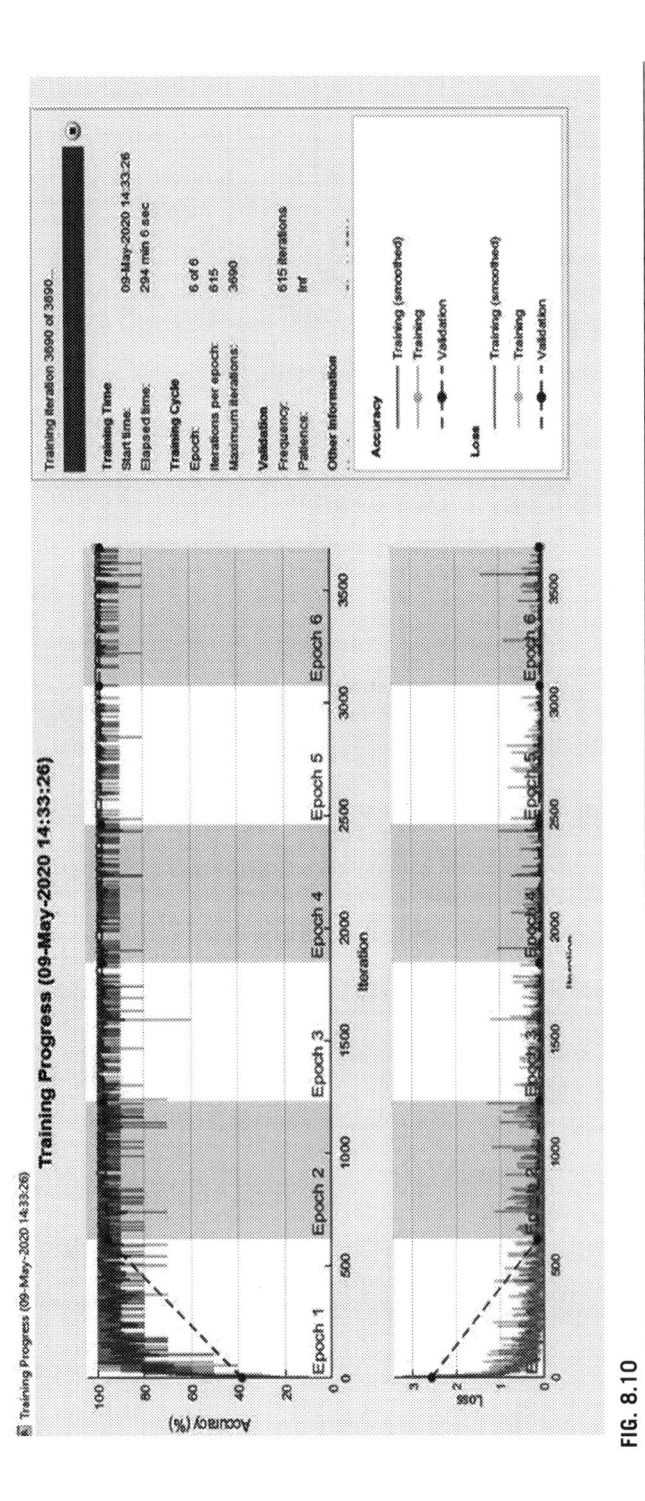

FIG. 8.10

Training progress of the retrained GoogLeNet convolutional neural network.

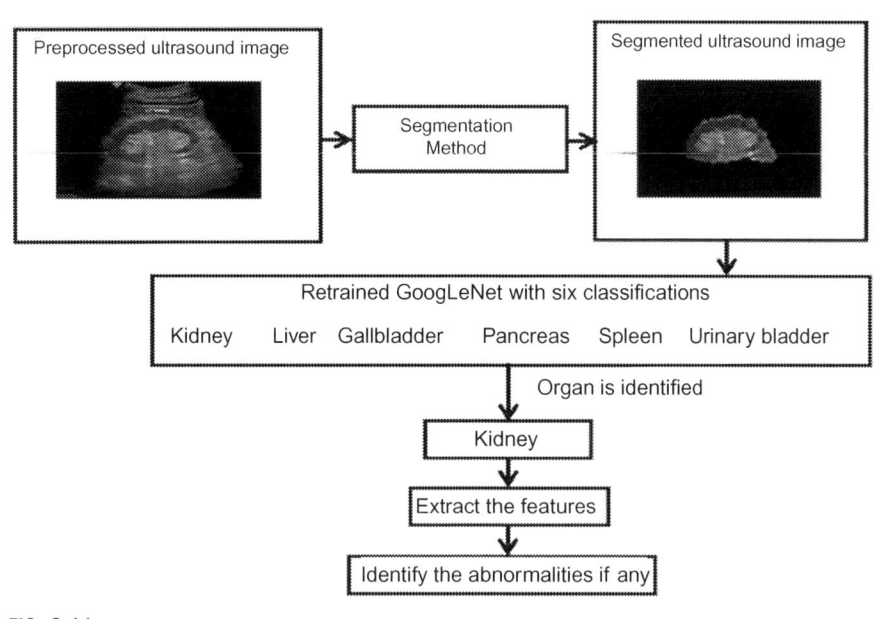

FIG. 8.11

Detailed representation of the proposed methodology.

Courtesy Irappa Madabhavi, Dr., Ultrasound Image, Kerudi Cancer and Heart Hospital, July 16, 2020,
Bagalkot, Karnataka.

Table 8.2 Performance of the proposed intraabdominal organ identification system.

SI. No.	Intraabdominal organ	Total number of testing images	Number of images correctly identified by network	Accuracy in %
1	Kidney	876	872	99.54
2	Liver	415	414	99.75
3	Gallbladder	183	182	99.45
4	Pancreas	346	346	100
5	Spleen	346	346	100
6	Urinary bladder	296	295	99.66

identifying liver abnormality, it was tested on 27 normal liver images, where it identified 24 images as normal. Similarly, it was tested on 94 fatty liver images, where it correctly identified 82 images as fatty liver. For identifying abnormalities in the gallbladder, it was tested on 30 normal gallbladder images, where it identified 21 images

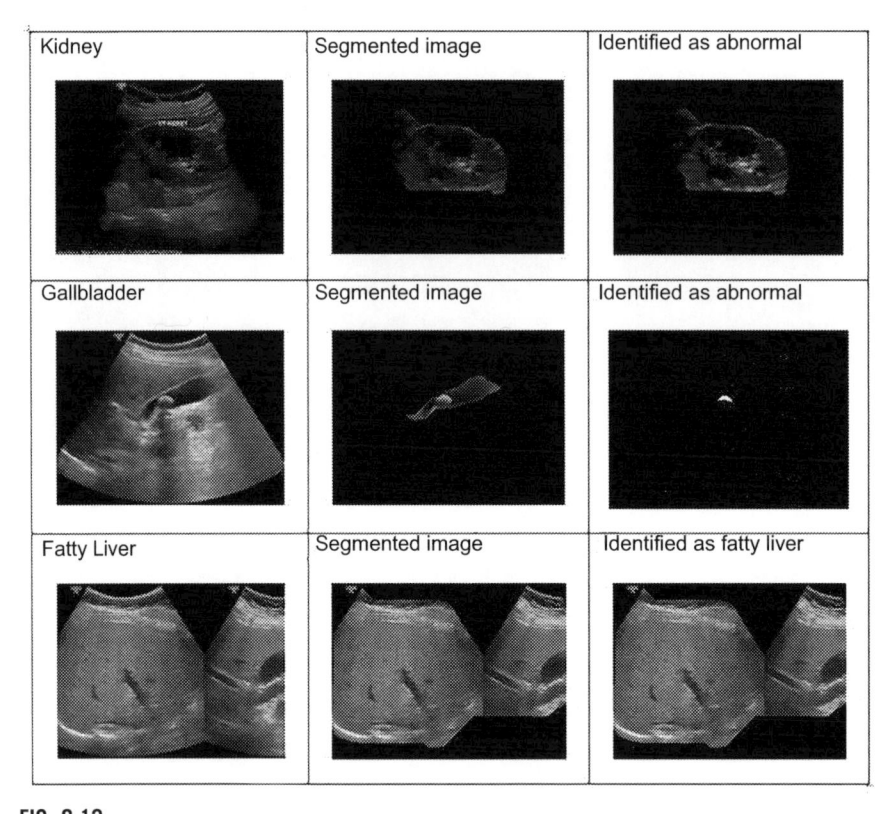

Kidney	Segmented image	Identified as abnormal
Gallbladder	Segmented image	Identified as abnormal
Fatty Liver	Segmented image	Identified as fatty liver

FIG. 8.12

Abnormal kidney, gallbladder, and liver ultrasound images.

Courtesy Irappa Madabhavi, Dr., Ultrasound Image, Kerudi Cancer and Heart Hospital, July 16, 2020, Bagalkot, Karnataka.

as normal. Similarly, it was tested on 25 gallbladder images with stones, where it correctly identified 16 images as abnormal.

8.6 Future scope and perceptive

Based on the available set of intraabdominal ultrasound images of the kidney, liver, pancreas, spleen, bladder, and gallbladder, the proposed clinical advisory system identifies the organ using a deep neural network and abnormalities related with organ are identified by using shape and texture features. In the kidney and gallbladder, abnormalities such as stones are identified. For liver abnormality, fatty liver is identified. The proposed system can be improved by collecting more abnormal intraabdominal ultrasound images.

8.7 Conclusion

ML and DL are unceasingly and perpetually evolving technologies of artificial intelligence (AI). Significant amounts of investments in AI research have been done in healthcare applications in recent years as compared to other sectors to help clinicians for early finding, identification, and treatment of illnesses as well as the prediction and prognosis of a particular disease. The proposed advisory system is one of the significant applications of a healthcare system that help the clinicians for early diagnosis. The proposed system uses DL techniques for identifying organs, and then shape and texture features are extracted for determining abnormalities, if any. The major challenges faced are characteristic artifacts in ultrasound imaging such as SNR and speckle noise having nonzero correlation over quite large distances, which cause difficulties in segmenting the intraabdominal organ. The identification of abdominal organs from ultrasound images using a soft computing technique such as neural networks requires a huge training set; even if a huge dataset is provided, it is unfamiliar to improve the network due to its opaqueness. Because of the incomplete volume effects, gray-level relationships of the adjacent organs, the contrast media affect, and the quite high dissimilarities of organ position and shape, it is a continuously challenging process for automatic identification of intraabdominal organs. Due to the lesser availability of data, identifying all abnormalities of intraabdominal organs is also a challenging task.

References

Akkus, Z., Cai, J., Boonrod, A., Zeinoddini, A., Weston, A. D., Philbrick, K. A., & Erickson, B. J. (2019). A survey of deep-learning applications in ultrasound: Artificial intelligence–powered ultrasound for improving clinical workflow. *Journal of the American College of Radiology, 16*(9), 1318–1328. https://doi.org/10.1016/j.jacr.2019.06.004.

Angadi, S. A., & Hatture, S. M. (2019). Face recognition through symbolic modeling of face graphs and texture. *International Journal of Pattern Recognition and Artificial Intelligence, 33*(12). https://doi.org/10.1142/S0218001419560081.

Arora, H., & Mittal, N. (2019). Image enhancement techniques for gastric diseases detection using ultrasound images. In *Proceedings of the 3rd international conference on electronics and communication and aerospace technology, ICECA 2019* (pp. 251–256). Institute of Electrical and Electronics Engineers Inc. https://doi.org/10.1109/ICECA.2019.8822148.

Bhavana, K., Varthamanan, Y., & Sonti, V. J. K. K. (2016). Enhancing and denoising of bowel gas using image fusion methods to upgrade ultra sound imaging systems. In *Proceedings of the 2016 IEEE international conference on wireless communications, signal processing and networking, WiSPNET 2016* (pp. 460–463). Presses Polytechniques et Universitaires Romandes. https://doi.org/10.1109/WiSPNET.2016.7566176.

Brattain, L. J., Telfer, B. A., Dhyani, M., Grajo, J. R., & Samir, A. E. (2018). Machine learning for medical ultrasound: Status, methods, and future opportunities. *Abdominal Radiology, 43*(4), 786–799. https://doi.org/10.1007/s00261-018-1517-0.

Chen, Y. C., Hong, D., Wu, C. W., & Mupparapu, M. (2019). The use of deep convolutional neural networks in biomedical imaging: A review. *Journal of Orofacial Sciences, 11*(1), 3–10. https://doi.org/10.4103/jofs.jofs_55_19.

Garg, A., & Khandelwal, V. (2016). Speckle noise reduction in medical ultrasound images using coefficient of dispersion. In *2016 international conference on signal processing and communication, ICSC 2016* (pp. 208–212). Institute of Electrical and Electronics Engineers Inc. https://doi.org/10.1109/ICSPCom.2016.7980577.

Gupta, M., & Garg, A. (2017). An efficient technique for speckle noise reduction in ultrasound images. In *2017 4th international conference on signal processing and integrated networks, SPIN 2017* (pp. 177–180). Institute of Electrical and Electronics Engineers Inc. https://doi.org/10.1109/SPIN.2017.8049939.

Hesamian, M. H., Jia, W., He, X., & Kennedy, P. (2019). Deep learning techniques for medical image segmentation: Achievements and challenges. *Journal of Digital Imaging, 32*(4), 582–596. https://doi.org/10.1007/s10278-019-00227-x.

Huang, Q., Zhang, F., & Li, X. (2018). Machine learning in ultrasound computer-aided diagnostic systems: A survey. *BioMed Research International.* https://doi.org/10.1155/2018/5137904.

Hussein, S., Kandel, P., Bolan, C. W., Wallace, M. B., & Bagci, U. (2019). Lung and pancreatic tumor characterization in the deep learning era: Novel supervised and unsupervised learning approaches. *IEEE Transactions on Medical Imaging, 38*(8), 1777–1787. https://doi.org/10.1109/TMI.2019.2894349.

Jabarulla, M. Y., & Lee, H. N. (2018). Speckle reduction on ultrasound liver images based on a sparse representation over a learned dictionary. *Applied Sciences (Switzerland), 8*(6). https://doi.org/10.3390/app8060903.

Kaur, P., Singh, G., & Kaur, P. (2018). A review of denoising medical images using machine learning approaches. *Current Medical Imaging Reviews, 14*(5), 675–685. https://doi.org/10.2174/1573405613666170428154156.

Kumar, E. S., & Bindu, C. S. (2019). Medical image analysis using deep learning: A systematic literature review. *Communications in Computer and Information Science, 985.*

Leong, S. S., Judy Westerhout, C., Chantre-Astaiza, A., Ramirez-Gonzalez, G., Arunkumar, N., Hoong See, M., … Fadzli, F. (2019). A novel algorithm for breast lesion detection using Textons and local configuration pattern features with ultrasound imagery. *IEEE Access, 7*, 22829–22842. https://doi.org/10.1109/ACCESS.2019.2898121.

Li, L., Ross, P., & Kruusmaa, M. (2013). Ultrasound image segmentation by Bhattacharyya distance with Rayleigh distribution. In *Signal processing—Algorithms, architectures, arrangements, and applications conference proceedings, SPA* (pp. 149–153).

Liu, S., Wang, Y., Yang, X., Lei, B., Liu, L., Li, S. X., … Wang, T. (2019). Deep learning in medical ultrasound analysis: A review. *Engineering, 5*(2), 261–275. https://doi.org/10.1016/j.eng.2018.11.020.

Marsousi, M., Plataniotis, K. N., & Stergiopoulos, S. (2014). Shape-based kidney detection and segmentation in three-dimensional abdominal ultrasound images. In *2014 36th annual international conference of the IEEE Engineering in Medicine and Biology Society, EMBC 2014* (pp. 2890–2894). Institute of Electrical and Electronics Engineers Inc. https://doi.org/10.1109/EMBC.2014.6944227.

Nikolić, M., Tuba, E., & Tuba, M. (2016). Edge detection in medical ultrasound images using adjusted Canny edge detection algorithm. In *24th telecommunications forum (TELFOR)* (pp. 1–4).

Precious, J. G., & Selvan, S. (2018). Detection of abnormalities in ultrasound images using texture and shape features. In *Proceedings of the 2018 international conference on current trends towards converging technologies, ICCTCT 2018*Institute of Electrical and Electronics Engineers Inc. https://doi.org/10.1109/ICCTCT.2018.8551174.

Zheng, Q., Furth, S. L., Tasian, G. E., & Fan, Y. (2019). Computer-aided diagnosis of congenital abnormalities of the kidney and urinary tract in children based on ultrasound imaging data by integrating texture image features and deep transfer learning image features. *Journal of Pediatric Urology*, 75.e1–75.e7. https://doi.org/10.1016/j.jpurol.2018.10.020.

Raghesh Krishnan, K., & Radhakrishnan, S. (2017). Hybrid approach to classification of focal and diffused liver disorders using ultrasound images with wavelets and texture features. *IET Image Processing*, *11*(7), 530–538. https://doi.org/10.1049/iet-ipr.2016.1072.

Shanmukhappa, A., & Sanjeevakumar, H. (2018). Hand geometry based user identification using minimal edge connected hand image graph. *IET Computer Vision*, 744–752. https://doi.org/10.1049/iet-cvi.2017.0053.

Sridevi, S., & Sundaresan, M. (2013). Survey of image segmentation algorithms on ultrasound medical images. In *Proceedings of the 2013 international conference on pattern recognition, informatics and mobile engineering, PRIME 2013* (pp. 215–220). https://doi.org/10.1109/ICPRIME.2013.6496475.

Suganya, R., & Rajaram, S. (2012). Classification of liver diseases from ultrasound images using a hybrid kohonen SOM and LPND speckle reduction method. In *IEEE international conference on signal processing* (pp. 1–6).

Van Sloun, R. J. G., Cohen, R., & Eldar, Y. C. (2020). Deep learning in ultrasound imaging. *Proceedings of the IEEE*, *108*(1), 11–29. https://doi.org/10.1109/JPROC.2019.2932116.

Zhang, X., Cheng, S., Ding, H., Wu, H., Gong, N., & Wang, J. (2016). Liver segmentation in ultrasound images based on FCM_I. In *Computational Intelligence and Communication Networks (CICN), international conference* (pp. 309–313). https://doi.org/10.1109/CICN.2016.67.

Xu, Z., Huo, Y., Park, J. H., Landman, B., Milkowski, A., Grbic, S., & Zhou, S. (2018). Less is more: Simultaneous view classification and landmark detection for abdominal ultrasound images. In *Lecture notes in computer science (including subseries Lecture notes in artificial intelligence and Lecture notes in bioinformatics): Vol. 11071* (pp. 711–719). Springer Verlag. https://doi.org/10.1007/978-3-030-00934-2_79.

Zhang, L., Zhu, H., & Yang, T. (2019). Deep neural networks for fatty liver ultrasound images classification. In *Proceedings of the 31st Chinese control and decision conference, CCDC 2019* (pp. 4641–4646). Institute of Electrical and Electronics Engineers Inc. https://doi.org/10.1109/CCDC.2019.8833364.

Zhang, X., Cheng, S., Ding, H., Wu, H., Gong, N., & Wang, J. (2017). Liver segmentation in ultrasound images based on FCM-I. In *Proceedings—2016 8th international conference on computational intelligence and communication networks, CICN 2016* (pp. 309–313). Institute of Electrical and Electronics Engineers Inc. https://doi.org/10.1109/CICN.2016.67.

Zhou, L. Q., Wang, J. Y., Yu, S. Y., Wu, G. G., Wei, Q., Deng, Y. B., … Dietrich, C. F. (2019). Artificial intelligence in medical imaging of the liver. *World Journal of Gastroenterology*, *25*(6), 672–682. https://doi.org/10.3748/wjg.v25.i6.672.

An improved time-frequency method for efficient diagnosis of cardiac arrhythmias

Sandeep Raj

Department of Electronics and Communication Engineering, Indian Institute of Information Technology Bhagalpur, Bhagalpur, India

9.1 Introduction

In the last few decades, a significant growth in the number of global mortalities has been reported in the statistics revealed by the World Health Organization (WHO). In 2008, nearly 17.3 million people lost their lives because of cardiac diseases; that is expected to reach approximately 23.3 million by 2030. Electrocardiography (ECG) has been a graph-based representation of cardiac activity and it serves as a noninvasive tool to diagnose cardiac abnormalities. It is widely used in clinical settings as a basic instrument for the identification of heart abnormalities. It gives significant data regarding the operational parts of the cardiovascular and heart frame. Surface tubes are positioned on an individual's limbs or chest for capturing and recording ECG. It is regarded as a descriptive signal of cardiovascular physiology, helpful for heart arrhythmia diagnosis. An ECG shape irregularity is generally referred to as arrhythmia. Arrhythmia is a normal word that varies from the usual sinus rhythm for any cardiovascular rhythm. Early diagnosis of cardiovascular disease can extend life by suitable therapy and improve the quality of life. Analyzing lengthy ECG records in a brief period is very hard for doctors. From the practical perspective, the study of the ECG sequence may need to be conducted over several hours for appropriate tests. Because the amount of ECG data is huge, its analysis is tiresome and complicated, and there is a high chance of losing the essential information. Therefore, the prevention of cardiac abnormality needs a strong computer-aided diagnosis (CAD) system.

Electrocardiogram (ECG) signals are considered irregular, unsystematic, and transient in nature, so it is very important to accept certain systems that possess those same guidelines in their quantitative scheme. Fourier transform (FT) is used to process stationary signals in the current usual methods, whereas wavelet analysis is useful for processing nonstationary signals. The classical wavelet transform (WT) property was possible due to its features describing the multiresolution observation

Demystifying Big Data, Machine Learning, and Deep Learning for Healthcare Analytics.
https://doi.org/10.1016/B978-0-12-821633-0.00012-X

185

and the ability to display local signal characteristics in time-frequency space (Erçelebi, 2004; Raj, 2018, 2020; Raj, Luthra, & Ray, 2015a; Raj & Ray, 2018b, 2018c; Saxena, Kumar, & Hamde, 2002). Because the WT investigation isn't self-versatile and is fairly founded on an appropriate determination of wavelet functions, the exactness of the conclusions is very constrained. Thus, for an adequate and accurate test, the fundamental characteristics of the ECG signal should be termed and the evaluation method should be selected accordingly.

Several studies have used different approaches to evaluate major aspects from an ECG signal; a few of them are presented below. Acir (2005) used amplitude values for making a feature set. Mahmoodabadi, Ahmadian, Abolhasani, Eslami, and Bidgoli (2005) primarily emphasized the extraction of the ECG function using wavelet transformation with multiresolution. Therefore, the hypotheses about the stationary signal are rendered in all these methods, so they lose the self-adaptability to data processing. The Hilbert-Huang Transform (HHT) is another research and data analysis tool developed by Huang et al. (1998). It produces specific physical data representations through irregular and nonstationary techniques. This method is focused on an empirically determined theory and generally consists of two areas: Empirical Mode Decomposition (EMD) and HSA. EMD is an adaptive process based completely on the amount of input data. EMD oversteps the limits of the classical discrete wavelet transform (DWT) method. The EMD approach dissolves any data input to distinctive intrinsic mode functions (IMFs) (Huang, 2005), which are used in the combined time-frequency-energy specification of the IMFs to perform HSA together. Moreover, the adaptive EMD approach remains an obvious choice for selecting a suitable set of features to provide a generic solution for adequate ECG beat disclosure. However, the classical EMD technique suffers from a drawback: it dissolves the input signals of multiple classes into an unequal number of modes. This report therefore uses the technique of variational mode decomposition (VMD) to define multiple classes of input ECG signals. Many feature extraction techniques are reported in the literature. These techniques extract different types of features such as time (De Chazal, O'Dwyer, & Reilly, 2004; De Chazal & Reilly, 2006; Ince, Kiranyaz, & Gabbou, 2009; Melgani & Bazi, 2008; Minami, Nakajima, & Toyoshima, 1999), frequency (Minami et al., 1999), time-frequency (Pourbabaee, Roshtkhari, & Khorasani, 2018), and nonlinear dynamism (Linh, Osowski, & Stodolski, 2003; Llamedo & Martinez, 2012; Raj, Luthra, & Ray, 2015; Raj, Praveen Chand, & Ray, 2015). These different kinds of features have their significance in their domain of analysis. The features are further applied as input to the machine learning (ML) stage and processed to report higher accuracy.

Artificial neural networks (ANN) (Yegnanarayana, 2009) are commonly used for identification due to their dominant self-determined recognition of random patterns and perceptual capacity. However, this has some disadvantages such as numerous local minima and computational complexity dependency on the input space dimension. A support vector machine (SVM) (Vapnik, 2000) is a fresh technique of computing with a basic geometric description and a sparse alternative to solve such limitations. SVM is superior to ANN because there is less chance of ANN to overfit.

A fresh set of features is built, utilizing EMD to guarantee that the suggested set of beats is classified more accurately. After the VMD (Raj & Ray, 2018a) process starts, five modes are produced to reflect a specific heartbeat with which to measure the frequency of the actual signal using the modes. In addition, for the first four modes, to efficiently differentiate between various categories of ECG beats, a constrained selection of statistical attributes with resultant. An SVM classifier model is instructed by selecting RBF for its kernel value for the categorization. The respective categorization precision is calculated as per the identification of the test information collection by getting the confusion matrix.

The present work aims for the development of an efficient method by combining the feature extraction and ML techniques for the automated and efficient recognition of cardiac signals in practical applications. As such, the dual tree complex wavelet transform (DTCWT) is employed to extract the time-frequency (TF) information from the subsequent input heartbeats and specific features are extracted to form a feature vector that characterizes a heartbeat. The feature vector characterizing every heartbeat is recognized using particle swarm optimization (PSO)-tuned SVMs. The developed methodology is tested in two evaluation schemes: category and subject-specific. Its validation is performed over the Physionet (Moody & Mark, 2001) for recognition into 16 categories of heartbeats. The performance of the method reported under both the evaluation schemes is compared with other methods to signify the superiority of the proposed methodology.

This chapter is organized according to the following. Section 9.2 provides the methods for extraction and feature classification. Section 9.3 provides the methodology proposed, along with the structure of the database and the experiment. Section 9.4 illustrates the outcome and performance analysis of this chapter, and Section 9.5 concludes the chapter.

9.2 **Methods**

This section discusses the existing method used for extracting features from the ECG data, dual tree wavelet transform (DTWT) (Physionet database), whereas a one-versus-one SVM model is designed to use these characteristics to distinguish these cardiac signals into 16 categories of arrhythmias.

9.2.1 **Dual tree wavelet transform**

The classical Fourier transform (FT) represents any input signal into infinitely oscillating sinusoidal basis functions, but it is incapable of providing the time instants of these functions. The discrete wavelet transform (DWT) replaces the limitation of FT using a set of basis functions varying locally known as wavelets. The wavelet functions are scaled and dilated versions of the mother wavelet $\psi(t)$ form an orthonormal basis for any input signal. In other words, any input signal $x(t)$ that is analog in nature and has finite energy can be decomposed using DWT. The WT preserves the time-frequency information of any input signal by estimating its frequency components at different instants of time. In the discrete domain, the DWT is implemented using a cascaded low-pass filter and a high-pass filter along with up and down sampling operations. These filters help to design wavelets and scaling functions in such a manner as to achieve desirable properties.

A recent enhancement in the classical DWT algorithm was made to achieve enhanced properties, which led to the development of DTCWT (Selesnick, Baraniuk, & Kingsbury, 2005). The DTCWT is implemented on any input signal $x(n)$ by employing two parallel DWTs, as shown in Fig. 9.1. Here, the two DWTs are critically sampled. The DTCWT is twice expansive, which generates an M number of coefficients to any input signal of length N where $M > N$. Here, twice the N length DWT coefficients are generated for an n length input signal. It must be noted that no enhanced properties can be achieved if the number of filters in the lower and upper DWTs is the same. Alternately, until the filters are specifically designed, then the subband signals of the lower DWT are considered imaginary and the upper DWT is considered the real part of the complex wavelet transform (Selesnick et al., 2005). Further, the upper DWT wavelet for specifically designed filters is an approximate Hilbert transform of the lower DWT wavelet. In contrast with the critically sampled DWT, this makes the DTCWT nearly shift-invariant. For k-dimensional signals, this property is achieved with only a redundant factor of 2^k. In other words, it is much more computationally efficient than the undecimated DWT.

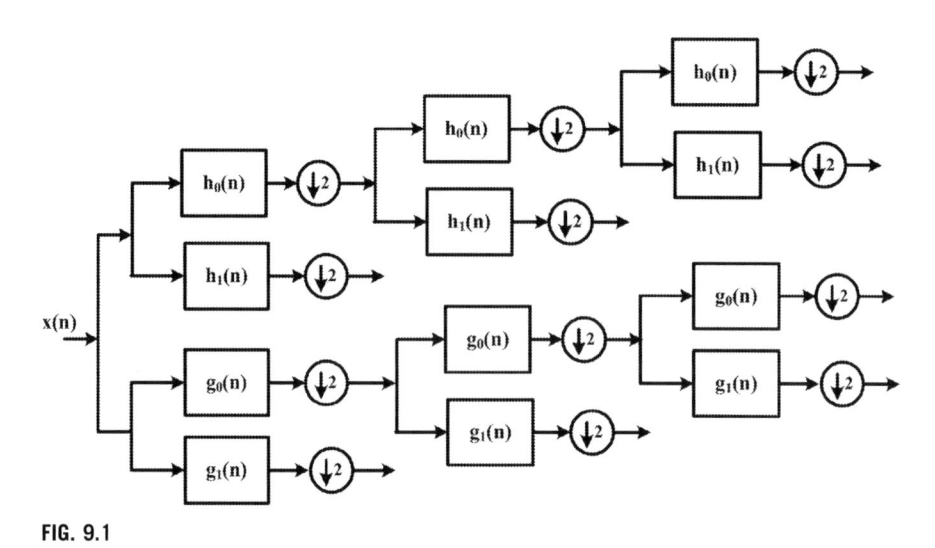

FIG. 9.1

Dataflow under the DTCWT algorithm.

9.2.2 **Support vector machines (SVMs)**

Vapnik's suggested SVM was designed for binary representation (Vapnik, 2000). However, a number of works are being done to efficiently introduce this to multiclass classification (Weston & Watkins, 1998). With certain priori-based nonlinear mapping, the SVM maps the input trends to higher dimensional function space. After that, a linear deciding surface is designed to identify the groups, so SVM classifies the input data of binary groups as a linear classifier all across the input space. Also, the SVM (Hsu & Lin, 2002; Raj & Ray, 2017) is capable of separating the nonlinear mapping of input patterns for which it uses kernel functions. These kernel functions project the nonlinear input patterns in the higher-dimensional space to distinguish among one another. The SVM method focuses on building a hyperplane $w^T x + b = 0$ in which w indicates the hyperplane parameter while b represents the bias word. Fig. 9.2 presents the hyperplane in the higher dimensional separating the binary classes. The limitation of constructing the optimized issue is that the margin should be max between the hyperplane and the support vector. Conditioning the SVM is presented as a quadratic optimization concern by implementing the goal and the restricted function (Vapnik, 2000).

SVMs (Hsu & Lin, 2002) are considerably better as a comparison mechanism compared to ANNs. They are less likely to overfit, they get an easy geometric understanding, and they provide a scarce alternative. It has also been seen that SVMs offer a larger capacity for generalization. The selection of the kernels for nonlinear identification in SVM purely depends on the nature of data. Therefore, it is essential to select the proper kernel function to achieve higher accuracy. Two frequently used techniques—one-against-all (OAA) and one-against-one (OAO)—are used against various binary classifiers. The OAA approach is the oldest MCSVM evaluation execution to build b binary SVM models in which b is the count of classes. In this strategy, the SVM routes the ith model of SVM through looking at every ith class dataset

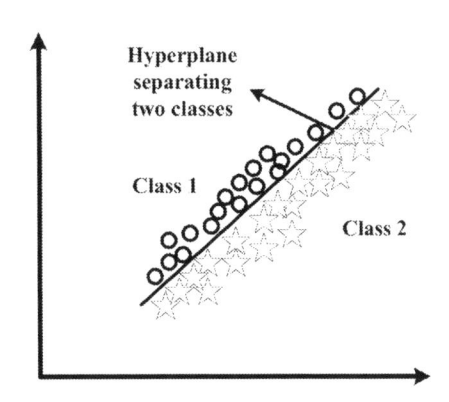

FIG. 9.2

Binary classification in SVM.

as one room and then the other dataset as another room. The OAO approach (Krebel, 1999) is an appropriate as well as effective methodology for applying the MCSVM. Such a technique shapes classifiers $k(k-1)/2$ in which every one of them is practiced on two-class information. The OAO technique is more appropriate for realistic use (Hsu & Lin, 2002) because it utilizes lower dual issues, causing them to stop quickly. In this study, the OAO strategy is used to classify MCSVM and objectively study the option of kernel function. The selection of RBF as a kernel is considered to give better performance.

9.2.3 PSO technique

In this chapter, particle swarm optimization (PSO) (Kennedy, 2010) is employed to gradually optimize the performance metrics of the developed classifier model. The PSO is a population-based stochastic technique inspired by the social behavior of bird flocking or fish schooling. It is widely used for optimization applications. In PSO, a problem is optimized iteratively and it involves higher computational complexity. It improves a particular solution in comparison to a given measure of quality. In PSO (Kennedy, 2010), the particles (variable) communicate with each other and participate in determining the global minimum value for QPP. The PSO provides a solution by moving the particles in the search-space. There are two components associated with every particle, that is, the position and velocity parameters. The position of each particle is influenced by its local best-known position, but it is also guided toward the best-known positions in the search-space. These are updated as better positions are found by other particles (Kennedy, 2010). It is expected that this step moves the swarm toward the best solutions. The various steps for implementing the PSO technique are depicted in Fig. 9.3.

9.3 Proposed methodology

In this chapter, the dual wavelet complex wavelet transform (DTCWT) is combined with the conventional SVM scheme for the identification of 16 classes of heartbeats. The proposed technique is validated over MIT-BIH arrhythmia data. The proposed methodology is presented in Fig. 9.4 and discussed subsequently.

9.3.1 Database

The database is formulated using all 47 recordings of patient data in this analysis for feature extraction and classification from the MIT-BIH directory for arrhythmias (Moody & Mark, 2001). The main motive for choosing this database is its broad use in research articles covering the same area. The logs and number of datasets used by the SVM classifier for training as well as testing are explored in Table 9.1. Here, the signals are collected with the ML-II model at a 360 Hz sampling rate and nearly 30 min duration. The specimen measurements with a specified notation of each peak

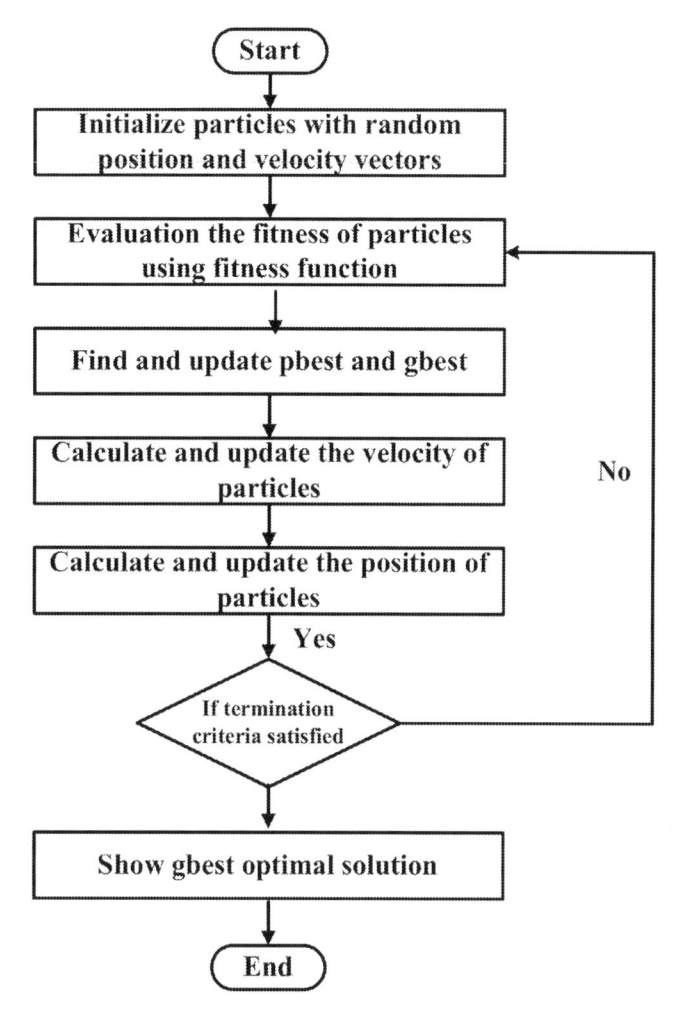

FIG. 9.3

Flowchart of the PSO method.

No permission required.

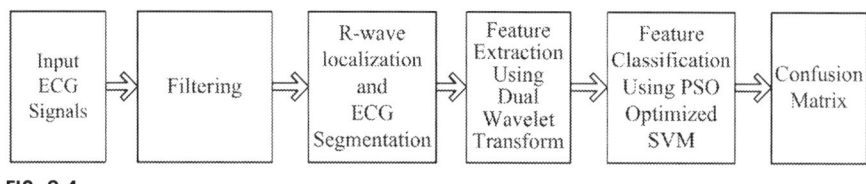

FIG. 9.4

Block diagram of the proposed method.

No permission required.

Table 9.1 Dataset.

Annotation	Heartbeat category	Train	Test	Total
N	Normal (NOR)	12,000	63,017	75,017
A	Atrial Premature Contraction (APC)	900	1646	2546
E	Ventricular Escape Beat (VE)	50	56	106
J	Nodal (Junctional) Premature Beat (NP)	40	43	83
Q	Unclassifiable Beat (UN)	16	17	33
F	Fusion of Ventricular and Normal Beat (VFN)	400	402	802
P	Paced Beat (PACE)	2500	4524	7024
a	Aberrated Atrial Premature Beat (AP)	70	80	150
!	Ventricular Flutter (VF)	200	272	472
V	Preventricular Contraction (PVC)	2500	4629	7129
j	Nodal (Junctional Escape Beat)	100	129	229
R	Right Bundle Branch Block (RBBB)	2600	4655	7255
x	Blocked Atrial Premature Beat (BAP)	90	103	193
f	Fusion of Paced and Normal Beat (FPN)	450	532	982
e	Atrial Escape Beat (AE)	8	8	16
L	Left Bundle Branch Block (LBBB)	2800	5272	8072
16 classes	Total	24,724	85,385	110,109

are used to separate sample beats in any of 16 classes of heartbeats (Moody & Mark, 2001). Once the beats are separated, the data for each class will be split into two equal sections.

9.3.2 Denoising

During the reception of ECG data using sensors, there are many types of unwanted noises that get introduced into the original signal, degrading its quality. The quality of ECG data greatly affects the performance of any automated diagnosis system. The heartbeat-related noise can include quantization noise, muscle artifacts, touch noise, electrosurgical noise, power line interference, and baseline drift. There is a need to eliminate these various types of noises to improve the performance of any diagnosis system by locating the fiducial points accurately. The signal-to-noise (SNR) ratio of any input gets improved in this stage of the classification system. Various filters and techniques are developed to eliminate these noises from the input signals. The raw ECG of record #105 from the Physionet database is shown in Fig. 9.5.

In this study, two median filters are used to eliminate the baseline wander within the heartbeats. A primary filter of size 200ms is employed for locating the QRS

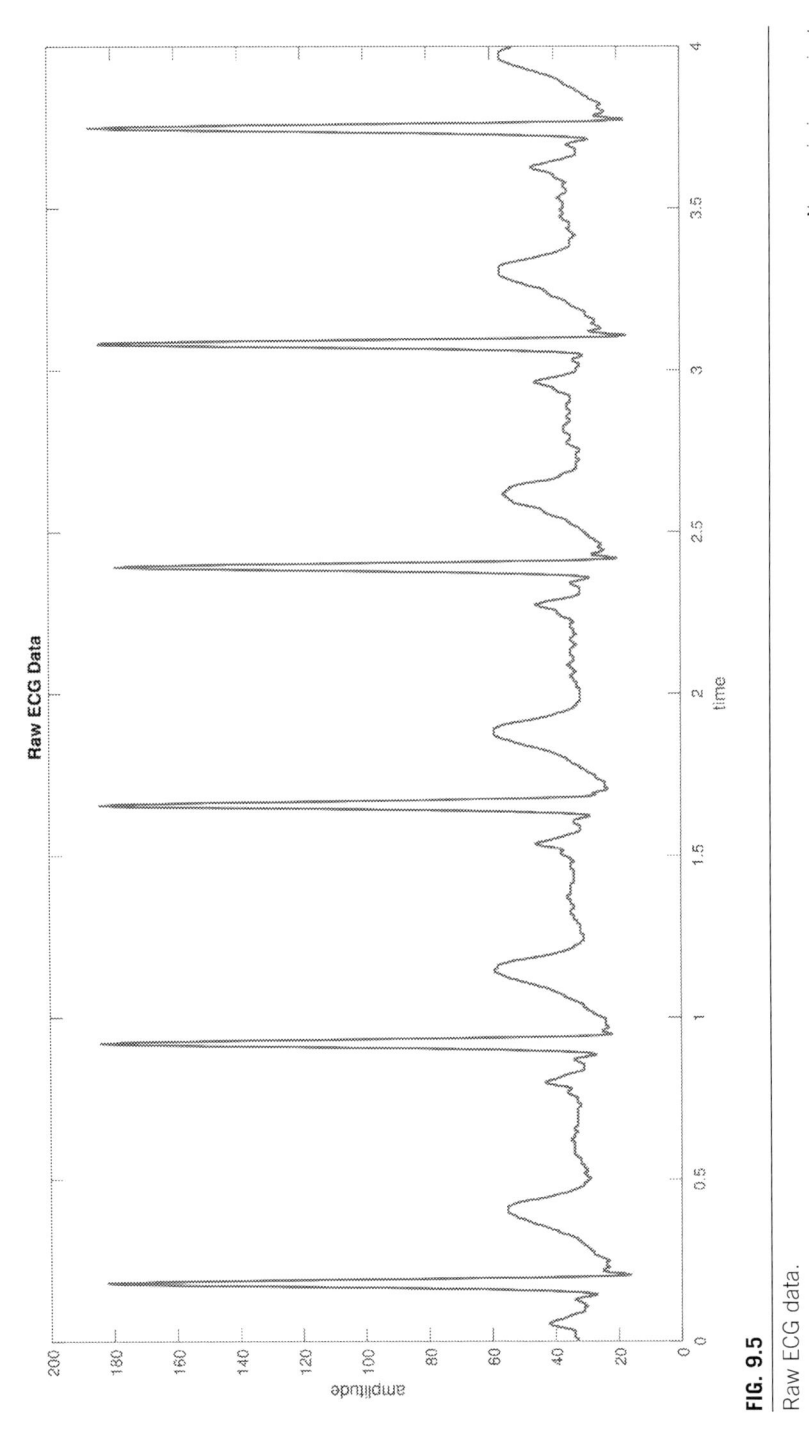

FIG. 9.5

Raw ECG data.

complex and P waves within the input signal. Further, a secondary filter of size 600 ms is used to demarcate the T wave in the heartbeat. As a result, the output of this stage obtained from the second filter is deducted from the raw heartbeat for eliminating the baseline wander from the input heartbeat. On the contrary, a 12-tapped low-pass filter with a cut-off of 35 Hz is used to remove the higher-frequency noise and power-line interferences from input cardiac signals. The output of this filter provides clean or filtered ECG, which is depicted in Fig. 9.6. That is further processed for the automated recognition of heartbeats.

9.3.3 QRS wave localization and windowing

This chapter presents the automatic recognition of arrhythmias depending on the localization of the QRS wave within the heartbeats. Therefore, it is essential in a cardiac diagnosis system to localize the R-peak effectively. Many studies have reported several techniques to demarcate the R-wave within the heartbeats (Raj, Ray, & Shankar, 2018). Among the various techniques reported in the literature, the standard Pan-Tompkins (PT) technique is chosen for R wave localization. The PT technique involves lower computational complexity and higher accuracy while being less noise-sensitive. The QRS waves localized using the PT technique are checked using the annotations file provided in the database. Fig. 9.7 presents the R-peak demarcation using the PT technique of record #100 of the Physionet database.

The segmentation stage involves the segmenting of each ECG signal from the Physionet data records. In the segmentation stage, the samples are considered as 65% after R peak and 35% before R peak. Therefore, a total of 256 samples is chosen to estimate the size of every cardiac signal. Further, it is ensured that no information is lost, starting from the P wave to the T wave of a particular cardiac signal.

9.3.4 Input representation

This chapter employs DTCWT (Selesnick et al., 2005) to carry out the distinction between signals of the same class by extracting the significant time-frequency (TF) characteristics for detecting dangerous cardiac abnormalities. Unlike DWT, DTCWT also provides the multiresolution analysis of any input data. In the implementation of DTCWT, two real filters are used that generates real and imaginary parts, that is, Trees A and B, respectively. Initially, the ECG signal is decomposed to five resolution scales. The output coefficients decomposed for the fourth- and fifth-scaled levels are computed and chosen as features. Then, the absolute values of these coefficients (detail coefficients) are computed at every scale. Further, one-dimensional fast Fourier transform (FFT) is performed over the chosen features. As a result, across the Fourier spectrum as output, a logarithm is taken. This shift-invariant characteristic of DTCWT provides an efficient representation of heartbeats. On the contrary, five kinds of other features are computed to represent the heartbeats in addition to DTCWT: (i) the AC power of the heartbeats, (ii) the kurtosis of the heartbeats, (iii) the skewness of the heartbeats, (iv) the mean of the heartbeat,

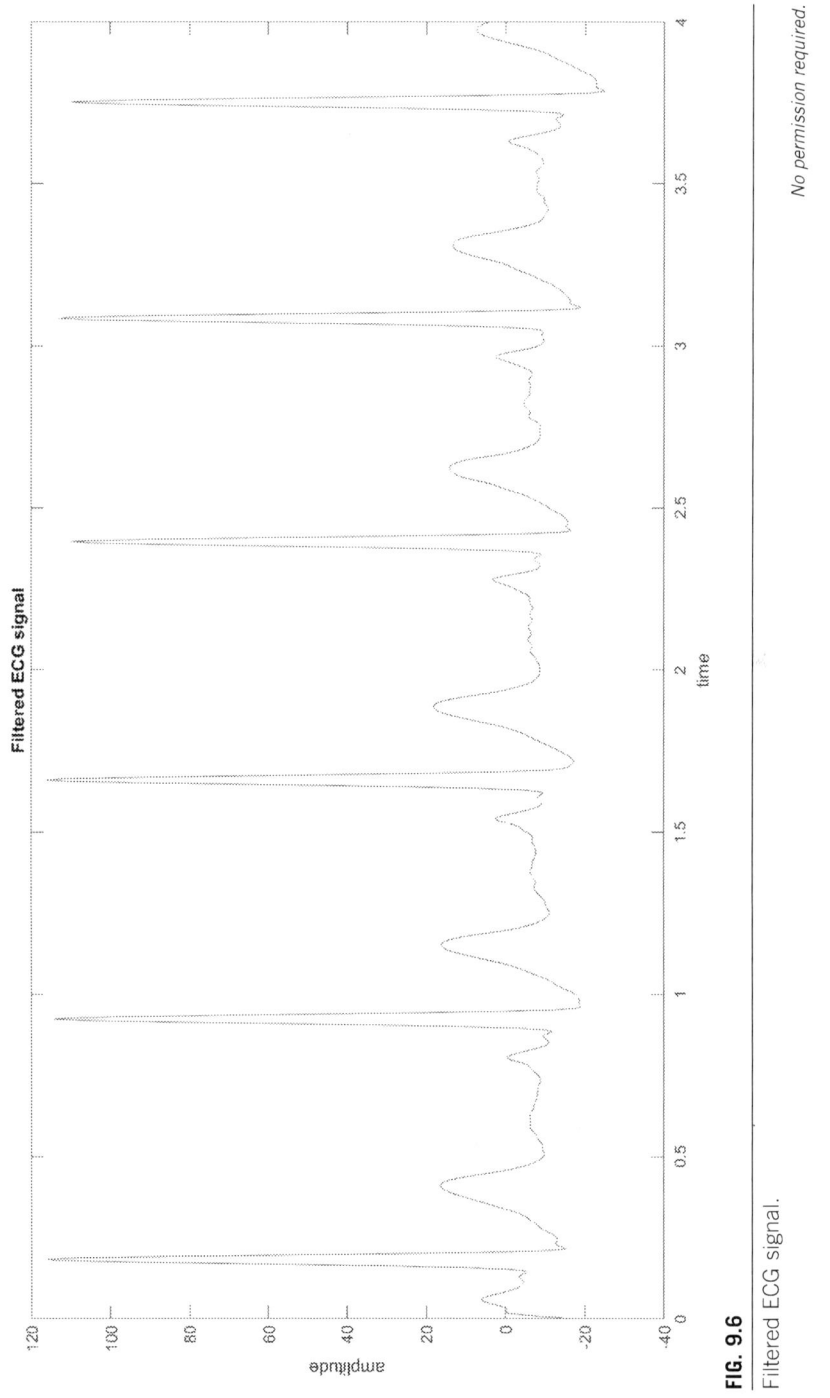

FIG. 9.6

Filtered ECG signal.

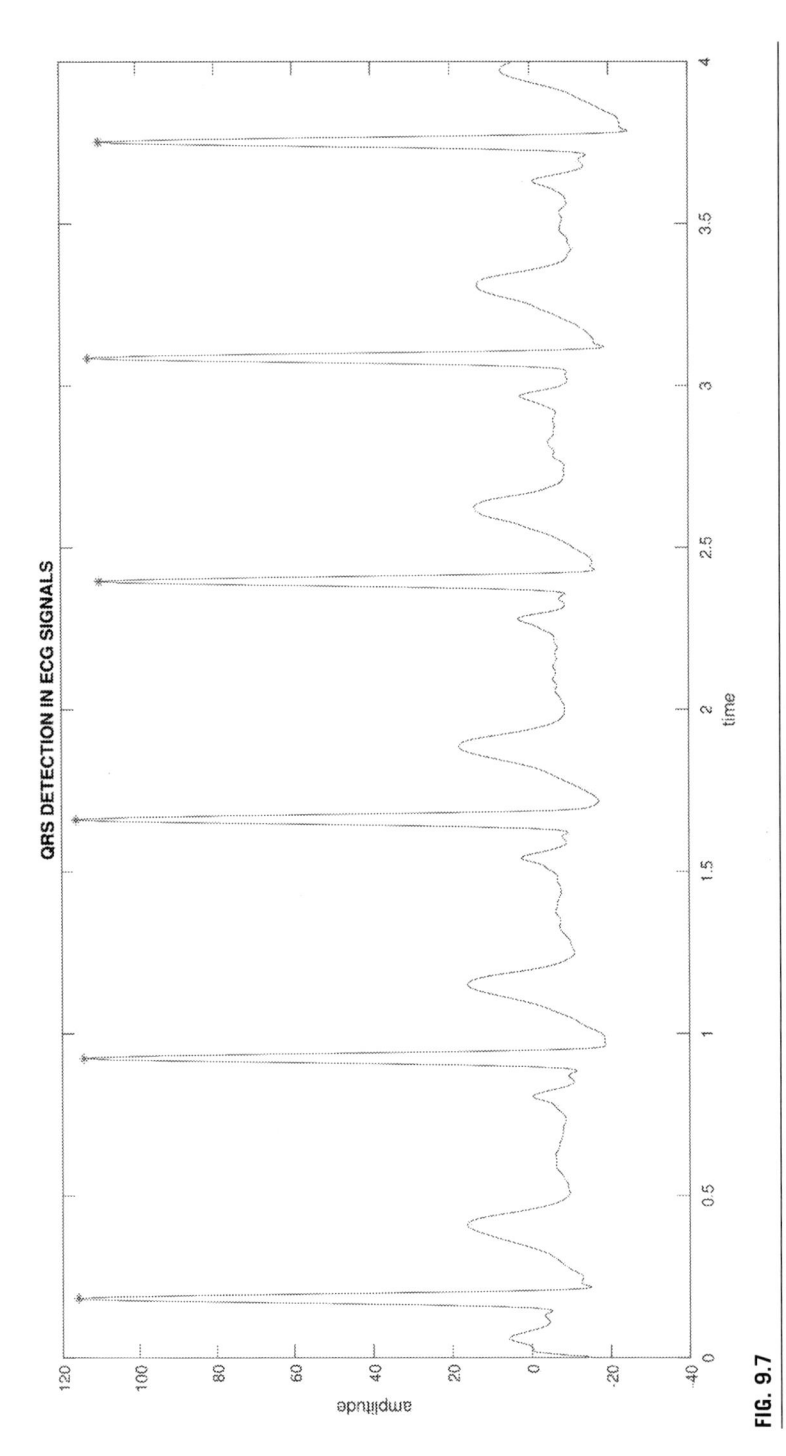

FIG. 9.7

QRS detection in ECG signals.

and (v) the central moment of the heartbeat. In total, 32 features using DTCWT are extracted and termed morphological features representing the different categories of heartbeats. Fig. 9.8 represents the complex wavelet designed specifically for the heartbeat.

The matrix featured enter data includes a substantial amount of useful data of its nature and characteristics. Fig. 9.9 shows the spectra of the normal ECG signal. Direct input data processing can increase the difficulty of the algorithm in the complexity of the classifier model. Therefore, it is essential to extract a significant amount of information as features from any input data to represent the heartbeat for classification purposes. These attributes are added to the classifier model as input, which is trained and tested; these features must be fed as inputs.

Weighted mean frequency (Feature I): The ECG beat data's Hilbert spectrum is calculated utilizing HHT throughout this chapter when measuring the mean frequency of an input ECG using both instant amplitude and instantaneous frequency to all modes of variation (with the exception of the remaining element). With m variational modes, it specifies the mean instant frequency, $IMF(j)$ of $c_j(t)$. The weight frequencies of different features are depicted in Fig. 9.10.

$$MIF(J) = \frac{\sum_{t=1}^{n} W(t)_j a(t)_j^2}{\sum_{t=1}^{n} a(t)_j^2} \tag{9.1}$$

From the original input signal, its frequency is determined by the following equation:

$$WMF = \frac{\sum_{j=1}^{n} MIF(j)\|a_j\|}{\sum_{j=1}^{m} \|a_j\|} \tag{9.2}$$

Absolute value (Feature II): Absolute IMF measures an IMF's absolute amplitude average value, for example,

$$AIMF = \frac{1}{N} \sum_{K=1}^{N} |X_k| \tag{9.3}$$

$$k = \frac{E(x-\mu)^4}{\sigma^4} \tag{9.4}$$

where μ is the mean of x, the standard deviation of x is represented by σ, and $E(t)$ speaks to the expected value of the quantity t.

vi. Skewness (Feature VI): This is a function of data asymmetry around the sample mean value. The skewness of a distribution is described as follows:

$$S = \frac{E(x-\mu)^3}{\sigma^3} \tag{9.5}$$

vii. Central Moment (Feature VII): A distribution's central time of order k is described as follows:

$$m_k = E(x-\mu)^k \tag{9.6}$$

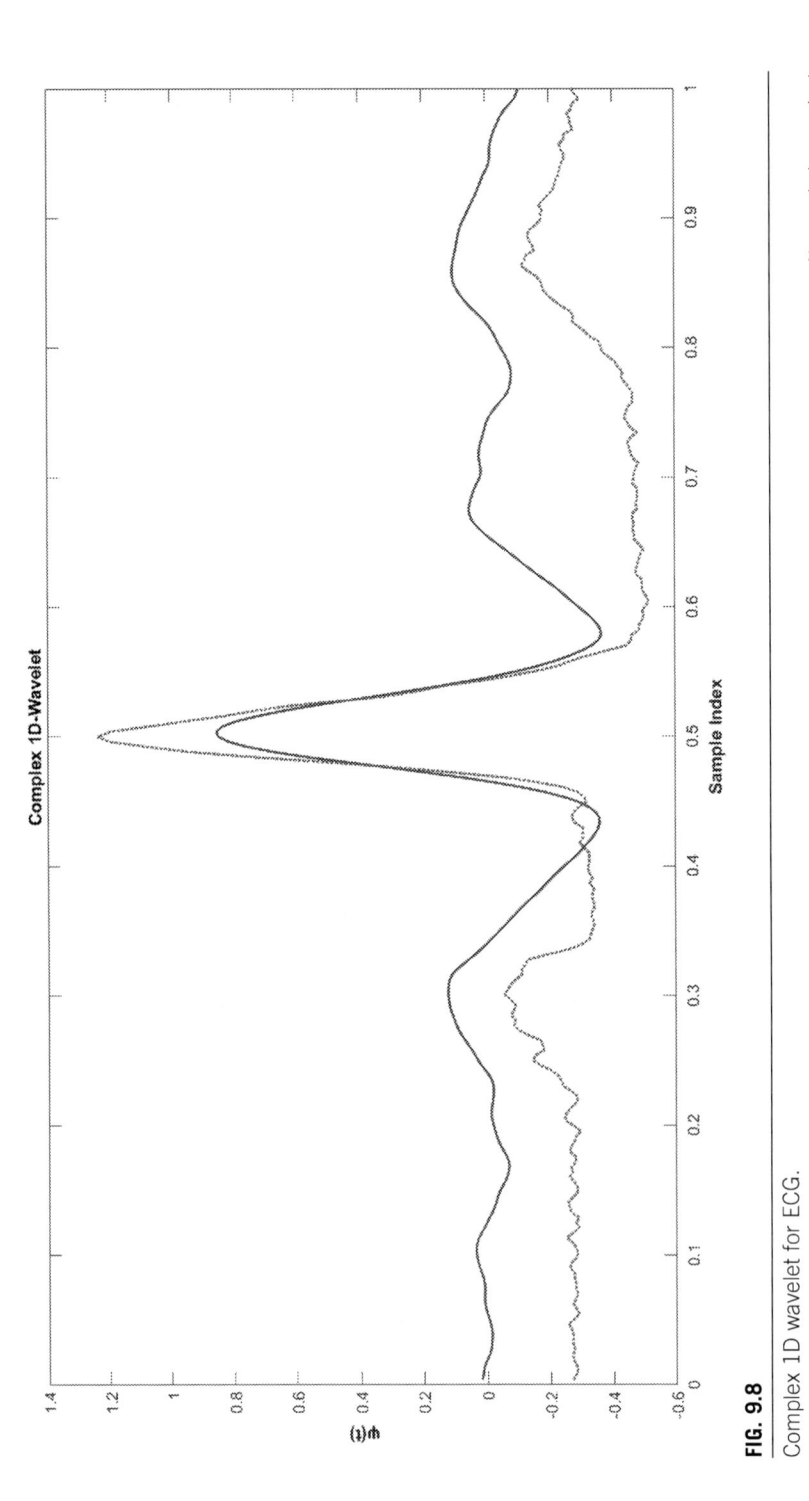

FIG. 9.8

Complex 1D wavelet for ECG.

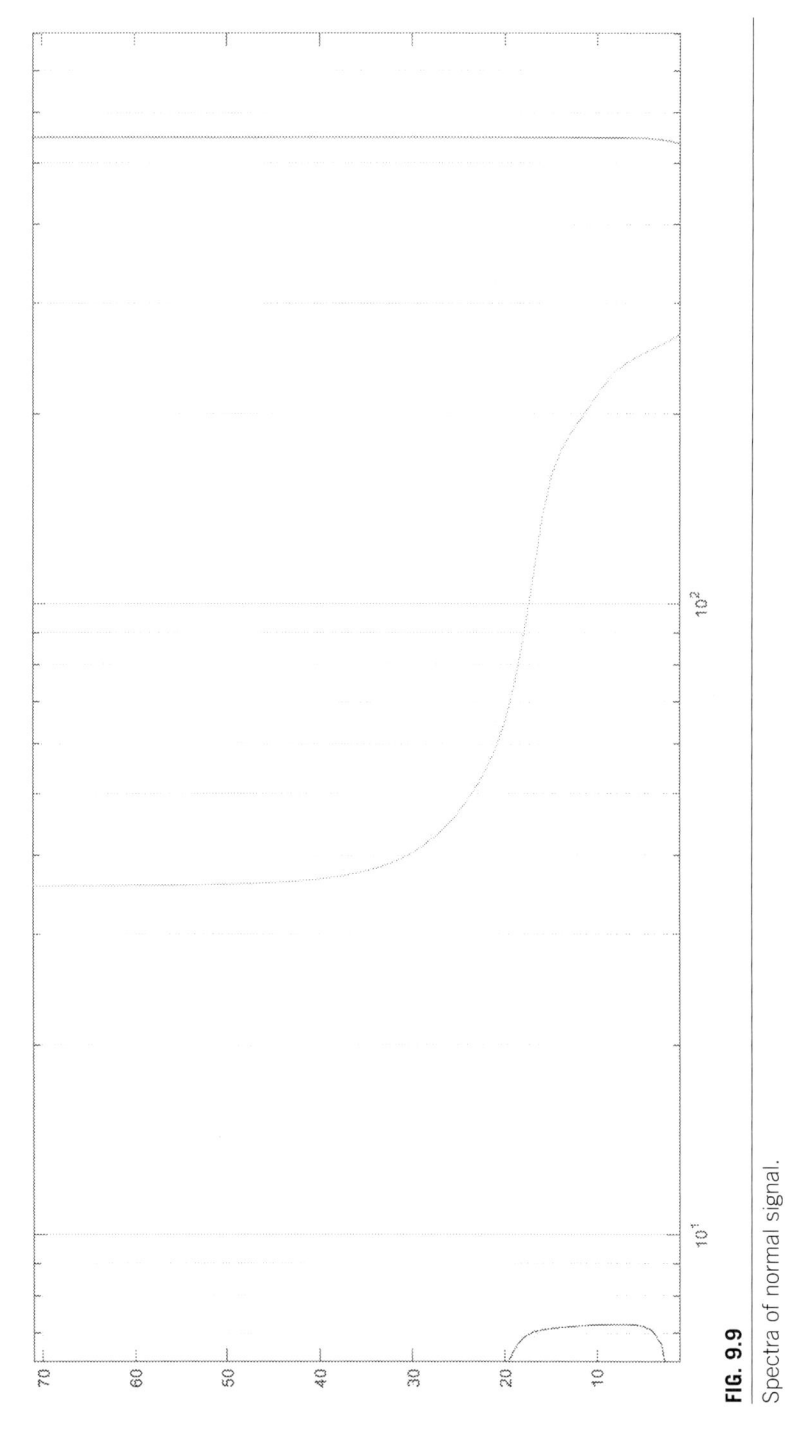

FIG. 9.9

Spectra of normal signal.

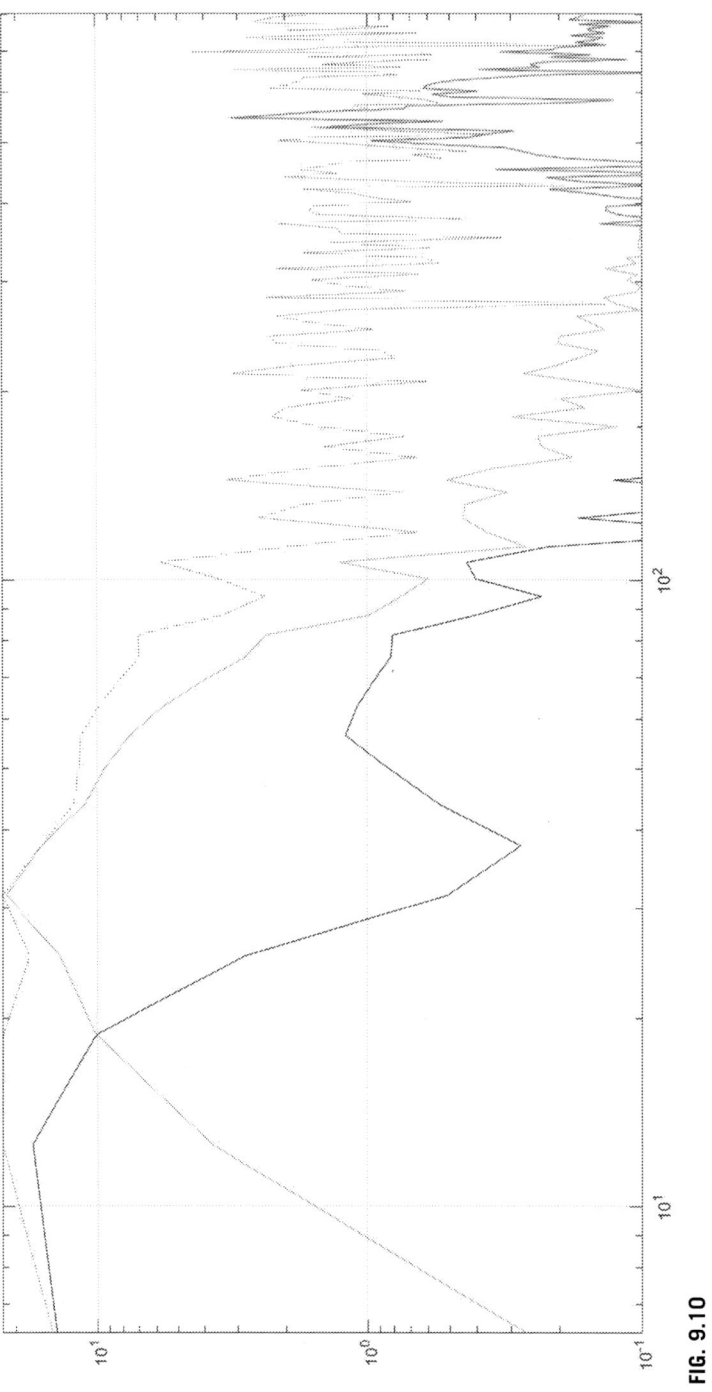

FIG. 9.10

Frequencies of different features.

Experimentally, it is observed in this analysis that the central third-order moment produces good outcomes.

9.3.5 Feature classification

In this chapter, the one-versus-one (OVO) strategy is implemented for recognition into 16 categories of cardiac inputs. This chapter deals with MCSVM grouping. Every class is compared to the remaining *(C-1)* classes, for *C* classes and no repetitions are allowed. In total, the strategy incorporates $C(C-1)/2$ number of categories of heartbeats being classified. Sample data is inserted consecutively into each block, where learning and validation occurs per each data point, keeping in mind just the two classes listed in the block. A division standard for the sorting process is considered to produce the performance of every binary SVM. During the testing phase, the majority rule was used to allocate the final category to every data point. Such a principle takes the majority of votes obtained by any specific class equal to every data point, which determines the assigned class in the input vector per point. Because SVM application requires quadratic programmers, they need to be configured for better results for whom a total of 1000 simulations being used at a time. It gets all possible combinations of the OAO group and instantiates the same amount of binary SVM configurations. After each binary SVM method is prepared and examined using the RBF kernel, the actual vote is performed for the categorized testing set of data. The categories from which they correspond are determined by listing the highest votes cast by each class in the selection system for each section of the test set of data. After the entire grouping method is established, the confusion matrix is established for each dataset of training so the subsequent identification precision is determined. There are various types of kernel functions available for recognition of the nonlinear distribution of input data. Table 9.2 summarizes the accuracies obtained from different kernels. From Table 9.2, the RBF kernel reported the best performance in comparison with other kernel functions in terms of accuracy.

During the training phase of OAO-SVM, the decisiveness of the support vectors is determined by the majority of the computing attempt performed on formulating the problem of quadratic optimization. We talked about the suggested features depicting the category of heartbeats and changes the appropriate information into the high

Table 9.2 Accuracies obtained for different kernels.

Kernel	Training performance (%)	Testing performance (%)
Linear	97.92	84.82
Radial basis function (RBF)	99.52	86.92
Multilayer perceptron (MLP)	98.76	87.23
Polynomial	98.91	86.62

Table 9.3 Accuracy reported by different combinations of features.

Feature subsets	Classification accuracy (%)
Feature I	81.64
Feature I+II	89.87
Feature I+II+III	91.59
Feature I+II+III+IV	96.45
Feature I+II+III+IV+V	97.37
Feature I+II+III+IV+V+VI	98.34
Feature I+II+III+IV+V+VI+VII	99.21

dimensional feature space improves the classification accuracy as well as gives important hints for separating the different categories of heartbeats. The selection of the kernel function is fully empirical; for training, the SVM classifier is therefore selected on an exploratory premise. However, it is concluded in the current research that the RBF kernel function produces better results. Therefore, the training of the classifier model is significant using the proper values of scaling factor σ and C parameters to achieve a higher performance. As listed in Table 9.2, the SVM classifier is implemented using the testing set of data of different records. A sample dataset is inserted onto the classifier when the classifier is programmed to determine the outcomes of the evaluation.

Table 9.3 summarizes the accuracies of the SVM classification scheme for different kinds of features extracted and concatenated together. The higher accuracy reported by the proposed methodology can appreciate the fact that each and every piece of data inserted in the characteristics mentioned is helpful for better results. When used collectively, all nine features achieve an accuracy of 99.21%, proving the full feature set's significance. The proposed methodology wide classification precision justifies the logic that the feature vector designed to represent the heartbeats are efficient which is also supported by the confusion matrix reported while performing the experiments.

The procedure used in implementing the PSO technique is reported in (Kennedy, 2010). The performance metrics employed along with their values are presented in Table 9.4. The PSO enables the classifier model to yield better values of parameters,

Table 9.4 Metric values employed in PSO implementation.

S. no.	PSO metrics	Values
1.	Swarm size (S)	100
2.	Inertia weights	0.6
3.	Maximum iterations	5000
4.	Position parameters	$X_{max} = -X_{min} = 2$
5.	Acceleration constants ($c1$ and $c2$)	2

No permission required.

which results in higher accuracy for multicategory recognition problems by evaluating the fitness of every particle. In addition, the cross-validation rule for k-folds is implemented on the trained as well as the tested data, as shown in Table 9.4.

9.3.6 **Performance metrics**

The efficiency of classification is decided by the calculation of positive predictivity (P_P), along with the accuracy (A_C), sensitivity (S_E), and F-score (F_s) metrics. For a particular category, the classification accuracy is specified as follows:

$$\eta = \frac{\text{Total beats} - \text{total misclassified beats}}{\text{Total beats}} \tag{9.7}$$

The classification accuracies for each class are actually similar to those referenced in the sensitivity parameters in Fig. 9.11 (presented in the next section) that resulted in achieving a classification performance with an accuracy of 99.21%. The computation of performance metrics such as sensitivity, specificity, and positive predictivity are calculated where the values of the TN, FP, TN, and FN parameters are derived via Table 9.5. Calculating the following parameters can determine the efficiency of the proposed methodology. It is possible to define sensitivity (S_E), specificity (S_P), and positive predictability (P_p) as $S_E = \text{TP}{\text{TP}+\text{FN}}$, $P_P = \text{TP}{\text{TP}+\text{FP}}$ and $F_s = 2\text{TP}{2\text{TP}+\text{FP}+\text{FN}}$ where TP denotes true positivity, TN denotes true negativity, FP denotes false positivity, and FN denotes false negativity. TP are those ECG signals correctly allocated to a specific category. As such, false negative (FN) relates to beats that must be categorized into the same class yet are misidentified into another category, whereas false positive measures the beats falsely allocated to that same category. True negative relates to those beats that relate to other classes and are categorized into the same category of another class. The percent of accurate detection or misrepresentation presents the method's overall performance.

Both the S_E and P_P metrics together show the capability of a measure for detecting the presence or absence of a particular condition or disease (i.e., the likelihood ratio). Physically, the S_E signifies subjects with a condition who are correctly classified by the test in percentage. S_P signifies the subjects without the condition who are correctly excluded by the test in percentage.

9.4 **Experiments and simulation performance**

The combination of the proposed features and the classification model is implemented using MATLAB software installed on personal computers (PC). The ECG signal, which is a series of minimal amplitude, is quite tough to deal with. Therefore, this study provides an automated diagnosis system as a generic solution for the recognition of 16 categories of cardiac signals. The validation of the proposed study is done on the Physionet database (Moody & Mark, 2001). A summary of the train and test data is presented in Table 9.1. The training of the developed method is performed over the trained data and the accuracy of the proposed diagnosis system is presented on the testing data. The accuracy of the proposed classification model is computed at a convergence state of the classifier. The performance of the developed methodology is shown in Fig. 9.11. The matrix is known as a confusion matrix. The row of the matrix highlights the actual or original category or number of signals whose labels are available in the database; this is also called ground truth. The column of the confusion matrix highlights the predicted categories of heartbeats recognized using the proposed methodology. In other words, the labels of the columns are the specific groups of beats while the inputs in the labels of the row reflect the number of beats related to the group of columns; this is marked as row illustrated class.

The performance of the proposed system is estimated on the testing data. In the class-based evaluation scheme, out of 86,113 tested cardiac signals, 85,432 cardiac beats recognized by the proposed diagnosis system are correct. As a result, an accuracy of 99.21% is achieved by the proposed methodology.

On the basis of Fig. 9.11, the various other performance metrics such as true positive, false negatives, and false positive are computed and reported in Table 9.5. Further, on the sensitivity of such performance metrics, the positive predictivity and f-score parameters are estimated at 99.21%, 99.1%, and 99.21%, respectively.

	Correctly Classified Instances: 85096 **Misclassified Instances: 1017**								**Accuracy: 98.82%** **Error Rate: 1.18%**								
							Ground Truth										
Class	N	L	R	A	V	P	u	!	F	x	j	f	E	J	e	Q	Σ
N	63178	41	0	286	139	0	27	0	71	0	15	7	0	0	0	0	63764
L	21	5211	0	0	15	0	0	0	0	0	0	0	0	0	0	0	5247
R	71	0	4628	11	6	0	0	0	0	0	0	0	0	0	0	0	4716
A	18	7	0	1609	4	0	0	13	4	0	0	0	0	0	0	0	1655
V	21	13	0	0	4576	0	0	5	19	0	0	0	0	0	0	0	4634
P	0	0	27	0	0	4501	0	0	0	0	0	38	0	0	0	0	4566
a	8	0	0	0	6	0	61	0	0	0	0	0	0	0	0	0	75
!	0	0	0	0	37	0	0	199	0	0	0	0	0	0	0	0	236
F	13	0	0	0	11	0	0	0	377	0	0	0	0	0	0	0	401
x	9	0	0	2	0	0	0	0	0	85	0	0	0	0	0	0	96
j	7	0	0	3	0	0	0	0	0	0	102	0	0	2	0	0	114
f	11	3	0	0	0	6	0	0	0	0	0	471	0	0	0	0	491
E	4	0	0	0	0	0	0	0	0	0	0	0	49	0	0	0	53
J	2	0	0	0	0	0	0	0	0	0	0	0	0	39	0	0	41
e	5	0	0	0	0	0	0	0	0	0	0	0	0	0	3	0	8
Q	3	0	0	0	2	0	0	0	0	0	0	4	0	0	0	7	16

(Left margin labels: Predicted Labels)

FIG. 9.11

Accuracy in class-oriented scheme.

Table 9.5 Performance metrics in class-oriented scheme.

Class	TP	FN	FP	S_E	P_P	F-score
N	63,178	586	193	99.08	99.69	99.39
L	5211	36	64	99.31	98.78	99.05
R	4628	88	27	98.13	99.41	98.77
A	1609	46	302	97.22	84.19	90.24
V	4576	58	220	98.74	95.41	97.05
P	4501	65	6	98.57	99.86	99.22
a	61	14	27	81.33	69.31	74.85
F	377	24	90	94.02	80.72	86.87
x	85	11	0	88.54	100	93.92
j	102	12	15	89.47	87.17	88.31
f	471	20	49	95.93	90.57	93.18
E	49	4	0	92.45	100	96.09
J	39	2	2	95.12	95.12	95.12
e	3	5	0	37.5	100	54.54
!	199	37	18	84.32	91.70	87.86
Q	7	9	0	99.08	99.69	60.87

Fig. 9.12 presents the variation of the sensitivity, positive predictivity, and F-score parameters of the 16 classes of heartbeats with M values. For example, the bars denote the value of the performance metrics for class 1, where class 1 is the normal heartbeat category. Likewise, bars from 2 to 16 denote the other 15 categories of heartbeats along the x axis. Along the y-axis, M is the measure of these out of 100.

9.4.1 Evaluation in patient-specific scheme

In this subsection, the proposed methodology is evaluated under a patient-specific scheme; the database description for performing the experiments is presented in Raj and Ray (2017). The accuracy achieved in an equally split dataset is presented in Table 9.7. The experiments are performed on 49,823 signals under an equally split scheme (S_1) and 101,352 signals under 22-fold cross-validation (S_2) schemes to estimate the accuracy of the proposed methodology. As a result, a total of 43,155 signals under the equally split scheme and 81,803 signals under the 22-fold cross-validation scheme are correctly recognized using the proposed diagnosis system. Accuracies of 86.61% and 88.37%, respectively, were reported. The error rates in these schemes were 13.31% in the equally split scheme and 11.63% in the 22-fold CV scheme.

In Tables 9.6 and 9.7, the performances reported for classes e and q are minimum in terms of accuracy in comparison to the performances obtained for other categories of cardiac signals. The reason behind the poor performance is that much less data

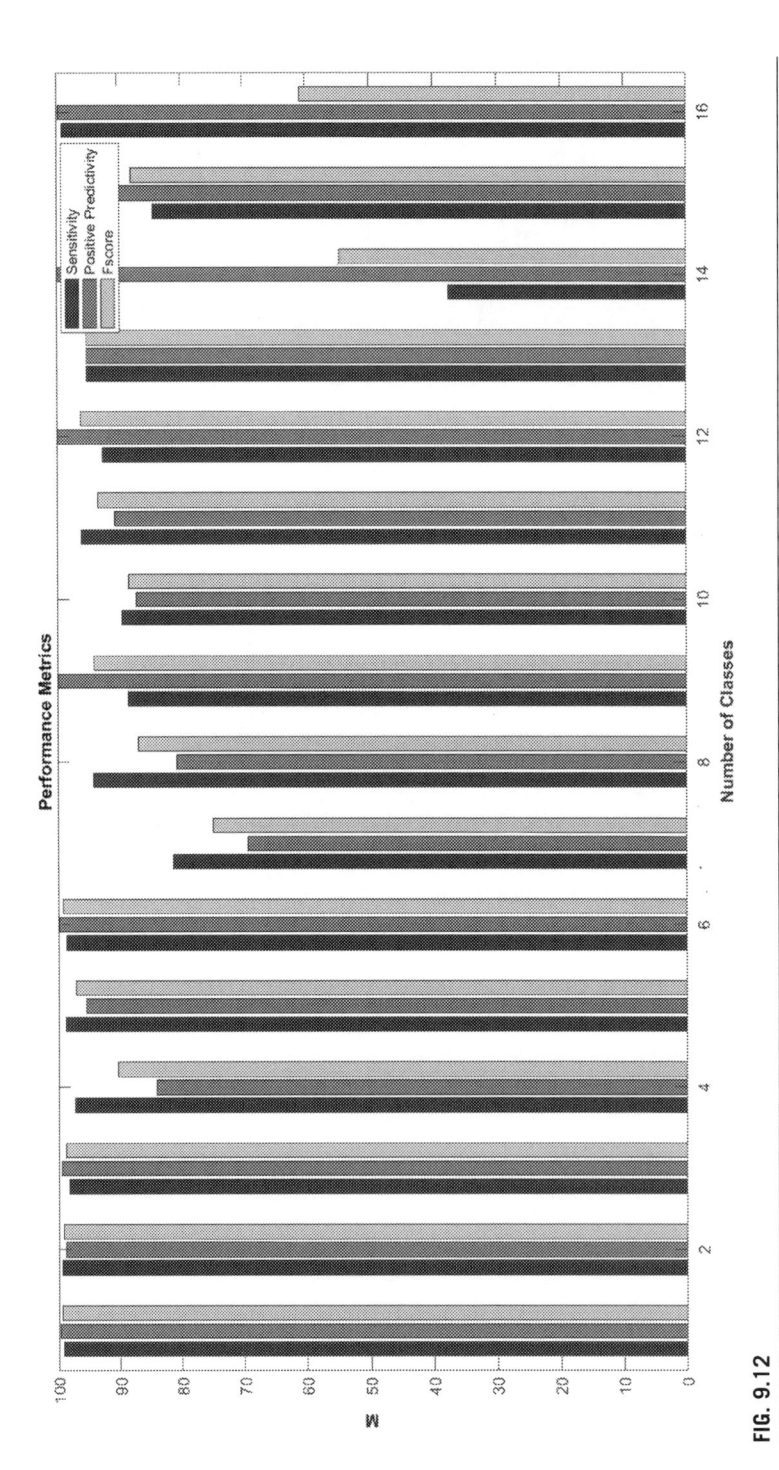

FIG. 9.12

Metrics plot in class-oriented scheme.

Table 9.6 Performance of proposed method in S_1 scheme.

	Category	n	s	v	f	q	Total
				Predicted labels			
GT	N	39,227	934	1268	2771	38	44,238
RR	S	485	1198	244	31	14	1972
OU	V	258	162	2644	145	11	3220
UT	F	191	3	107	84	1	386
NH	Q	3	0	1	1	2	7
D	Total	40,164	2297	4264	3032	66	49,823

have been trained in the e and q classes. In this chapter, the experiments are carried out over all the records of standard MIT-BIH arrhythmia data for these two classes.

Once the confusion matrix is formulated, for each class of heartbeat the following clinimetric parameters are computed subsequently: sensitivity (S_E), positive predictivity, (P_P) and F-score (Fs). These metrics can only be calculated after determining the other parameters such as true positive (T_P), false negative (F_N), and false positive (F_P). All these parameters and metrics are estimated for each category of heartbeat as per the AAMI standard; these are presented in Tables 9.8 and 9.9. In the subject-oriented scheme, Table 9.9 presents the F_s, P_P, S_E metrics calculated for each category of cardiac signal. They are reported to be an overall 86.61% each in the S_1 strategy and 88.37% each in the S_2 strategy.

Figs. 9.13 and 9.14 present the variations of the sensitivity, positive predictivity, and F-score parameters for five categories of heartbeats as per the AAMI recommendation with V_L. In these figures, the bars denote the value of the performance metrics for different classes. Here, 1–5 represent the five heartbeat categories, that is, n, s, f, v, q

Table 9.7 Accuracy reported using CV scheme.

	Category	n	s	v	f	q	Total
				Predicted labels			
GT	N	81,156	997	3611	4222	97	90,083
RR	S	729	1701	517	19	6	2972
OU	V	401	247	6407	407	18	7480
UT	F	339	6	153	301	3	802
NH	Q	3	1	3	5	2	15
D	Total	82,628	2952	10,691	4954	126	101,352

Table 9.8 Performance metrics under S_1.

TP	FP	FN	S_E	P_P	F_s
39,227	937	5011	88.67	97.66	92.95
1198	1099	774	60.75	52.15	56.12
2644	1620	576	82.11	62.00	70.65
84	2948	302	21.76	2.77	4.91
2	64	4	28.571	3.03	5.47
43,112	6711	6711	86.61	86.61	86.61

Table 9.9 Performance metrics under S_2.

TP	FP	FN	S_E	P_P	F_s
81,156	1472	8927	90.09	98.21	93.97
1701	1251	1271	57.23	57.62	57.42
6407	4284	1073	85.65	59.92	70.51
301	4653	501	37.53	6.07	10.458
2	124	12	14.28	1.58	2.85
89,414	11,784	11,784	88.37	88.37	88.37

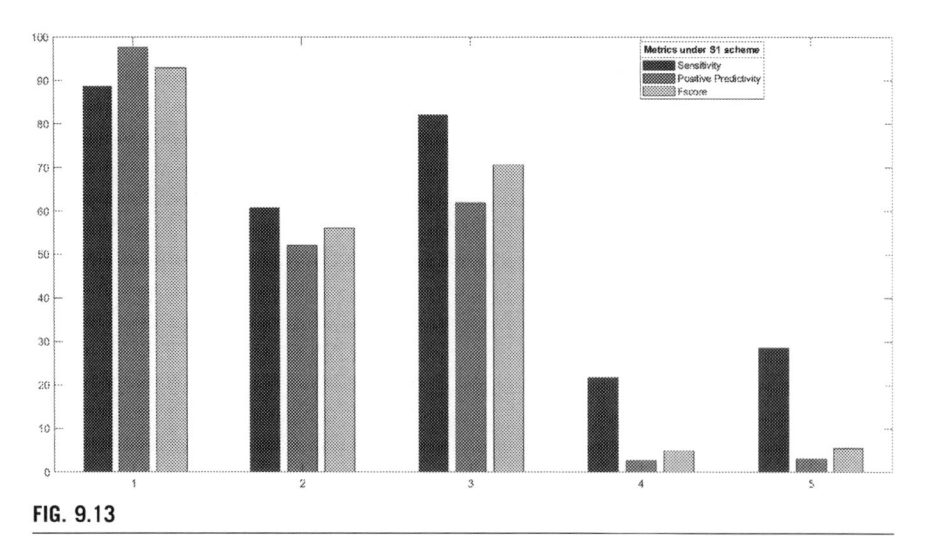

FIG. 9.13

Variation in metrics for different classes under S1 scheme.

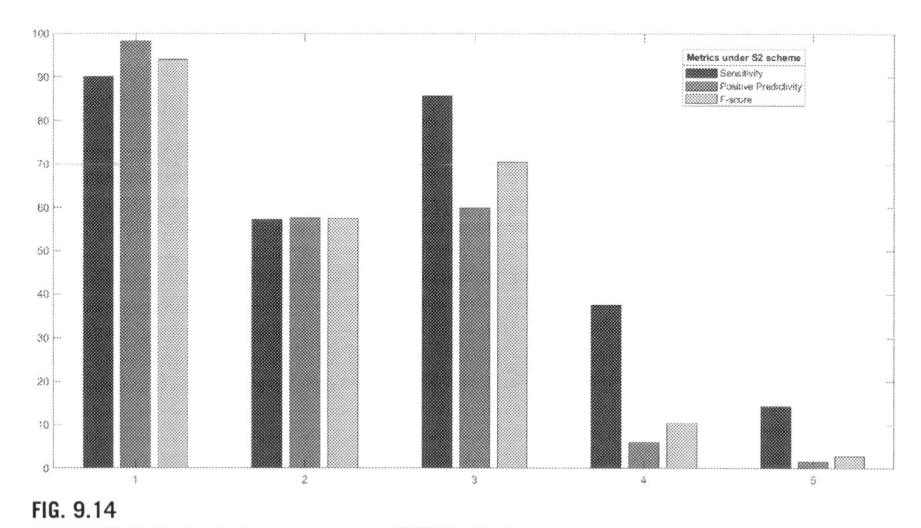

FIG. 9.14

Variation in metrics for different classes under S2 scheme.

No permission required.

along the x-axis. Along the y-axis, V_L is the measure or value of these parameters out of 100.

9.4.2 Advantages of proposed method

Table 9.10 presents the prediction efficiency of the proposed approach compared to that of previous studies reported for an accurate identification of different classes of heartbeats. In Raj (2019), the authors computed the features based on the morphology and R to R interval characteristics. Those are concatenated together to shape the feature vector representing the heartbeats. Then, the feature vector representing the ECG is categorized by a linear discriminant classification scheme. The accuracy of the classification achieved with this method is 85.90%. In Ince et al. (2009), the authors investigated DWT along with principal component analysis to distinguish five categories of cardiac signals using SVM and multidimensional PSO as evaluation techniques to obtain accuracies of 95% and 93%, respectively. Osowski and Linh (2001) categorized a total of seven classes with an accuracy of 96.06%. For three category classification models, an accuracy of 98% was reported in Martis et al. (2013). In, the authors used wavelet transform (WT) to extract significant characteristics from the cardiac signals and identified the back propagation neural network with an accuracy of 97.40% to categorize eight classes of heartbeats. In Raj, Maurya, and Ray (2015), the authors proposed comparable research performed in real time for four classes by implementing the rule-based method; this achieved a 97.96% accuracy. In this chapter, the suggested methodology produces better outcomes in terms of 99.21% classification accuracy than the presented methods, as illustrated in Table 9.10.

Table 9.10 Comparison with other methodologies.

Literature	Feature extraction	Classification model	Classes	A_c (%)
Raj, Maurya, and Ray (2015b)	R to R wave	Morphological Rules	4	97.96
Ince et al. (2009)	WT+PCA	Multidimensional PSO	5	95.58
Llamedo and Martinez (2012)	RR-interval and its derived features	Linear discriminant analysis and expectation maximization	3	98.0
Martis, Acharya, Mandana, Ray, and Chakraborty (2013)	Bio-spectrum + PCA	SVM with RBF kernel	5	93.48
De Chazal and Reilly (2006)	Morphology and heartbeat interval	LDA	5	85.90
Hu, Palreddy, and Tompkins (1997)	Time-domain approach	Mixture of experts	2	94.00
Osowski and Linh (2001)	HOSA	Hybrid fuzzy neural network	7	96.06
Raj, Luthra, and Ray (2015)	DWT using FFT	BPNN	8	97.40
Proposed	**DTCWT +Time domain features**	**PSO-SVM**	**16**	**99.21**

A_C, accuracy; BPNN, *back propagation neural networks*; DWT, *discrete wavelet transform*; FFT, *fast Fourier transform*; LDA, *linear discriminant analysis*; PCA, *principal component analysis*; PSO, *particle swarm optimization*; RBF, *radial basis function*; SVM, *support vector machines*; WT, *wavelet transform*.

9.5 Conclusion and future scope

This chapter introduces a unique technique for the classification of an arrhythmia beat by using a strategy consisting of DTWT-SVM that is simulated and evaluated in MATLAB. Compared to the other current methods, the suggested method is considered highly efficient, as it is reported an accuracy of 99.21% in category (S_1) and 88.37% in subject-specific schemes (S_2). A third validation set is used to optimize the feature matrix, which further decreases the computational burden of the training phase and the recognition system. It is also possible to train the SVM model developed by using the cross-validation strategy to decide the selection of kernel functions. The validation of the proposed methodology is performed over Physionet data to identify 16 generic kinds of heartbeats. These heartbeats are taken from the different records of subjects available in the database. The developed method is efficient in diagnosing cardiac diseases and can be employed in hospitals to provide computer-aided diagnosis solutions for cardiac healthcare.

In the future, the proposed work can be implemented on a microcontroller platform to develop a proof-of-concept prototype and validated over the data of patients in a real-time scenario. Further, the work can be extended to include a greater number of categories of heart abnormalities. In addition to the current work, new computationally efficient techniques with higher accuracy can be developed to provide a more generic solution to the state-of-the-art techniques.

References

Acir, N. (2005). Classification of ECG beats by using a fast least square support vector machines with a dynamic programming feature selection algorithm. *Neural Computing and Applications, 14*(4), 299–309. https://doi.org/10.1007/s00521-005-0466-z.

De Chazal, P., O'Dwyer, M., & Reilly, R. B. (2004). Automatic classification of heartbeats using ECG morphology and heartbeat interval features. *IEEE Transactions on Biomedical Engineering, 51*(7), 1196–1206. https://doi.org/10.1109/TBME.2004.827359.

De Chazal, P., & Reilly, R. B. (2006). A patient-adapting heartbeat classifier using ECG morphology and heartbeat interval features. *IEEE Transactions on Biomedical Engineering, 53*(12), 2535–2543. https://doi.org/10.1109/TBME.2006.883802.

Erçelebi, E. (2004). Electrocardiogram signals de-noising using lifting-based discrete wavelet transform. *Computers in Biology and Medicine, 34*(6), 479–493. https://doi.org/10.1016/S0010-4825(03)00090-8.

Hsu, C. W., & Lin, C. J. (2002). A comparison of methods for multiclass support vector machines. *IEEE Transactions on Neural Networks, 13*(2), 415–425. https://doi.org/10.1109/72.991427.

Hu, Y. H., Palreddy, S., & Tompkins, W. J. (1997). A patient-adaptable ECG beatclassifier using a mixture of experts approach. *IEEE Transactions on Biomedical Engineering, 44*(9), 891–900.

Huang, N. E. (2005). Introduction to the hilbert huang transform and its related mathematical problems. In *Vol. 5. Hilbert-Huang transform and its applications* (pp. 1–26). World Scientific. https://doi.org/10.1142/9789812703347_0001.

Huang, N. E., Shen, Z., Long, S. R., Wu, M. C., Shih, H., Zheng, Q., & Liu, H. (1998). The empirical mode decomposition and the Hilbert spectrum for nonlinear and non-stationary time series analysis. *Proceedings of the Royal Society of London. Series A: Mathematical, Physical and Engineering, 454*(1971), 903–995. https://doi.org/10.1098/rspa.1998.0193.

Ince, T., Kiranyaz, S., & Gabbou, M. (2009). A generic and robust system for automated patient-specific classification of ECG signals. *IEEE Transactions on Biomedical Engineering, 56*(5), 1415–1426. https://doi.org/10.1109/TBME.2009.2013934.

Kennedy, J. (2010). *Particle swarm optimization* (pp. 760–766). Springer US. https://doi.org/10.1007/978-0-387-30164-8_630.

Krebel, U. H. G. (1999). Pairwise classification and support vector machines. In *Advances in kernel methods: Support vector learning* (pp. 255–268). Cambridge, MA, USA: MIT Press.

Linh, T. H., Osowski, S., & Stodolski, M. (2003). On-line heart beat recognition using hermite polynomials and neuro-fuzzy network. *IEEE Transactions on Instrumentation and Measurement, 52*(4), 1224–1231. https://doi.org/10.1109/TIM.2003.816841.

Llamedo, M., & Martinez, J. P. (2012). An automatic patient-adapted ECG heartbeat classifier allowing expert assistance. *IEEE Transactions on Biomedical Engineering, 59*(8), 2312–2320. https://doi.org/10.1109/TBME.2012.2202662.

Mahmoodabadi, S. Z., Ahmadian, A., Abolhasani, M. D., Eslami, M., & Bidgoli, J. H. (2005). ECG feature extraction based on multiresolution wavelet transform. In *Vol. 7. Annual international conference of the IEEE engineering in medicine and biology – Proceedings* (pp. 3902–3905).

Martis, R. J., Acharya, U. R., Mandana, K. M., Ray, A. K., & Chakraborty, C. (2013). Cardiac decision making using higher order spectra. *Biomedical Signal Processing and Control, 8* (2), 193–203. https://doi.org/10.1016/j.bspc.2012.08.004.

Melgani, F., & Bazi, Y. (2008). Classification of electrocardiogram signals with support vector machines and particle swarm optimization. *IEEE Transactions on Information Technology in Biomedicine, 12*(5), 667–677. https://doi.org/10.1109/TITB.2008.923147.

Minami, K. I., Nakajima, H., & Toyoshima, T. (1999). Real-time discrimination of ventricular tachyarrhythmia with Fourier-transform neural network. *IEEE Transactions on Biomedical Engineering, 46*(2), 179–185. https://doi.org/10.1109/10.740880.

Moody, G. B., & Mark, R. G. (2001). The impact of the MIT-BIH arrhythmia database. *IEEE Engineering in Medicine and Biology Magazine, 20*(3), 45–50. https://doi.org/10.1109/51.932724.

Osowski, S., & Linh, T. H. (2001). ECG beat recognition using fuzzy hybrid neural network. *IEEE Transactions on Biomedical Engineering, 48*(11), 1265–1271. https://doi.org/10.1109/10.959322.

Pourbabaee, B., Roshtkhari, M. J., & Khorasani, K. (2018). Deep convolutional neural networks and learning ECG features for screening paroxysmal atrial fibrillation patients. *IEEE Transactions on Systems, Man, and Cybernetics: Systems, 48*(12), 2095–2104. https://doi.org/10.1109/TSMC.2017.2705582.

Raj, S. (2018). *Development and hardware prototype of an efficient method for handheld arrhythmia*. Ph.D. Thesis, IIT Patna.

Raj, S. (2019). A real-time ECG processing platform for telemedicine applications. In *Advances in telemedicine for health monitoring: Technologies, design, and applications*. London, UK: IET.

Raj, S. (2020). An efficient IoT-based platform for remote real-time cardiac activity monitoring. *IEEE Transactions on Consumer Electronics*, 106–114. https://doi.org/10.1109/tce.2020.2981511.

Raj, S., Luthra, S., & Ray, K. C. (2015). Development of handheld cardiac event monitoring system. *IFAC-PapersOnLine, 28*(4), 71–76. https://doi.org/10.1016/j.ifacol.2015.07.010.

Raj, S., Maurya, K., & Ray, K. C. (2015). A knowledge-based real time embedded platform for arrhythmia beat classification. *Biomedical Engineering Letters, 5*(4), 271–280. https://doi.org/10.1007/s13534-015-0196-9.

Raj, S., Praveen Chand, G. S. S., & Ray, K. C. (2015). ARM-based arrhythmia beat monitoring system. *Microprocessors and Microsystems, 39*(7), 504–511. https://doi.org/10.1016/j.micpro.2015.07.013.

Raj, S., & Ray, K. C. (2017). ECG signal analysis using DCT-based DOST and PSO optimized SVM. *IEEE Transactions on Instrumentation and Measurement, 66*(3), 470–478. https://doi.org/10.1109/TIM.2016.2642758.

Raj, S., & Ray, K. C. (2018a). Application of variational mode decomposition and ABC optimized DAG-SVM in arrhythmia analysis. In *Vol. 2018. 2017 7th International symposium on embedded computing and system design, ISED 2017* (pp. 1–5). Institute of Electrical and Electronics Engineers Inc. https://doi.org/10.1109/ISED.2017.8303935.

Raj, S., & Ray, K. C. (2018b). Automated recognition of cardiac arrhythmias using sparse decomposition over composite dictionary. *Computer Methods and Programs in Biomedicine, 165,* 175–186. https://doi.org/10.1016/j.cmpb.2018.08.008.

Raj, S., & Ray, K. C. (2018c). Sparse representation of ECG signals for automated recognition of cardiac arrhythmias. *Expert Systems with Applications, 105,* 49–64. https://doi.org/10.1016/j.eswa.2018.03.038.

Raj, S., Ray, K. C., & Shankar, O. (2018). Development of robust, fast and efficient QRS complex detector: A methodological review. *Australasian Physical and Engineering Sciences in Medicine, 41*(3), 581–600. https://doi.org/10.1007/s13246-018-0670-7.

Saxena, S. C., Kumar, V., & Hamde, S. T. (2002). Feature extraction from ECG signals using wavelet transforms for disease diagnostics. *International Journal of Systems Science, 33* (13), 1073–1085. https://doi.org/10.1080/00207720210167159.

Selesnick, I. W., Baraniuk, R. G., & Kingsbury, N. C. (2005). The dual-tree complex wavelet transform. *IEEE Signal Processing Magazine,* 123–151. https://doi.org/10.1109/MSP.2005.1550194.

Vapnik, V. N. (2000). *The nature of statistical learning theory* (2nd ed.). New York: Springer-Verlag. https://doi.org/10.1007/978-1-4757-3264-1.

Weston, J., & Watkins, C. (1998). Multi-class support vector machines. In *Technical report.* Citeseerx: Royal Holloway, University of London.

Yegnanarayana, B. (2009). *Artificial neural networks.* Prentice Hall of India, PHI Learning Pvt. Ltd.

Local plastic surgery-based face recognition using convolutional neural networks

10

Roshni Khedgaonkar[a], Kavita Singh[a], and Mukesh Raghuwanshi[b]

[a]*Computer Technology, Yeshwantrao Chavan College of Engineering, Nagpur, India,* [b]*Computer Engineering, G.H. Raisoni College of Engineering and Management, Pune, India*

10.1 Introduction

Nowadays, face recognition has become a generally used application in different areas of social networks, for example, Facebook, Instagram, and Twitter, to name a few. Different utilizations of face recognition are authentication, login systems, day care, crime-controlling bodies, commercial areas, and so forth. Moreover, face recognition is also been used as a biometric identification technology. As per a recent survey, the performance of face recognition is influenced by aging, pose variation, facial expression, occlusion, cosmetics, and illumination variations. This makes face recognition a most challenging and ongoing research field of computer vision. On the other hand, in addition to all these factors, an emergent issue is facial plastic surgery, which occurs as a result of medical surgical procedures. In general, there are two forms of facial plastic surgery: global plastic surgery and local plastic surgery. Global plastic surgery changes the complete facial structure. This kind of surgery is mostly suggested in cases where functional losses have to be cured, such as major accidents, acid attacks, or chemical assaults. In global plastic surgery, people undergo various other types of surgeries such as rhytidectomy (face lift), liposhaving, skin resurfacing, dermabrasion, etc.

Aside from global plastic surgery, local plastic surgery just changes a portion of the facial features rather than complete facial image reconstruction. Examples include evacuating birth inconsistencies or irregularities that have shaped the face throughout the years, fixing imperfections from some wounds on the face, and so forth. Rhinoplasty, blepharoplasty, forehead surgery, cheek implants, otoplasty, lip augmentation, and craniofacial changes are various kinds of local facial plastic surgeries. In any case, because of innovations in the medical field, it has increasingly become a cosmetic application. Frequently, celebrities undergo plastic surgery to look better; this will be used to change the facial structure in numerous angles. Common people can now also afford such treatment. Therefore, the presence of individuals with changed facial structure has increased in recent years. Now, it has become a progressively unpredictable issue.

There are various approaches for face recognition that act as a crucial aspect of biometric identification. Some of them are discussed below. The first classification is the algebraic characteristics method, which predominantly incorporates principal component analysis (PCA) (Tian, 2013; Xiaoqian & Huan, 2010), linear discriminant analysis (LDA) (Hua & Jie, 2001; Kim, Kim, & Bang, 2002), and the hidden Markov method (HMM) (Bobulski, 2016). Another, nearby component descriptor method, there are normal LBP features (Lei, Kim, Park, & Ko, 2014; Zhai, Wang, Gan, & Xu, 2015), Gabor features (Kishore, Rana, Manikantan, et al., 2014), and the histogram of oriented gradient (HOG) characteristics (Ghorbani, Targhi, & Dehshibi, 2016). In addition, sparse representation is consider, for example, HMM (Chan & Kittler, 2010; Singh, Zaveri, & Raghuwanshi, 2012). In any case, there are clear downsides to these approaches; for instance, the accurate recognition rate of PCA will be diminished significantly with changes in light and attitude. Sparse representation is a technique that requires fixing the alignment of input images. This is not promising for applications that require practical knowledge. To address the solution to these problems, recently, a new research direction in the field of machine learning (ML) has been introduced with remarkable successes in the area of deep learning (DL). A convolutional neural network (CNN) is a characteristic DL technique that has extraordinary predominance in image processing due to its special construction of local weight sharing. Specifically, the multidimensional input vector can be given as an input into the network. This network escapes feature extraction and reduction in the data reconstruction difficulty of the classification task. Effective applications of CNNs are character recognition (Ahranjany, Razzazi, & Ghassemian, 2010), face recognition (Syafeeza, Khalil-Hani, Liew, & Bakhteri, 2014), human pose estimation (Cheng, 2005), and object detection (Toshev & Szegedy, 2014; Xiaoqian & Huan, 2010).

10.2 Overview of convolutional neural network

CNNs were first proposed by LeCun, Kavukcuoglu, and Farabet (2010). They have been successfully used in computer vision tasks such as hand-composed digit recognition (LeCun et al. (1990). CNNs have of late become prevalent in pattern classification and face recognition. A CNN is a kind of artificial neural network (ANN) inspired by the visual recognition of objects by animals and human beings. This is utilized for various applications consisting of system recommender (Postorino & Sarne, 2011) video and image recognition (Ciresan, Meier, Masci, Gambardella, & Schmidhuber, 2011), and natural processing of languages.

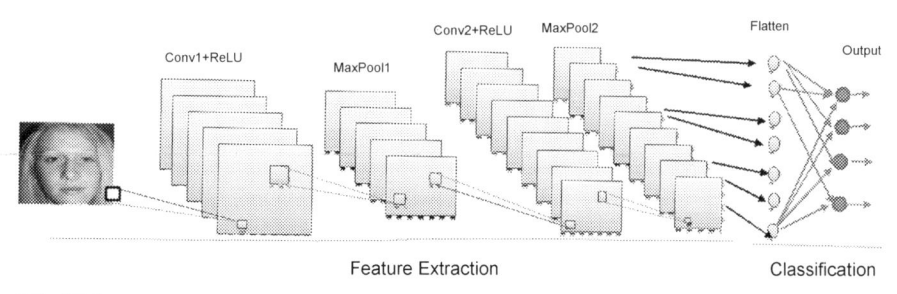

FIG. 10.1

The general architecture of CNN.

The CNN (Kumar, 2017) model makes the presumption that the inputs are images. It allows encoding specific properties into the architecture. Neurons in CNNs are three-dimensional (3D) channels that activate depending on the inputs. They are connected distinctively to a small region, called the responsive field (Rojas, 2013), of a previous neuron's activations. They process a convolution task between the connected inputs and their interior parameters. They also got activated relying on their output and a nonlinearity function (Karpathy, 2019). A typical CNN architecture can be seen in Fig. 10.1. The CNN structure contains convolutional, pooling, rectified linear unit (ReLU), and fully connected layers.

10.2.1 Convolutional layer

The convolutional layer is considered an essential block of the CNN. In a CNN, it is crucial to understand that the layers' parameters and channel are comprised of a set of learnable channels or neurons. These channels have a small receptive field. In the feed forward process, every individual channel goes over the dimensions of the input, thus calculating the dot product from the filter (kernel) pixels and the input pixels. The result of this calculation is a two-dimensional feature map. Through this, the system learns channels made when it detects some specific sort of feature at a spatial location within the feature map. Every one of the neurons calculates convolutions with small portions in LeCun et al. (2010), as shown in Eq. (10.1).

$$y_i = b_i + \sum_{x_i \in x} W_{ij} {}^* x_i \qquad (10.1)$$

Where $y_i \in Y$, $i = 1,2,\ldots D$. D is the depth of the convolutional layer. Each filter W_{ij} is a 3D matrix of size $[F \times F \times CX]$. Its size is determined by a selected receptive field (F), and its feature-map input's depth (CX). An individual neuron in the next layer is associated with certain neurons in the previous layer, called the receptive field (Indolia, Goswami, Mishra, & Asopa, 2018). The local region features from the input image are extracted utilizing a receptive field. A neuron's receptive field is related to a specific portion in the previous layer and frames a weight vector. Fig. 10.2 depicts the receptive fields of neurons in the next layer. For example, if the receptive field is

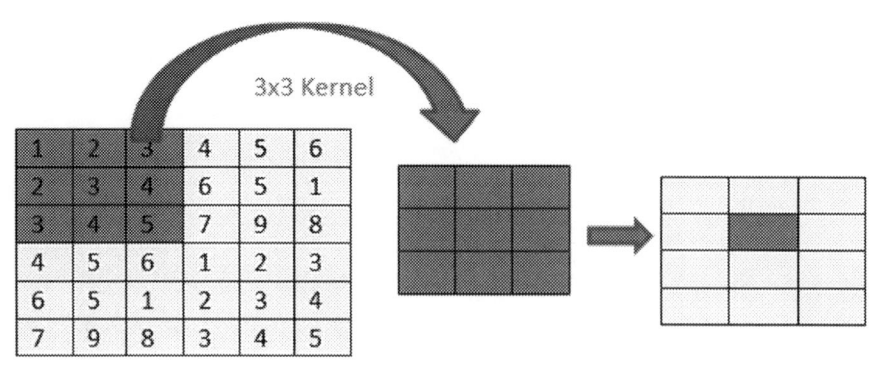

FIG. 10.2

Receptive field of a particular neuron.

three pixels and the feature-map input is a $[32 \times 32 \times 3]$ RGB image, then the size is [3,3,3]. This improves the system's performance by convolutional layer by utilizing identical neurons for every pixel in the layer.

10.2.2 Pooling layers

Pooling layers are mainly authorized for dimensionality reduction, which reduces the input dimensions by performing the downsampling operation. The reductions perform less computation as the information advances to the next pooling layer, and it also neutralizes overfitting. The most well-known schemes utilized in the layer networks are max pooling and average pooling. The max pooling example with operation is shown in Fig. 10.3. It gets max pixel value from the given arrangement of input. In Boureau, Ponce, and Lecun (2010), the most popular pooling layers are max pooling and average pooling. Max pooling can give an outcome in a quicker convergence of information. The image analysis is the most commonly used application of max pooling.

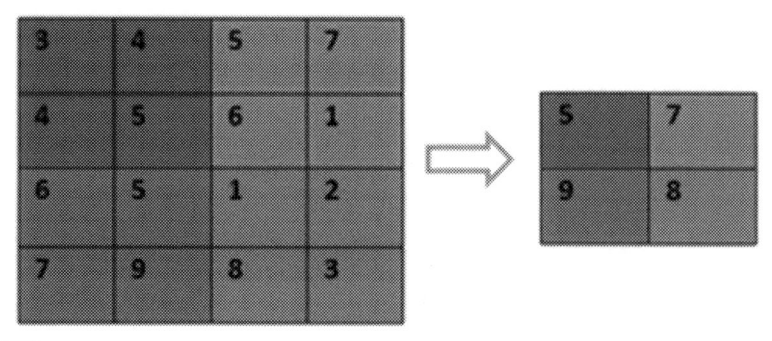

FIG. 10.3

Max pooling operation performed by choosing a 2×2 window.

In addition to average pooling and max pooling, a pooling layer has other variations: stochastic pooling (Wu & Gu, 2015), spatial pyramid pooling (He, Zhang, Ren, & Sun, 2015), and def pooling (Tang et al., 2015).

10.2.3 Fully connected layers

The neurons of the FC layer are connected to nearly all neurons in the previous layers. This gives full associations. FC mainly translates the two-dimensional (2D) feature map to a one-dimensional (1D) feature vector. In respect to this, the 1D feature vector acts as a classifier for classification purposes (Krizhevsky, Sutskever, & Hinton, 2012).

10.2.4 Activation functions

The commonly used activation functions are of a nonlinear nature. The nonlinear activation functions have various categories. **Sigmoid** activation predicts the probability with the range between 0 and 1. This functions in the conventional ML algorithms. It is represented in Eq. (10.2) (Moraga, 1995).

$$f(x) = \frac{1}{1 + e^{-x}} \tag{10.2}$$

Next, the tanh activation function (LeCun, Bengio, & Hinton, 2015) is also like the logistic sigmoid, but the difference is that the range may vary from -1 to 1 as expressed in Eq. (10.3).

$$f(x) = \frac{e^x - e^{-x}}{e^x + e^{-x}} \tag{10.3}$$

The utilization of **Rectified Linear Unit (ReLU)** has substantiated itself superior to the previous, due to simplest computation of partial derivative of ReLU. The saturating nonlinearities such as sigmoid are slower than nonsaturating nonlinearities such as ReLU, as represented in Eq. (10.4) (Krizhevsky et al., 2012). The ReLU is the most used activation in every CNN or deep learning method.

$$f(x) = \begin{cases} 0, & x < 0 \\ x, & x \geq 0 \end{cases} \tag{10.4}$$

ReLU doesn't allow gradients to disappear. Still, the effectiveness of ReLU declines because of the large gradient operator. This makes modified weight causes the neuron deactivate, it leads to Dying ReLU. A dying ReLU is an acceptable factor that is frequently affected. This factor can be settled utilizing **Leaky ReLU** (Maas, Hannun, & Ng, 2013), as described in Eq. (10.5).

$$f(x) = \begin{cases} x, & x \geq 0 \\ \alpha x, & x < 0 \end{cases} \tag{10.5}$$

Where α is a constant.

10.2.5 **CNN architectures**

There are various CNN architectures. The details of each architecture are mention as follows:

10.2.5.1 *LeNet*

LeNet was introduced by LeCuN in 1998 (Lecun et al., 1995). It is well known because of its chronicled significance, as it was the first CNN. This CNN indicated a classical performance on hand digit recognition tasks. It can order digits while not being influenced by low alterations or the change of position and rotation. LeNet is a feed-forward neural network that establishes five substituting layers of convolution with pooling, trailed by two fully connected layers, as shown in Fig. 10.4. It has a seven-layer structure with 60,000 trainable parameters. The shortcoming of a conventional multilayered fully connected neural network is that it is computationally expensive (Gardner & Dorling, 1998). LeNet broke the bordering pixels that are associated with one another and are dispersed among the complete image. In this way, with very few parameters, convolution found an effective mode to extract the same features at various locations. It is mostly use to read ZIP codes, digits, etc.

10.2.5.2 *AlexNet*

LeNet (Lecun et al., 1995) started the historical backdrop of deep CNNs. However, around that time, CNN was constrained to hand digit recognition tasks and did not show considerable performance on various categories of images. AlexNet, proposed by Krizhevsky et al. (2012), is measured as the first deep CNN architecture. This architecture indicated weighty outcomes for image classification and recognition tasks. In addition to this, it also improved the learning ability of the CNN by making it more profound and by applying a few parameter optimization strategies. The architecture of AlexNet is shown in Fig. 10.5. In mid-2000, the hardware capacity was limited to learn the ability of deep CNN networks. So, to benefit from the illustrative limit of deep CNNs, AlexNet was prepared. It is formed on two NVIDIA GTX 580 GPUs to conquer the deficiencies of the hardware. AlexNet contain an eight-layer

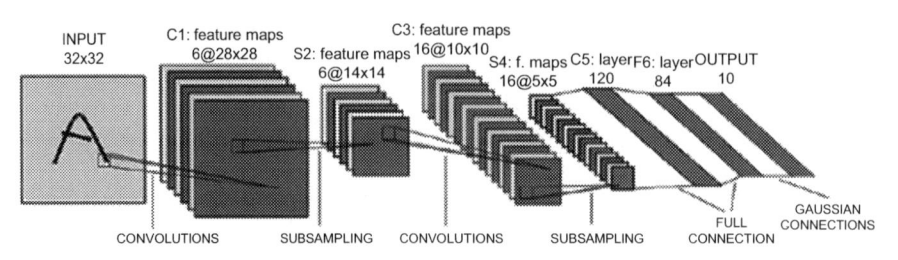

FIG. 10.4

LeNet consists of a 7-layered architecture with 60K parameters.

From Syed et al. (c. 2018). Convolutional neural network architectures for image classification and detection. *https://www.researchgate.net/publication/328102687_Convolutional_neural_network_architectures_for_ image_classification_and_detection.*

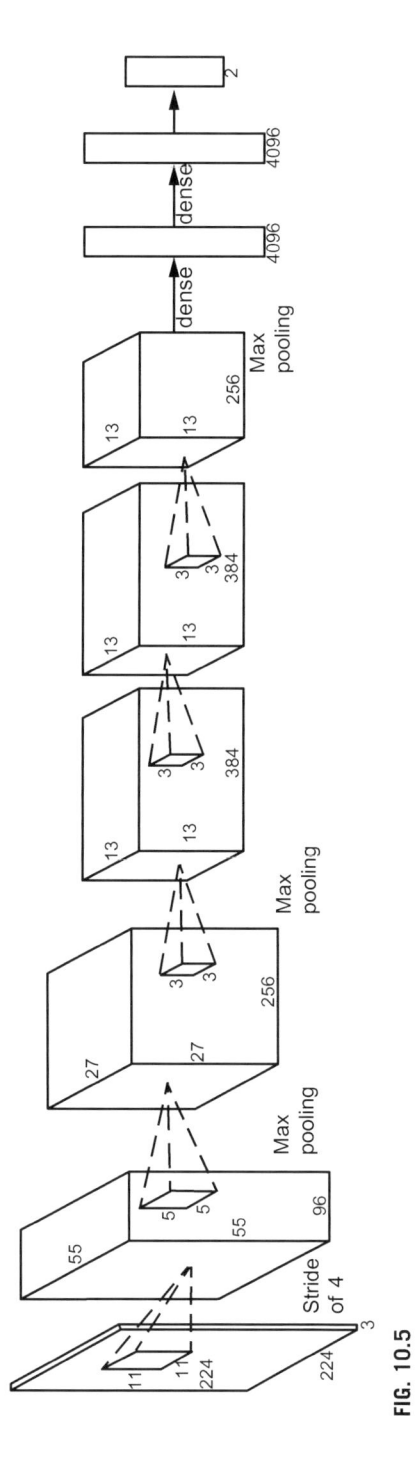

FIG. 10.5

AlexNet architecture.

From Syed, M. G., McKenna, S., & Trucco, E. (c. 2018). Convolutional neural network architectures for image classification and detection. https://www.researchgate.net/ publication/328102687_Convolutional_neural_network_architectures_for_image_classification_and_detection.

architecture to increase the depth. This eight-layer architecture forces the model to learn more robust features. Moreover, the ReLU acts as a nonsaturating activation function (Hochreiter, 1998). ReLU relieves the cause of the vanishing gradient to some extent and improves the convergence rate.

10.2.5.3 ZFNet

Matthew Zeiler and Rob Fergus were the champions of ILSVRC 2013. It became referred to as the ZFNet (short for Zeiler and Fergus Net). The lack of understanding constrained the presentation of deep CNNs on complex images. In 2013, Zeiler and Fergus introduced an intriguing multilayer deconvolutional NN (DeconvNet), which became known as ZfNet (Hu et al., 2018). ZfNet was created to arrange execution quantitatively. The representation of system action was to screen CNN execution by deciphering neuron initiation. The feature perception proposed by ZfNet was tentatively approved on AlexNet utilizing DeconvNet, which demonstrated that a couple of neurons were dynamic. Interestingly, different neurons were placed in the first two layers of the network. Also, it indicated that the layer showed associating ancient rarities. In light of these discoveries, Zeiler and Fergus balanced the CNN topology and streamlined the parameters. Zeiler and Fergus expanded CNN learning by decreasing both the channel size and step to hold the greatest number of highlights in the initial two convolutional layers. This rearrangement in CNN topology brought about execution improvement, which recommended that highlight perception can be used for the distinguishing proof of plan deficiencies and for the convenient alteration of parameters. Fig. 10.6 shows an example of ZFNet.

10.2.5.4 VGG

The effective utilization of CNN in image recognition tasks has quickened the exploration in architectural design. In such a manner, Zisserman (2015) presented a straightforward and successful CNN architecture, called VGG, that was measured in layer design. To represent the depth capacity of the network, VGG had 19 deep layers compared to AlexNet and ZfNet (Krizhevsky et al., 2012). ZfNet introduced the small size kernel aid to improve the performance of the CNNs. This was a cutting edge network of 2013-ILSVRC competition in 2013. In view of these discoveries, VGG followed the 11×11 and 5×5 kernels with a stack of 3×3 filter layers. It then tentatively showed that the immediate position of the kernel size (3×3) could activate the weight of the large-size kernel (5×5 and 7×7).

To gain the advantage of low computational complexity, a small size kernel is the best choice with a reduction in the number of parameters. These discoveries set another pattern in research to work with a small-size kernel in CNN. Fig. 10.7 shows the VGG architecture. VGG demonstrated great outcomes for both image classification and localization problems. VGG was in second place in the 2014-ILSVRC competition. It became more popular due to its homogenous strategy, simplicity, and increased depth. The principle impediment related to VGG was the utilization of 138 million parameters. This make it computationally costly and hard to use on low-asset frameworks (Khan, Sohail, Zahoora, & Qureshi, 2020).

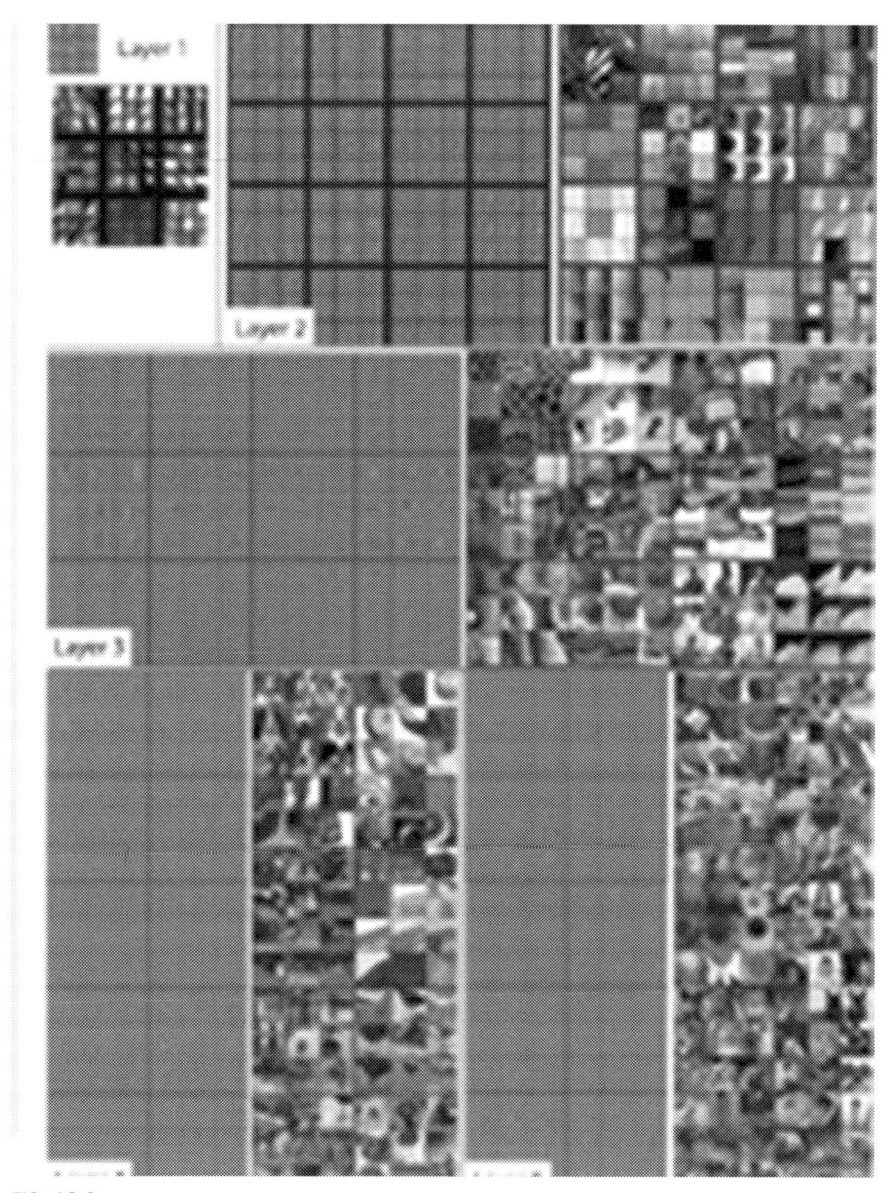

FIG. 10.6

Example of ZFNet.

From Syed, M. G., McKenna, S., & Trucco, E. (c. 2018). Convolutional neural network architectures for image classification and detection. *https://www.researchgate.net/publication/328102687_Convolutional_neural_network_architectures_for_image_classification_and_detection.*

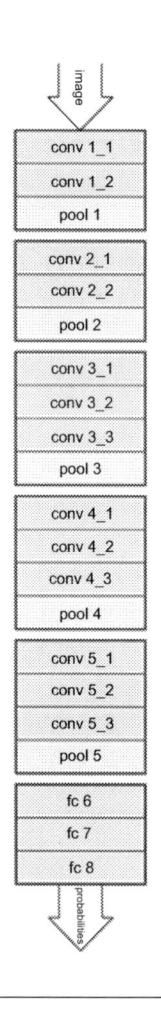

FIG. 10.7

VGG architecture with 19 layers.

From Syed, M. G., McKenna, S., & Trucco, E. (c. 2018). Convolutional neural network architectures for image classification and detection. *https://www.researchgate.net/publication/328102687_Convolutional_neural_network_architectures_for_image_classification_and_detection.*

10.2.5.5 GoogLeNet

In the 2014-ILSVRC competition, GoogLeNet won first position, also called inception-V1. The main concern behind the GoogLeNet architecture was to undertake high accuracy with reduced computational cost (Toshev & Szegedy, 2014). It presented the new idea of an inception block in CNN. This inception block is applied to split and merge techniques to join multiscale convolutional transformations. The architecture of the inception block is shown in Fig. 10.8. In a network-in-network architecture, Lin, Chen, and Yan (2013) proposed replacing every layer with microneurons. This block summarizes kernels of various sizes ($3 \times 3, 5 \times 5$) to catch spatial data at various scales. It is comprised of coarse grain and a fine grain level.

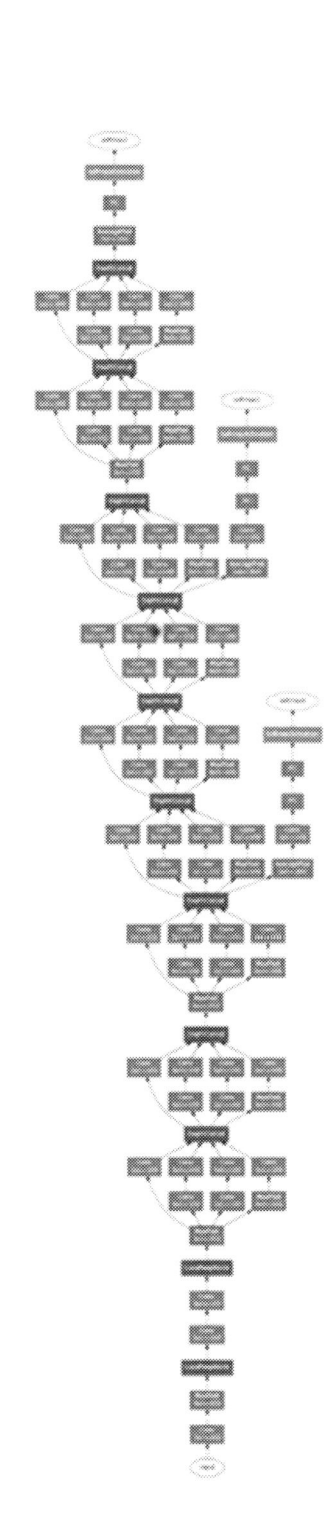

FIG. 10.8

GoogLeNet's architecture with a 22-layer deep CNN.

In GoogLeNet, a 1×1 convolution is used as a measurement decrease model to ease computation. It has increased the size of the depth and width to deal with the reduction of the computation bottleneck. GoogLeNet comprises a deep CNN with 22 layers and a decreased parameter size from 60 million, as in case of AlexNet, to 4 million.

10.2.5.6 ResNet

ResNet was introduced by He, Zhang, Ren, and Sun (2016), and it was assumed to be an extension of deep networks. ResNet upset the CNN architectural race by presenting the idea of residual learning in CNNs. The ResNet architecture proposed a 152-layer deep CNN, and won the 2015-ILSVRC competition. The architecture of ResNet is shown in Fig. 10.9. ResNet was more than 20 times as deep compared to AlexNet and 8 times as deep in structure against VGG. ResNet was less computationally complex than the network discussed earlier. ResNet with 50 layers/101 layers/152 layers achieved reduced errors on image classification than 34 layers in a normal neural network, as suggested by He et al. after experimentation. The successful execution of ResNet on image recognition and localization tasks indicated that illustrative depth plays a significant role for some visual identification and recognition.

10.3 Literature survey

The face recognition algorithms extract different facial features to recognize the face image of a specific person. For instance, an algorithm may analyze the corresponding size, position, and current state of the eyes, nose, mouth, and forehead. These features are scanned for images with equivalent features (Bonsor & Johnson, 2001). Various face algorithms standardize the training of face images and afterward suppress the face value, finally storing this face value in the image. This is valuable for face recognition. One of the past successful systems (Brunelli & Poggio, 1993) depends on template matching techniques (Brunelli, 2009). This is useful to a lot of significant facial features, providing a compressed face representation.

Face recognition algorithms can be expressed by two main methodologies: the geometric-based approach and the template-based approach, which takes distinguishing features and a statistical approach. These algorithms may be divided into two common classifications: holistic and feature-based models. The previous endeavors tried to perceive the face completely while the element-based ones partition them into segments. These segmented partitions are used to analyzes the spatial location of the features (Zhang & Fu, 2010). A commonly used recognition algorithm is principal component analysis (PCA). It is a dimensionality reduction technique that changes correlated features into a large number of uncorrelated features. These are called rule segments using eigen faces, linear discriminant analysis, elastic bunch graph matching using the Fisher face algorithm, the hidden markov model, and dynamic matching of neurons.

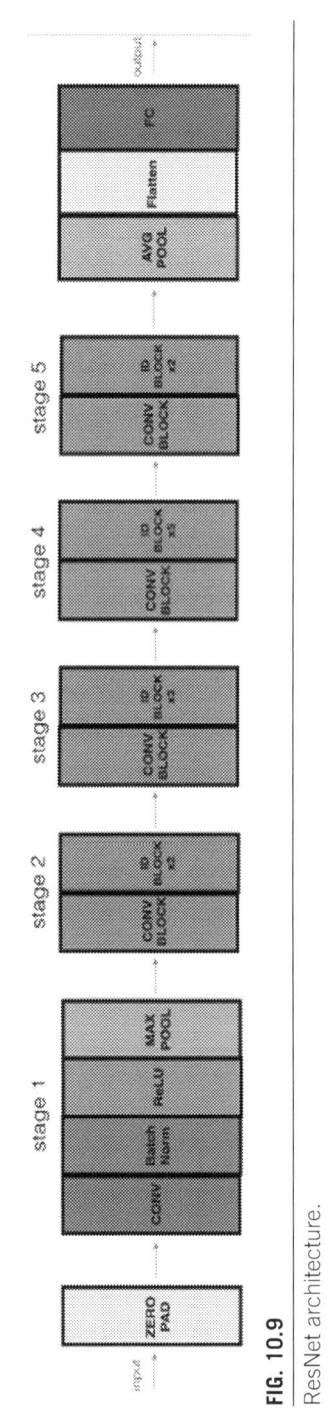

FIG. 10.9

ResNet architecture.

As per the FRVT (White, 1977) face recognition performance is affected by different factors such as aging; variations in age are a main issue in face recognition. As age grows, the facial appearance is also affected, which leads to a decline in recognition. It is a tedious task to accumulate facial images of an individual at different ages to train the system. Next, face recognition systems consider to be the most complex problem in varying poses. The pose problem involves facial changes that occur due to changes in head movement in a 3D vertical coordinate system. Deep rotation perpendicular to two directions of the image plane may cause partial damage of facial information. Another challenge of face recognition is occlusion, which is a serious problem for face image acquisition in a noncooperative situation.

If the object under observation repeatedly wears glasses, scarves, and similar items, the face image is incomplete; this is an occlusion problem (Khadatkar, Khedgaonkar, & Patnaik, 2016; Khedgaonkar, Raghuwanshi, & Singh, 2018). Moreover, illumination is an age-old problem in the area of face recognition. The shadows created due to improper illumination will heighten or weaken the existing facial features. The face recognition system may choose the wrong identity of a person due to changes in the degree of illumination, even with the same pose and expression. All the above mentioned problems make face recognition most challenging. On the other hand, in addition to all these factors, an emergent issue is facial plastic surgery. Normally, facial plastic surgery is applied to enhance the facial appearance, such as removing scars and birthmarks as well as correcting flaws. In earlier days, only celebrities were likely to have facial plastic surgery. Nowadays, numerous people can have facial plastic surgery due to its affordable cost (Khedgaonkar, Singh, & Gawande, 2011; Singh, Khedgaonkar, & Gawande, 2011).

Local plastic surgery-based face recognition system permanently changes certain portion of the face due to various reasons as mentioned above. Among the different types of local facial plastic surgeries, rhinoplasty, blepharoplasty and lip augmentation are the most popular types. The first time the issue of local plastic surgery-based face recognition was addressed was by Richa Singh et al., from IIIT Delhi (Singh, Vatsa, & Noore, 2009). The authors addressed the local plastic surgery-based face recognition problem with classical face recognition algorithms, then applied it on a standard plastic surgery face database (PSD). In the presented work, the author discussed the effectiveness of face recognition using facial plastic surgery with six distinct face recognition algorithms: PCA, Fisher discriminant analysis (FDA), circular local binary pattern (CLBP), local feature analysis (LFA), speeded-up robust features (SURF), and 2D log polar Gabor transform (GNN). These algorithms showed considerable accuracy on the PSD database. Later, other researchers also addressed the problem of plastic surgery face recognition. Researchers in Lakshmiprabha and Majumder (2012) extended the work of plastic surgery face recognition with local region analysis. Shape local binary texture feature (SLBT) allows the cascading of particular features with the combination of Gabor and LBP features.

Further, Maria De Marsico et al. presented a PCA technique for correct recognition of a fact that has undergone facial plastic surgery using region-based approaches (De Marsico, Nappi, Riccio, & Wechsler, 2015). Nevertheless, the PCA technique was unsuccessful at covering the simplest invariance and it is found to be most tedious for evaluating the covariance matrix. De Marsico, Nappi, Riccio, and Wechsler (2016) evaluated the performance of two integrative methods, FARO and FACE. These methods are form of fractals and a localized version of a correlation index in sight of facial plastic surgery. The FARO and FACE compare positively in contrast to PCA and LDA. Liu, Shan, and Chen (2013) applied the concept of Gabor patch classifiers through rank-order list fusion (GPROF). This technique is applied on the face database of plastic surgery. It achieved much higher accuracy than the standard face recognition algorithm for various types of facial plastic surgeries. Bansal and Shetty (2018) computed the Fisher vector encodings of various features such as the nose, eyes, mouth, lip, etc. These features were extracted from prefacial plastic surgical images and postfacial plastic surgical images. Bhatt, Bharadwaj, Singh, and Vatsa (2013) proposed a multiobjective evolutionary granular algorithm to match preplastic surgery and postplastic surgery facial images. In addition, Sable, Talbar, and Dhirbasi (2019) presented the EV-SIFT technique for both the volume and contrast information. The EV-SIFT method achieved a considerable recognition rate when applied on plastic surgery face databases for various kinds of surgeries. Recently, the author (Khedgaonkar, Singh, Raghuwanshi, & Sonsare, 2019) also discussed naïve Bayes and neural network classifiers for local plastic surgery-based face recognition on the plastic surgery face database (PSD). Table 10.1 shows the analysis of different types of facial plastic surgery with recognition rates.

Table 10.1 Analysis of different types of facial plastic surgery with recognition rates.

Sl. no.	Face recognition techniques	Recognition rate (%) of local Facial plastic surgery			Ref.
		Rhinoplasty	Blepharoplasty	Lip augmentation	
1	PCA	21.4	24.3	12.0	Singh et al. (2009)
	FDA	22.1	25.0	12.9	
	GF	31.4	34.7	12.3	
	LFA	23.3	27.6	12.7	
	LBP	32.6	36.4	18.2	
	GNN	37.3	40.7	19.1	
2	GPROF	81.77	89.52	–	Lei, Kim, Par, and Ko (2016)
3	FARO	38	45	–	De Marsico et al. (2016)
	FACE	63	60	–	
4	EV-SIFT	99	98	96	Sable et al. (2019)
5	Naïve Bayes	52	47.71	36	Khedgaonkar et al. (2019)

The main objective of all the presented research was to achieve improved computational time and higher accuracy. Nonetheless, better accuracy is still a challenging factor for local plastic surgery-based face recognition. In addition to the above-mentioned face recognition algorithm, a neural network can be used. An artificial neural network (ANN) is inspired by the biological neural networks that constitute animal brains (Yung-Yao, Yu-Hsiu, Chia-Ching, Ming-Han, & I-Hsuan, 2019). These kinds of systems "learn" to perform tasks.

The utmost aim of the ANN approach was to resolve issues in a similar fashion to a human brain. But over time, it deviates from biology and grabs attention by performing particular tasks. The diversity of work comprises machine translation, computer vision, speech recognition, social network filtering, and medical diagnosis by ANNs. The nonlinearity feature of the network attracts neural networks. Hence, the process of feature extraction is considered to be more effective compared to linear methods. The face recognition by approximation spaces is stated in Raghuwanshi and Singh (2009) with a combination of rough set theory with neural networks. One of the first ANN techniques used for face recognition was a single-layer adaptive network. For face detection, multilayer perceptron (Sung & Poggio, 1995) and the CNN have been applied. The performance of the face recognition algorithm improves rapidly with the advancement of computer science and the development of the DL learning technique. Currently, deep CNN-based methods on face recognition are outpacing traditional ones with hand-crafted features. DL is cognitive learning that obtains high-level features. It follows two basic strategies in face recognition methods based on DL. First, the higher-dimension feature vectors can be extracted using a deep CNN. Next, the PCA or metric-learning method is executed to minimize the dimensions of the extracted feature vectors from CNN. This is more efficient and distinct, and it grabs more discriminative low-dimension feature vectors. Then, the faces of various identities are represented by these vectors.

In 2012, Huang, Lee et al. for the first time used the LFW database for recognition using DL techniques. They achieved 87% accuracy (Huang, Lee, & Learned-Miller, 2012) by applying an unsupervised feature learning method. Recently, various international projects have successfully applied DL techniques for face recognition, such as DeepFace (Taigman, Yang, Ranzato, & Wolf, 2014), DeepID (Sun, Wang, & Tang, 2014), FaceNet (Schroff, Kalenichenko, & Philbin, 2015), etc. The above -mentioned algorithms are built on massive amounts of training data. It allows learning face features by a DL algorithm even when the facial expression, illumination of the face, and head pose are unchanged. Among the above algorithms, FaceNet gives an overview of face recognition methods. Schroff et al. (2015) achieved 99.63% accuracy when using the LFW database. One of the probable future research directions would be to apply ML techniques on proteomics analysis, which is very well addressed by P.M. Sonsare et al., in Sonsare and Gunavathi (2019).

10.4 Design of deep learning architecture for local plastic surgery-based face recognition

Face recognition suffers from performance issues due to numerous challenges, including illumination variations, expression variations, pose variations, occlusion, makeup spoofing, and aging. All these issues occur during taking facial images and the variations in facial images do not remain permanent. In addition, a new challenge has emerged: facial plastic surgery. Facial plastic surgery is roughly characterized as global plastic surgery and local plastic surgery. Global plastic surgery is the most considered type of surgery where functional injury is cured. Local plastic surgery is used only when the local region of the face is damaged. Global plastic surgery and local plastic surgery were already discussed in the introduction. We are more focused on local plastic surgeries. Of those, we will evaluate the performance on three kind of surgeries: rhinoplasty (nose surgery), blepharoplasty (eyelid surgery), and lip augmentation (liposhaving).

As already discussed in the literature many authors have used various face recognition techniques on these types of surgery. As per the literature, performance issues still exist. However, due to local plastic surgery, the face structure changes and therefore the facial features also change. Because a change in facial structure is not in one's control, feature extraction for facial representation under local plastic surgery becomes a more complex task. Under such circumstances, considering plastic surgery face recognition a two-step (feature extraction and recognition) problem makes it a more challenging issue. This motivates us to propose local plastic surgery-based face recognition using a CNN.

10.4.1 Proposed CNN model

CNNs are a natural way to learn neural networks for treating images. CNNs learn features that are shift-invariant. It automatically extracts the high-level features, unlike the hand-crafted features in state-of- the-art algorithms. Those features are then fed to a classifier for classification purpose. Fig. 10.10 shows the proposed CNN model. The proposed CNN model has been used to extract important features and utilize them for classification purposes.

The general process comprises the same traditional steps: image capture, feature extraction, and classification. Feature extraction is getting the desired feature values using a CNN rather than the extraction of hand-crafted features. In this model, we used a nine-layer CNN architecture. The person identification is done by the classification step. This work includes the binary classification problem. The overall process of the proposed CNN model consists of various layers, as described below.

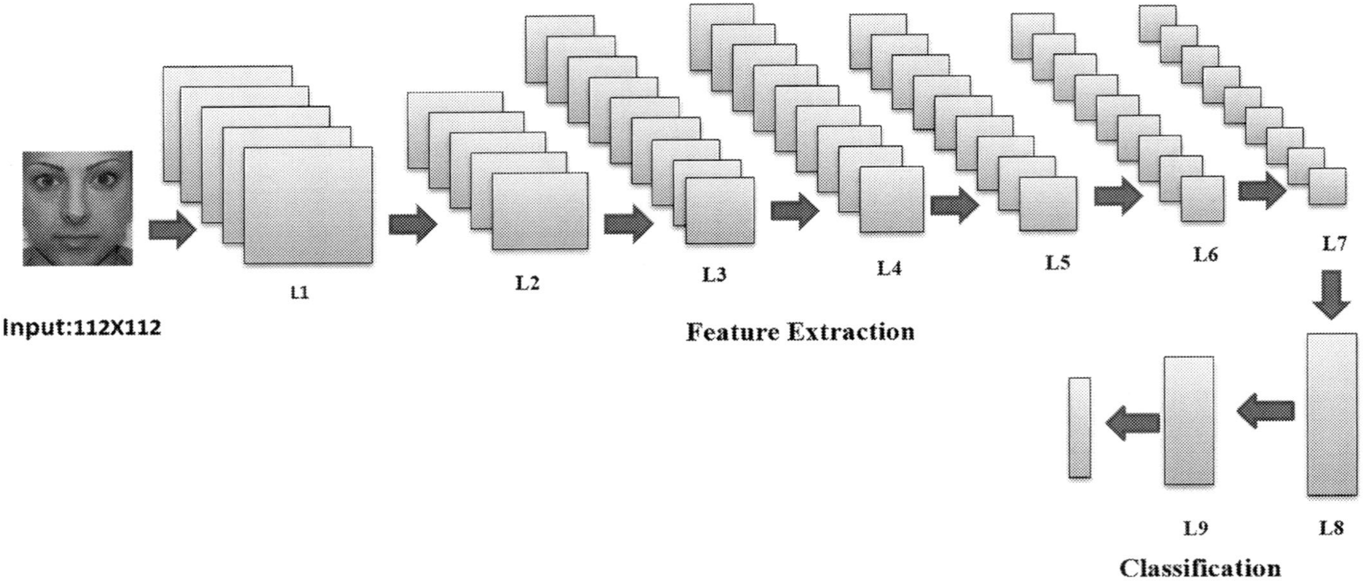

FIG. 10.10

Proposed CNN model.

10.4.1.1 Convolution layer

The contribution of every convolution layer, similar to the CNN, is the yield of the upper layer; it is tangled by a few convolution kernels. The convolution kernels are utilized over and over in each field of the entire region, and the convolution result comprises a feature map of the input image. The convolution layer contains the weight matrix w and bias b. In this chapter, the sizes of the convolution kernels are 5 × 5 and 3 × 3.

The mathematical representation of the convolution layer (Syafeeza et al., 2014) is expressed in Eq. (10.6).

$$x_j^l = f\left(\sum\nolimits_{i \in M_j^{l-1}} x_i^{l-1} k_{ij}^l + b_j^l \right) \tag{10.6}$$

Where l represents the layer, k is the convolution kernel, b is the bias, i denotes the indices of the input feature map, and j denote the indices of the output feature map. M_j is the feature map. f denotes the activation function as ReLU (Hinton, 2010). The ReLU activation function is mathematically expressed in Eq. (10.7).

$$f(x) = \max(0, x) \tag{10.7}$$

It generates 0 output if the input is less than 0; otherwise it gives raw output.

10.4.1.2 Pooling layer

The pooling layer is a commonly used layer to reduce the dimension of the feature map and retain the most valuable information. It is a nonlinear downsampling method. In the network, each feature map that has been placed into the pooling layer is sampled, and the number of resultant feature maps is unaltered. However, the size of each feature map will be smaller. In this chapter, the pooling layer is sampled with the maximum value. The sampling size is 2 × 2 with a stride of 2.

10.4.1.3 Fully connected layer

For the network, after a few persistent levels of convolution layers and pooling layers, for the most part, there will be various fully connected layers close to the output layer. Furthermore, these fully associated layers structure a multilayer perceptron (MLP), which assumes the job of a classifier.

In this chapter, we use two fully connected layers, each of which is connected to all the neurons in the previous layer. The mathematical expression of the layer (Ziming & Fu, 2010) is expressed in Eq. (10.8)

$$y_{pj} = f\left(\sum\nolimits_{i=1}^{n} x_i^{l-1} w_{ji}^l + b_j^l \right) \tag{10.8}$$

where n is the number of neurons in the preceding layer $l - 1$, w_{ji}^l is the weight for the connection from neuron i in layer $l - 1$ to neuron j in layer l, and b_j^l is the bias of neuron j in layer l and represents the activation function of layer l.

10.4.1.4 Parameter tuning

The convolution layer's parameters are comprised of a lot of learnable filters (kernel). Each kernel is a small dimension (width and height), yet stretches out through the full depth of the input dimensions. The whole arrangement of kernels in each convolution layer is 32 filters and each one of these filters creates a 2D activation map. We have stacked these activation maps beside the size of depth and generated the output dimensions, as tabulated in the table of the parameter tuning of the proposed approach.

The connectivity between neurons of various layers is called the hyperparameter (receptive field). These hyperparameters control the size of the output dimension by **depth, stride**, and **padding, where depth** corresponds to the number of kernels. First, a 2D convolutional layer takes as input the original image, then various neurons alongside activate the depth dimension. **Stride** is actually slide the filter (kernel). A stride of 1 makes the kernel shift one pixel at a time. A stride of two shifts the kernel 2 pixels ahead. In this, we have used a stride of 1 for convolution layers and 2 for max pooling layers, which is used to reduce the dimension of the 2D face image. Next, padding is used to preserve the size of the input dimensions. We have considered zero padding for all the layers so as to keep the equal size of the input and output dimensions.

The output dimension (O) of the convolution layer is computed as $(W2 \times H2 \times D2)$ where $W2,H2,D2$ are the functions of the input size as width, height, and depth, respectively. $(W1 \times H1 \times D1)$ are the dimensions of the input image. The dimensions of the kernel (size) of the convolutional layer $(K),(S)$ represent the stride and the amount of padding used (P), and this is mathematically expressed in Eqs. (10.9)–(10.11).

$$O_{W2} = \frac{(W1 - K + 2P)}{S} + 1 \tag{10.9}$$

$$O_{H2} = \frac{(H1 - K + 2P)}{S} + 1 \tag{10.10}$$

$$D2 = K_n \tag{10.11}$$

Where K_n is the number of kernels.

The input image is of size $(112 \times 112 \times 3)$. On the first convolutional layer it used a kernel size $K = 3 \times 3$, a stride $S = 1$ with zero padding $P = 0$ and $K_n = 32$. The output size of the convolution layer is expressed in Eqs. (10.12)–(10.15).

$$O_{W2} = \frac{(112 - 3 + 2 \times 0)}{1} + 1 = 108 \tag{10.12}$$

$$O_{H2} = \frac{(112 - 3 + 2 \times 0)}{1} + 1 = 108 \tag{10.13}$$

$$O_{W2} = \frac{(W1 - K + 2P)}{S} + 1 \tag{10.14}$$

$$O_{H2} = \frac{(H1 - K + 2P)}{s} + 1 D2 = K_n \tag{10.15}$$

Similarly, the max pooling layer is used after the convolution layer and its output dimension is computed in Eqs. (10.16)–(10.18) as follows:

$$O_{W2} = \frac{(W1 - K)}{S} + 1 \tag{10.16}$$

$$O_{H2} = \frac{(H1 - K)}{S} + 1 \tag{10.17}$$

$$D2 = D1 \tag{10.18}$$

As per Table 10.2, we have considered parameters such as the input face image of dimensions ($108 \times 108 \times 32$). The stride (S) is 2, the kernel size (K) is 3, and the depth of the input face image ($D1$) is 32. After applying the mathematical formation of the max pooling layer as expressed in the above equations, we have reached the output with dimensions ($54 \times 54 \times 32$), this leads to the fiunctionality of dimensionality reduction.

Likewise, the series of convolution layers and max pooling layers is used in a cascaded way to generate the high-level feature in terms of the feature map. The feature map thus generated is converted to a feature vector by a fully connected layer. Fully connected layers act as a regular neural network because the neurons in a fully

Table 10.2 Summary of parameters.

Layers	Input size	Output size	Number of kernels	Kernel size, stride, padding	Parameter
Input	112 × 112 × 3	112 × 112 × 3	–	–	–
Conv1	112 × 112 × 3	108 × 108 × 32	32	5 × 5, 1, 0	2432
MaxPool1	108 × 108 × 32	54 × 54 × 32	32	2 × 2, 2, 0	0
Conv2	54 × 54 × 32	52 × 52 × 32	32	3 × 3, 1, 0	9248
MaxPool2	52 × 52 × 32	26 × 26 × 32	32	2 × 2, 2, 0	0
Conv3	26 × 26 × 32	24 × 24 × 32	32	3 × 3, 1, 0	9248
MaxPool3	24 × 24 × 32	12 × 12 × 32	32	2 × 2, 2, 0	0
Conv4	12 × 12 × 32	10 × 10 × 32	32	3 × 3, 1, 0	9248
FC1	10 × 10 × 32	128	–	–	409,728
FC2	128	1	–	–	129

connected layer have connected to various activations in the previous layer. Here, in the first fully connected layer (FC1), 128 features are considered. With a sigmoid activation function, it generates a binary output by the second fully connected layer (FC2).

In addition, the learnable parameter is computed for two main layers, namely the convolution layer and the fully connected layer. The computation of parameter learning of convolution layers requires a kernel size ($Width_K \times Height_K$) and total kernels K_n. Thus, the total learnable parameter in a convolution layer is expressed in Eq. (10.19).

$$P_{CONV} = \left(\left(Width_K \times Height_K \times Depth_{Prev\_Layer} \right) + 1 \right) \times (K_n) \qquad (10.19)$$

Where 1 is used as the bias term.

Fully connected layers generate the highest number of parameters because each neuron in a fully connected layer is connected to another neuron in a network. The computation of the learnable parameter for a fully connected layer requires a number of neurons in the current layer (N) and in the previous layer (M). Thus, the total learnable parameter in a fully connected layer is given in Eq. (10.20).

$$P_{FC} = N_{Current\_Layer} \times M_{Previous\_Layer} + bias \qquad (10.20)$$

Where bias is connected to each neuron. So, there are N biases.

10.5 Experimental setup

In the proposed model, we have designed a 2D convolutional deep neural network for local plastic based-face recognition. The experimentation is carried out by considering three types of facial plastic surgeries: rhinoplasty, blepharoplasty, and lip augmentation. The experimentation is done on two different datasets: the Plastic Surgery Face Database (PSD) of the Image Analysis and Biometrics Lab from IIIT Delhi (Singh, 2010) and the American Society of Plastic Surgeons face database (ASPS) (*American Society of Plastic Surgeon Face Database*, 2019). All experiments are conducted in TensorFlow using the Keras library where the weights in the proposed architecture are initialized using the default setting of Keras. The performance of the proposed CNN model is evaluated on accuracy.

10.6 Database description

The proposed CNN model utilizes face images from those two surgical face datasets. From each database, 80% of face samples are used as the training set and the remaining are used as the testing set. Both databases consist of presurgical face images and postsurgical face images of variable size. All the images in the PSD databases are color images with variations in skin tone and expressions. In addition, the ASPS database consists of color images with variations in skin tone, changes in facial expression, and variations in poses. First, we performed the experimentation with the PSD database. The PSD face database consists of 541 local plastic surgical face images of variable size. Out of 541 face images, 329 sample face images are rhinoplasty, 197 sample face images are of blepharoplasty, and 15 sample face images are of lip augmentation. The subdatabases under each category were divided into a training set and a testing set. A total of 263 sample face images were randomly selected for training and the remaining 66 images were used for testing in rhinoplasty. Similarly, a total of 157 sample face images of blepharoplasty were selected for training and 40 sample face images were selected for testing. Likewise, 12 sample face images of lip augmentation were selected for training and the remaining images were used for testing. In addition to the PSD database, the performance of the proposed algorithm was also evaluated on ASPS. According to Michele Nappi (De Marsico et al., 2016), among the one of the renowned organizations for plastic surgery in the world, demonstrating 94% of all board-certified plastic surgeons in the United States. ASPS consists of more than 10,000,000 face images, including actual surgical procedures of 250,000 face images of rhinoplasty, 200,000 face images of blepharoplasty, and 130,000 face images of lip augmentation. Out of these, the proposed algorithm was performed on 680 samples of rhinoplasty, 650 samples of blepharoplasty, and 130 samples of lip augmentation. A training set included 544 samples, 520 samples, and 104 samples of rhinoplasty, blepharoplasty, and lip augmentation, respectively. The remaining face samples are selected for testing. Sample face images for including PSD and ASPS databases are shown in Figs. 10.11 and 10.12, respectively.

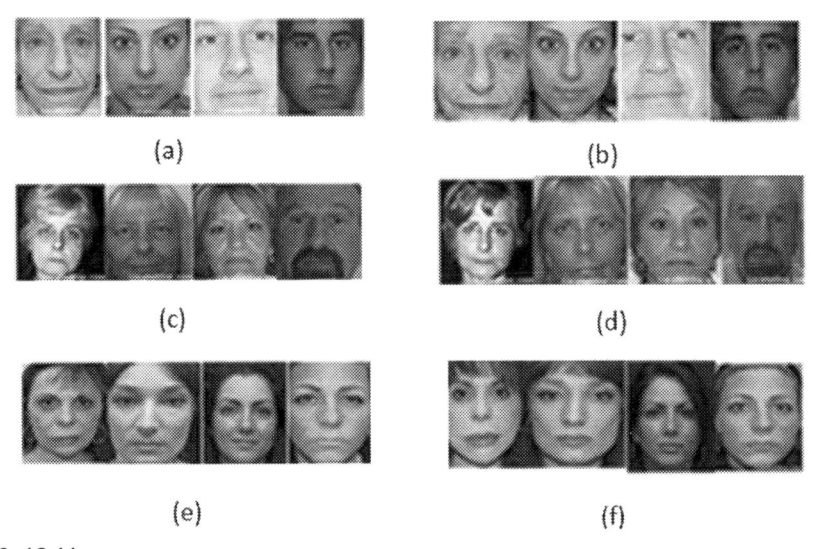

(a) (b)

(c) (d)

(e) (f)

FIG. 10.11

Sample face images from Plastic Surgery Face Database. (A, C, E) Nose job, eyelid, and lip-augmentation before plastic surgery, (B, D, F) nose job, eyelid, and lip-augmentation after plastic surgery.

From Singh, R. (c. 2010). Plastic Surgery Face Database. *Indraprastha Institute of Information Technology Delhi.*

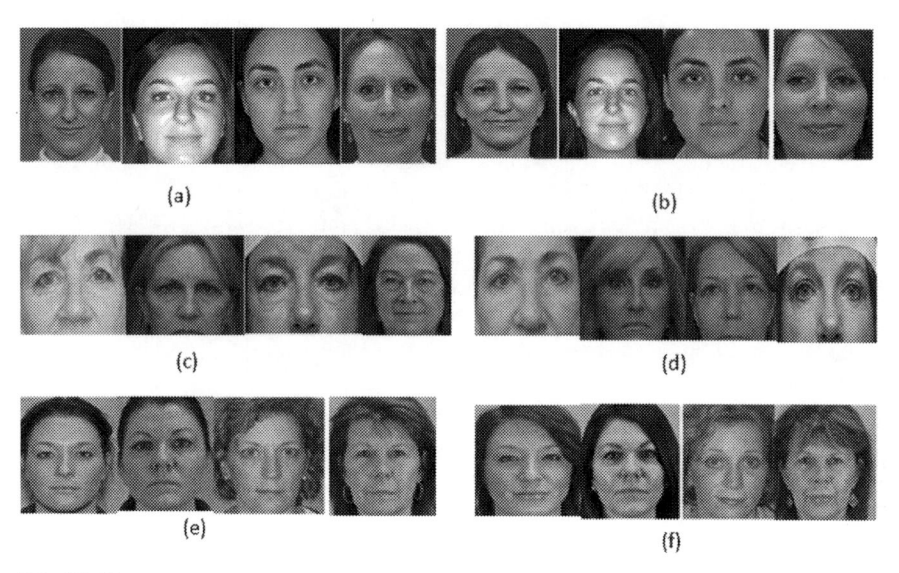

FIG. 10.12

Sample images from American Society of Plastic Surgeon Face database. (A, C, E) Nose job, eyelid, and lip-augmentation before plastic surgery, (B, D, F) nose job, eyelid, and lip-augmentation after plastic surgery.

From American Society of Plastic Surgeon Face Database. *(c. 2019).*

10.7 **Results**

We have investigated and evaluated the performance of the face recognition algorithm on the PSD and ASPS databases. We have considered three types of surgery for experimentation purpose namely, Rhinoplasty, Blepharoplasty, and Lip-augmentation for experimentation.

The proposed algorithm has a binary_crossentropy loss function that shows the sum of all losses and the optimizer algorithm as adam, which was good for CNN. Because we use the binary_crossentropy loss, we need to set the class mode as binary labels with two classes, presurgical face images and postsurgical face images. From each of these classes, 80% of the face images were used for training and the remaining 20% were used for testing for both the PSD and ASPS databases. The train and test set added a 32 batch size where the steps-per-epoch and validation steps were set to 63 and 10, respectively. However, as a result of testing the model it achieved good accuracy. The line graph as depicted in Fig. 10.13 shows the result after applying the proposed CNN model to the types of local plastic surgery, which is also given in Table 10.3. The performance of the proposed CNN model is evaluated on the two different databases the three types of local plastic surgery in terms of accuracy.

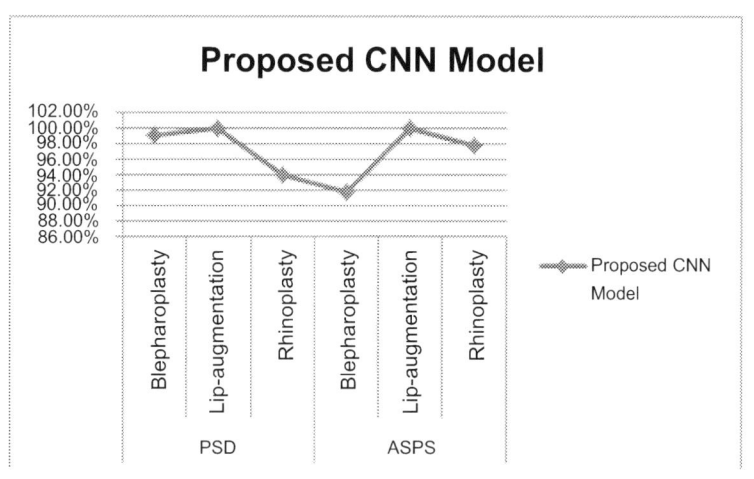

FIG. 10.13

The graph representation of the proposed CNN model on types of local plastic surgery.

No Permission Required.

Table 10.3 Performance analysis of the proposed CNN model on two databases for various facial plastic surgeries.

Types of databases	Facial local plastic surgery	Proposed CNN model
PSD	Blepharoplasty	99.09%
	Lip augmentation	100%
	Rhinoplasty	93.96%
ASPS	Blepharoplasty	91.57%
	Lip augmentation	100%
	Rhinoplasty	97.71%

It achieved 99.09% and 91.57% accuracies for blepharoplasty. Similarly, for rhinoplasty it achieved 93.96% and 97.71% accuracies. It achieved 100% accuracy for lip augmentation for both databases.

This evaluation was based on the performance on frontal faces with natural expressions under proper illumination. This chapter also shows the comparison of the proposed CNN model with related work by Singh et al. (2009), and also compares the result of author Khedgaonkar et al. (2019). Table 10.4 presents a statistical analysis in the comparison of the performance with the algorithm presented by Singh et al. (2009) and Khedgaonkar et al. (2019) in terms of the PSD database. The statistical analysis was performed on the level of accuracy it achieved. The statistics shown in the table give a lower accuracy in the range of 12%–38% for PCA,

Table 10.4 Comparative analysis of the proposed CNN model with other face recognition models.

Facial local plastic surgery	PCA	FDA	GF	LFA	LBP	GNN	Naïve Bayes	Neural network	Proposed CNN model
Blepharoplasty	24.30%	25%	34.70%	27.60%	36.40%	40.70%	55.71%	61.44%	99.09%
Lip augmentation	12%	12.90%	12.30%	12.70%	18.20%	19.10%	46%	50%	100%
Rhinoplasty	21.40%	22.10%	31.40%	23.30%	32.00%	37.30%	60%	71%	93.96%

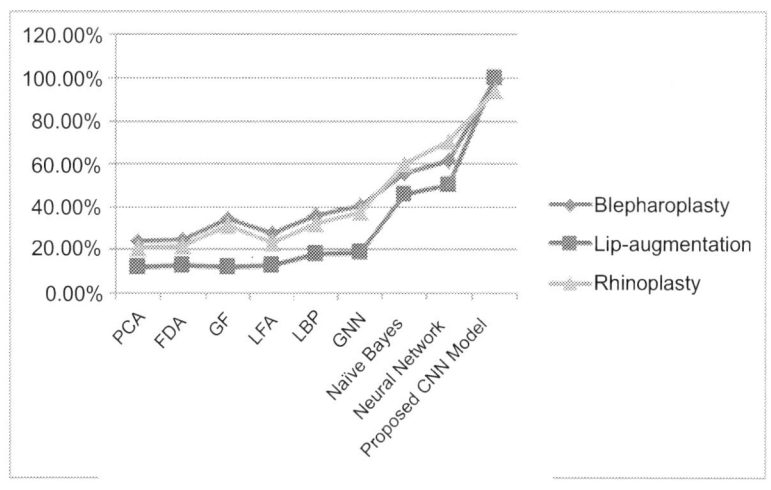

FIG. 10.14

Graphical representation of the comparison of different face recognition algorithms with the proposed CNN model.

No Permission Required.

FDA, GF, LFA, LBP, and GNN. After that, naïve Bayes gives better accuracy as compared to the previous methods just mentioned of between 40% and 72%, but again this was not a promising figure. Contrary to this, the proposed CNN model achieved higher accuracy. This statistics are placed graphically in Fig. 10.14.

10.8 Conclusion and future scope

As per the recent survey, the performance of face recognition is affected by different factors such as aging, pose variation, facial expression, occlusion, and illumination variation. This makes face recognition a most challenging field in the area of computer vision. In addition to all the aforementioned factors, a growing issue is facial plastic surgery, which occurs as a result of a medical surgical procedure. Our main focus is to address local plastic surgery face recognition, which changes certain portions of the face permanently. Furthermore, various state-of-the-art face recognition algorithms are used to extract hand-crafted facial features. This motivates us to propose local plastic surgery -based face recognition using a CNN.

In this chapter, we have presented an experimental evaluation of the performance of the proposed CNN model. The overall performances were obtained using a number of training images and test images. This proposed CNN model was evaluated by using types of surgeries such as rhinoplasty, blepharoplasty, and lip augmentation on two types of databases, PSD and ASPS. It uses a nine-layered CNN model for feature extraction. For recognition, the extracted features were fed to a fully connected neural network classifier. The performance measures were analyzed by accuracy. It achieved accuracy ranges from 90% to 100% for the three types of surgeries when applied on two different types of face databases. It achieved considerable accuracy because when surgical face recognition is performed, even though there is a permanent change in one local region of the face, the other region of the face retains its features, which may help in the recognition of the correct face. Moreover, the proposed nine-layered CNN model was compared with other face recognition algorithm such as, PCA, FDA, GF, LFA, LBP, and GNN. In addition, it also compares the results of naïve Bayes and the neural network classifier for the three different types of surgery. We demonstrated that the performance of rhinoplasty was 93.96%, blepharoplasty achieved 99.09% accuracy, and lip augmentation achieved 100% accuracy. From these statistics, we have concluded that the proposed CNN model outperformed the other face recognition algorithms.

In future work, we will be introducing generative adversarial networks (GANs). In this mechanism, we would generate large numbers of images to build a large dataset. In the proposed model, we have utilized datasets that were somewhat controlled and CNN worked better on the large dataset. So, this large dataset generated by GAN improves the performance of face recognition to a great extent. Moreover, we would significantly improve the performance of face recognition by incorporating the use of other classifiers for classification tasks such as fuzzy support vector machines.

References

Ahranjany, S. S., Razzazi, F., & Ghassemian, M. H. (2010). A very high accuracy handwritten character recognition system for Farsi/Arabic digits using convolutional neural networks. In *Proceedings 2010 IEEE 5th international conference on bio-inspired computing: Theories and applications, BIC-TA 2010* (pp. 1585–1592). https://doi.org/10.1109/BICTA.2010.5645265.

Lecun, Y., Jackel, L. D., Bottou, L., Cortes, C., Denker, J. S., Drucker, H., … Vapnik, V. (1995). Learning algorithms for classification: A comparison on handwritten digit recognition. In *Neural networks: The statistical mechanics perspective* (pp. 261–276). World Scientific.

American Society of Plastic Surgeon Face Database. (2019).

Bansal, A., & Shetty, N. P. (2018). Matching before and after surgery faces. *Procedia Computer Science, 132*, 141–148. Elsevier B.V. https://doi.org/10.1016/j.procs.2018.05.175.

Bhatt, H. S., Bharadwaj, S., Singh, R., & Vatsa, M. (2013). Recognizing surgically altered face images using multiobjective evolutionary algorithm. *IEEE Transactions on Information Forensics and Security, 8*(1), 89–100. https://doi.org/10.1109/TIFS.2012.2223684.

Bobulski, J. (2016). Face recognition method with two-dimensional HMM. In *Vol. 403. Advances in Intelligent Systems and Computing* (pp. 317–325). Springer Verlag. https://doi.org/10.1007/978-3-319-26227-7_30.

Bonsor, K., & Johnson, R. (2001). *How facial recognition systems work. Vol. 4*. HowStuffWorks. com. Np.

Boureau, Y. L., Ponce, J., & Lecun, Y. (2010). A theoretical analysis of feature pooling in visual recognition. In *ICML 2010—Proceedings, 27th international conference on machine learning* (pp. 111–118).

Brunelli, R. (2009). Template matching techniques in computer vision: Theory and practice. In *Template matching techniques in computer vision: Theory and practice* (pp. 1–338). John Wiley and Sons. https://doi.org/10.1002/9780470744055.

Brunelli, R., & Poggio, T. (1993). Face recognition: Features versus templates. *IEEE Transactions on Pattern Analysis and Machine Intelligence*, *15*(10), 1042–1052. https://doi.org/10.1109/34.254061.

Chan, C. H., & Kittler, J. (2010). Sparse representation of (multiscale) histograms for face recognition robust to registration and illumination problems. In *Proceedings—International conference on image processing, ICIP* (pp. 2441–2444). https://doi.org/10.1109/ICIP.2010.5651933.

Cheng, D. (2005). Controllability of switched bilinear systems. *IEEE Transactions on Automatic Control*, *50*(4), 511–515. https://doi.org/10.1109/TAC.2005.844897.

Ciresan, D. C., Meier, U., Masci, J., Gambardella, L. M., & Schmidhuber, J. F. (2011). High performance convolutional neural networks for image classification. In *Proceedings of the twenty-second international joint conference on artificial intelligence* (pp. 1237–1242).

De Marsico, M., Nappi, M., Riccio, D., & Wechsler, H. (2015). Robust face recognition after plastic surgery using region-based approaches. *Pattern Recognition*, *48*(4), 1261–1276. https://doi.org/10.1016/j.patcog.2014.10.004.

De Marsico, M., Nappi, M., Riccio, D., & Wechsler, H. (2016). Robust face recognition after plastic surgery using local region analysis. *Image Analysis and Recognition*, 191–200.

Gardner, M. W., & Dorling, S. R. (1998). Artificial neural networks (the multilayer perceptron)—A review of applications in the atmospheric sciences. *Atmospheric Environment*, *32*(14–15), 2627–2636. https://doi.org/10.1016/S1352-2310(97)00447-0.

Ghorbani, M., Targhi, A. T., & Dehshibi, M. M. (2016). HOG and LBP: Towards a robust face recognition system. In *The 10th international conference on digital information management, ICDIM 2015* (pp. 138–141). Institute of Electrical and Electronics Engineers Inc. https://doi.org/10.1109/ICDIM.2015.7381860.

He, K., Zhang, X., Ren, S., & Sun, J. (2015). Spatial pyramid pooling in deep convolutional networks for visual recognition. *IEEE Transactions on Pattern Analysis and Machine Intelligence*, *37*(9), 1904–1916. https://doi.org/10.1109/TPAMI.2015.2389824.

He, K., Zhang, X., Ren, S., & Sun, J. (2016). Deep residual learning for image recognition. In *Vol. 2016. Proceedings of the IEEE computer society conference on computer vision and pattern recognition* (pp. 770–778). IEEE Computer Society. https://doi.org/10.1109/CVPR.2016.90.

Hinton, G. E. (2010). Rectified linear units improve restricted Boltzmann machines. In *International conference on machine learning* (pp. 807–814). Omnipress.

Hochreiter, S. (1998). The vanishing gradient problem during learning recurrent neural nets and problem solutions. *International Journal of Uncertainty, Fuzziness and - Knowledge-Based Systems*, *6*(2), 107–116. https://doi.org/10.1142/S0218488598000094.

Hu, Y., Wen, G., Luo, M., Dai, D., Ma, J., & Yu, Z. (2018). *Competitive inner-imaging squeeze and excitation for residual network*.

Hua, Y., & Jie, Y. (2001). A direct LDA algorithm for high-dimensional data—With application to face recognition. *Pattern Recognition*, 2067–2070. https://doi.org/10.1016/s0031-3203(00)00162-x.

Huang, G. B., Lee, H., & Learned-Miller, E. (2012). Learning hierarchical representations for face verification with convolutional deep belief networks. In *Proceedings of the IEEE computer society conference on computer vision and pattern recognition* (pp. 2518–2525). https://doi.org/10.1109/CVPR.2012.6247968.

Indolia, S., Goswami, A. K., Mishra, S. P., & Asopa, P. (2018). Conceptual understanding of convolutional neural network—A deep learning approach. *Procedia Computer Science, 132*, 679–688. Elsevier B.V. https://doi.org/10.1016/j.procs.2018.05.069.

Karpathy. (2019). *CS231n convolutional neural networks for visual recognition.*

Khadatkar, A., Khedgaonkar, R., & Patnaik, K. S. (2016). Occlusion invariant face recognition system. In *IEEE WCTFTR 2016—Proceedings of 2016 world conference on futuristic trends in research and innovation for social welfare*Institute of Electrical and Electronics Engineers Inc. https://doi.org/10.1109/STARTUP.2016.7583985.

Khan, A., Sohail, A., Zahoora, U., & Qureshi, A. S. (2020). A survey of the recent architectures of deep convolutional neural networks. *Artificial Intelligence Review.* https://doi.org/10.1007/s10462-020-09825-6.

Khedgaonkar, R., Raghuwanshi, M. M., & Singh, K. R. (2018). Patch-based face recognition under plastic surgery. In *ICSCCC 2018—1st international conference on secure cyber computing and communications* (pp. 364–368). Institute of Electrical and Electronics Engineers Inc. https://doi.org/10.1109/ICSCCC.2018.8703270.

Khedgaonkar, R. S., Singh, K. R., & Gawande, S. P. (2011). Identifying resemblance in local plastic surgical faces using near sets for face recognition. In *Proceedings—2011 international conference on communication systems and network technologies, CSNT 2011* (pp. 589–592). https://doi.org/10.1109/CSNT.2011.126.

Khedgaonkar, R. S., Singh, K. R., Raghuwanshi, M. M., & Sonsare, P. M. (2019). Face recognition using probabilistic model for locally changed face. *International Journal of Innovative Technology and Exploring Engineering, 8*(9), 2400–2406. https://doi.org/10.35940/ijitee.i7633.078919.

Kim, H. C., Kim, D., & Bang, S. Y. (2002). Face recognition using LDA mixture model. In *Vol. 16, Issue 2. Proceedings—International Conference on Pattern Recognition* (pp. 486–489).

Kishore, B., Rana, V. L., Manikantan, K., et al. (2014). Face recognition using Gabor-feature-based DFT shifting. In *9th International Conference on Industrial and information systems (ICIIS)* (pp. 1–8). IEEE.

Krizhevsky, A., Sutskever, I., & Hinton, G. E. (2012). ImageNet classification with deep convolutional neural networks. In *Vol. 2. Advances in neural information processing systems* (pp. 1097–1105). ACM.

Kumar, R. N. (2017). Language processing. In *Machine learning and cognition in enterprises* (pp. 65–73). Berlin: Springer.

Lakshmiprabha, N. S., & Majumder, S. (2012). Face recognition system invariant to plastic surgery. In *International conference on intelligent systems design and applications, ISDA* (pp. 258–263). https://doi.org/10.1109/ISDA.2012.6416547.

LeCun, Y., Boser, B. E., Denker, J. S., Henderson, D., Howard, R. E., Hubbard, W. E., & Jackel, L. D. (1990). Handwritten digit recognition with a back-propagation network. In *Proceedings of the Advances in Neural Information Processing Systems* (pp. 396–404).

LeCun, Y., Kavukcuoglu, K., & Farabet, C. (2010). Convolutional networks and applications in vision. In *ISCAS 2010–2010 IEEE international symposium on circuits and systems: Nano-bio circuit fabrics and systems* (pp. 253–256). https://doi.org/10.1109/ISCAS.2010.5537907.

LeCun, Y., Bengio, Y., & Hinton, G. (2015). Deep learning. *Nature, 521*, 436–444. https://doi.org/10.1038/nature14539.

Lei, L., Kim, D. H., Park, W. J., & Ko, S. J. (2014). Face recognition using LBP Eigenfaces. *IEICE Transactions on Information and Systems, E97-D*(7), 1930–1932. https://doi.org/10.1587/transinf.E97.D.1930.

Lei, L., Kim, D.-H., Par, W.-J., & Ko, S.-J. (2016). Face recognition using LBP Eigenfaces. *IEICE TRANSACTIONS on Information and Systems, 97*(7), 1930–1932.

Lin, M., Chen, Q., & Yan, S. (2013). *Network in network* (pp. 1–10).

Liu, X., Shan, S., & Chen, X. (2013). Face recognition after plastic surgery: A comprehensive study. In *Vol. 7725, Issue 2. Lecture notes in computer science (including subseries lecture notes in artificial intelligence and lecture notes in bioinformatics)* (pp. 565–576). https://doi.org/10.1007/978-3-642-37444-9_44.

Maas, A., Hannun, A., & Ng, A. (2013). Rectifier nonlinearities improve neural network acoustic mode. In *International conference on machine learning.*

Moraga, J. H. (1995). The influence of the sigmoid function parameters on the speed of back-propagation learning. In *Natural to artificial neural computation. IWANN 1995. Lecture notes in computer science* (pp. 195–201).

Postorino, M. N., & Sarne, G. M. L. (2011). A neural network hybrid recommender system. *Frontiers in Artificial Intelligence and Applications, 226*, 180–187. https://doi.org/10.3233/978-1-60750-692-8-180.

Raghuwanshi, M. M., & Singh, K. R. (2009). Face recognition with rough-neural network: A rule based approach. In *International workshop on machine intelligence research* (pp. 123–129).

Rojas, R. N. N. (2013). *A systematic introduction.* Berlin, New York: Springer-Verlag.

Sable, A. H., Talbar, S. N., & Dhirbasi, H. A. (2019). Recognition of plastic surgery faces and the surgery types: An approach with entropy based scale invariant features. *Journal of King Saud University—Computer and Information Sciences, 31*(4), 554–560. https://doi.org/10.1016/j.jksuci.2017.03.004.

Schroff, F., Kalenichenko, D., & Philbin, J. (2015). FaceNet: A unified embedding for face recognition and clustering. In (Vols. 07-12). *Proceedings of the IEEE computer society conference on computer vision and pattern recognition* (pp. 815–823). IEEE Computer Society. https://doi.org/10.1109/CVPR.2015.7298682.

Singh, K. R., Khedgaonkar, R. S., & Gawande, S. P. (2011). *A new approach to local plastic surgery face recognition using near sets* (pp. 71–75).

Singh, K. R., Zaveri, M. A., & Raghuwanshi, M. M. (2012). Recognizing faces under varying poses with three states hidden Markov model. In *Vol. 2. CSAE 2012—Proceedings, 2012 IEEE international conference on computer science and automation engineering* (pp. 359–363). https://doi.org/10.1109/CSAE.2012.6272792.

Singh, R. (2010). *Plastic surgery face database.* Indraprastha Institute of Information Technology Delhi.

Singh, R., Vatsa, M., & Noore, A. (2009). Effect of plastic surgery on face recognition: A preliminary study. In *2009 IEEE conference on computer vision and pattern recognition, CVPR 2009* (pp. 72–77). https://doi.org/10.1109/CVPR.2009.5204287.

Sonsare, P. M., & Gunavathi, C. (2019). Investigation of machine learning techniques on proteomics: A comprehensive survey. *Progress in Biophysics and Molecular Biology, 149*, 54–69. https://doi.org/10.1016/j.pbiomolbio.2019.09.004.

Sun, Y., Wang, X., & Tang, X. (2014). Deep learning face representation from predicting 10,000 classes. In *Proceedings of the IEEE computer society conference on computer vision and pattern recognition* (pp. 1891–1898). IEEE Computer Society. https://doi.org/10.1109/CVPR.2014.244.

Sung, K.-K., & Poggio, T. (1995). Learning human face detection in cluttered scenes. In *Computer analysis of image and patterns* (pp. 432–439). Springer.

Syafeeza, A. R., Khalil-Hani, M., Liew, S. S., & Bakhteri, R. (2014). Convolutional neural network for face recognition with pose and illumination variation. *International Journal of Engineering and Technology*, *6*(1), 44–57. http://www.enggjournals.com/ijet/docs/IJET14-06-01-041.pdf.

Taigman, Y., Yang, M., Ranzato, M., & Wolf, L. (2014). DeepFace: Closing the gap to human-level performance in face verification. In *Proceedings of the IEEE computer society conference on computer vision and pattern recognition* (pp. 1701–1708). IEEE Computer Society. https://doi.org/10.1109/CVPR.2014.220.

Tang, X., Wang, X., Zeng, X., Qiu, S., Luo, P., Tian, Y., … Loy, C. C. (2015). DeepID-Net: Deformable deep convolutional neural networks for object detection. In (Vols. 07-12). *Proceedings of the IEEE computer society conference on computer vision and pattern recognition* (pp. 2403–2412). IEEE Computer Society. https://doi.org/10.1109/CVPR.2015.7298854.

Tian. (2013). Face recognition using a hybrid algorithm based on improved PCA. In *Information technology and computer application engineering: International conference on information technology and computer application engineering.*

Toshev, A., & Szegedy, C. (2014). Human pose estimation via deep neural networks. In *IEEE conference on computer vision and pattern recognition (CVPR)* (pp. 1653–1660).

White, R. (1977). Job evaluation: Report. *Nursing Mirror*, *145*(13), 31.

Wu, H., & Gu, X. (2015). Max-pooling dropout for regularization of convolutional neural networks. In *Vol. 9489. Lecture notes in computer science (including subseries lecture notes in artificial intelligence and lecture notes in bioinformatics)* (pp. 46–54). Springer Verlag. https://doi.org/10.1007/978-3-319-26532-2_6.

Xiaoqian, D., & Huan, W. A. H. (2010). Comparative study of several face recognition algorithms based on PCA. In *The third international symposium computer science and computational technology.*

Yung-Yao, C., Yu-Hsiu, L., Chia-Ching, K., Ming-Han, C., & I-Hsuan, Y. (2019). Design and implementation of cloud analytics-assisted smart power meters considering advanced artificial intelligence as edge analytics in demand-side management for smart homes. *Sensors*, 2047. https://doi.org/10.3390/s19092047.

Zhai, Y., Wang, X., Gan, J., & Xu, Y. (2015). Towards practical face recognition: A local binary pattern non frontal faces filtering approach. In *Vol. 9428. Lecture notes in computer science (including subseries lecture notes in artificial intelligence and lecture notes in bioinformatics)* (pp. 51–59). Springer Verlag. https://doi.org/10.1007/978-3-319-25417-3_7.

Zhang, Z., & Fu, S. (2010). Profiling and analysis of power consumption for virtualized systems and applications. In *Conference proceedings of the IEEE international performance, computing, and communications conference* (pp. 329–330). https://doi.org/10.1109/PCCC.2010.5682290.

Ziming, Z., & Fu, S. (2010). Profiling and analysis of power consumption for virtualized systems and applications. In *IEEE 29th international performance computing and communications conference* (pp. 329–330).

Zisserman, K. S. (2015). *Very deep convolutional networks for large-scale image recognition. Vol. 75* (pp. 398–406). ICLR.

Machine learning algorithms for prediction of heart disease

11

Rashmi Rachh[a], Shridhar Allagi[b], and Shravan B.K.[a]

[a]*Department of Computer Science and Engineering, Visvesvaraya Technological University, Belagavi, India,* [b]*Department of Computer Science and Engineering, KLE Institute of Technology, Hubballi, India*

11.1 Introduction

The advent of social networking sites, the Internet of Things (IoT), and many other recent technological advancements has led to a deluge of data. This tremendous amount of data is termed big data. Traditionally, retaining large quantities of data wasn't economically feasible. Also, it was inefficient and expensive to analyze large datasets that were unstructured and incomplete, with many noisy features (Michalik, Štofa, & Zolotová, 2014a, 2014b). Traditionally unavailable opportunities are now a reality, with new and more revealing insights extracted from various sources and technologies such as cloud computing, artificial intelligence, and machine learning (ML).

Smart analysis of data streams has become a key area of research as the number of applications requiring such processing has increased. This smart analysis is bound to become pervasive as a necessary ingredient of technological progress.

The rapid embracement of cloud computing has made it possible to provide unlimited storage of data. The data deluge from various sources combined with other technological advancements such as artificial intelligence (AI), ML, advances in hardware such as GPUs, and algorithmic inventions such as deep learning (DL) have a very significant role to play in learning from the data (Auto, 2019).

AI and ML have relevance to almost all fields ranging from business to healthcare to agriculture and many more. The digitization of diagnostics and diagnoses in the healthcare industry has led to an abundance of available data. This abundance of data has led to the possibility of using ML techniques for disease detection. Disease prediction using ML techniques is an area that is very important, as a clinical diagnosis by doctors can lead to the possibility of human errors as there are a large number of attributes to be taken into consideration.

The anatomy of the chapter is as follows: Section 11.1 provides a brief introduction to ML, including types and related terms. In Section 11.2, a brief review of the related work and how these works serve as a baseline to work described in this chapter are discussed. In Section 11.3, we discuss the workflow of the models used in our experimentation. The experimental set-up used for this work is described in Section 11.4. Base supervised ML algorithms used for this case study are discussed in Section 11.5. In Section 11.6, ensemble learning models are discussed. Section 11.7 provides a comprehensive analysis and discussion of results as well as a visualization of the performances of various algorithms. The chapter concludes with Section 11.8 with a summary of the overall work.

11.1.1 Introduction to ML

In this section, a brief introduction to ML and its types is provided. There are various definitions of ML in the literature; most of them hint at providing machines with the ability to "learn" from data, without being explicitly programmed. Unlike in traditional programming, where the input to the computer is data and the program produces the output, in ML, the input to the machine is a lot of data as well as the expected output. In this case, a machine is required to find patterns from this data and, in turn, learn from this data. It is also expected to come up with set of rules or generate an algorithm.

Thus, instead of programming explicitly step by step, machines are made to go through a learning process to come up with a set of rules or algorithms. Hence, the name of ML. The traditional program takes data and rules as input; the rules are applied to the input data to produce the output. In the case of ML, the data and output are given to the machine, and it comes up with rules, patterns, or models that it sees in the input data. This phase is called the training phase. There is another phase where we make the model take this unseen or new data and get output called inference or prediction. The inference or prediction phase is similar to the traditional programming paradigm; however, the training phase is something new that is not so common in traditional programming.

11.1.2 Types of ML

There are four types of ML: supervised, unsupervised, semisupervised, and reinforcement learning.

11.1.2.1 Supervised learning

In this type, the machine is provided with examples that are labeled with their desired outputs. These learning algorithms use patterns to predict outcomes for unseen data. Supervised learning is comparable to learning in the presence of a supervisor or teacher. Because the correct answers are known, the algorithm iteratively makes predictions on the training data. When an acceptable level of performance is achieved, learning can be stopped. Thus, the goal of supervised learning is to learn the mapping (the rules) between a set of inputs and outputs (Brownlee, 2019). Supervised learning can be further grouped into two categories: classification and regression.

Classification: This can be treated as the task of approximating a mapping function (f) from input variables (x) to discrete output variables (y). The output variables are called labels or categories. The mapping function predicts the class or category for a given observation. The instances need to be classified into one of two or more classes. The variables can be real-valued or discrete. If the number of classes is two, then it is called a binary or two-class classification problem. For more than two classes, the problem is called a multiclass classification problem (Yadav, Jadhav, Nagrale, & Patil, 2020a, 2020b).

Regression: In this type, the algorithms are used to predict continuous values such as price, weight, age, etc. Some of the popular supervised learning algorithms are decision trees (DT), naïve Bayes (NB) classifier, K-nearest neighbor, linear regression, and logistic regression (LR).

11.1.2.2 Unsupervised learning

Unsupervised learning algorithms try to group unsorted information based on similarities, patterns, and differences without any prior training of data. The data provided in unsupervised algorithms are unlabeled, that is, without being provided with the correct label or answer. Unsupervised learning algorithms can be used to organize complex and seemingly unrelated data in potentially meaningful ways. Some of the variants of unsupervised learning are clustering, association, and data compression.

Unsupervised learning algorithms can be used for anomaly detection and recommender systems. Some of the popular unsupervised algorithms are Apriori algorithms, principal component analysis (PCA), singular value decomposition (SVD), and K-means, which is used for clustering.

11.1.2.3 Semisupervised learning

This category of algorithms uses an approach somewhere between supervised and unsupervised learning and use labeled as well as unlabeled data for training. Semisupervised learning is used where the amount of labeled data is very small, but there is a huge amount of unlabeled data. Usually, semisupervised learning algorithms are preferred when there is not enough labeled data to produce an accurate model, and there are no resources from which more data can be obtained. In such a scenario, semisupervised techniques can be used to increase the size of the training data. Some of the applications of semisupervised learning are in speech analysis and Internet content classification. Some of the semisupervised algorithms are self-training, generative methods, graph-based methods, and semisupervised support vector machine (SVM).

11.1.2.4 Reinforcement learning

This type of learning algorithm observes by interacting with the environment and takes actions accordingly so that the reward is maximized or the risk is minimized. The reinforcement learning algorithm (called the agent) continuously learns from the environment in an iterative fashion. In order to maximize the performance, the algorithm or software agent automatically determines the ideal behavior for a given context. Reinforcement learning is used in applications wherein dialog agents can learn from user interactions so that it can be improved over time. Some of the areas are gameplay, robotics, and text summarization engines. Popular reinforcement learning algorithms are Q-learning, temporal difference (TD), and deep adversarial networks.

11.2 Literature review

Health is the most important aspect of human life, and this has been very well understood by all in these tough times dominated by the coronavirus. Health systems have always depended on machines for diagnosis. Of late, people have started using ML models and DL models for Q-Learning diagnosis and prediction of diseases, as they would complement the traditional diagnosis techniques. A cardiovascular disease prediction system has also been studied from the ML and DL perspectives by many individuals and groups over the past few years because of the emerging importance of ML algorithms in disease prediction and prevention. This section discusses various studies spanning varying studies carried out on the use of ML and DL techniques in healthcare and related works.

Purushottam, Saxena, and Sharma (2016) tried to develop an efficient heart disease prediction system using ML and data mining techniques. They assumed that all doctors are not specialists, and even with specialists, not all are of the same skill set. Thus, they proposed that by using rule-based methods of data mining, an efficient method of heart disease prediction can be formed that will augment the skills of doctors, even specialists. Their techniques include pruned rules, original rules, classified rules, rules without duplicates, sorted rules, and Polish rules. Latha and Jeeva (2019) explored ensemble classification techniques to forecast the disease at a premature stage. The dataset was used from the UCI ML repository. The ensemble algorithms explored in this paper are boosting, bagging, majority voting (MV), and stacking. The peak accuracy improvement was obtained using the MV ensemble technique.

Gandhi and Singh (2015) used data mining techniques for heart disease prediction. They used neural networks, DT, and NB classifiers to classify the medical data and to forecast heart disease. These techniques are used to search for unseen patterns in the data that can be used to help hospitals in decision making. The work carried out by Santhana Krishnan and Geetha (2019) explored ML techniques to classify the medical dataset from the UCI repository to predict heart disease using supervised learning algorithms such as DT and NB. The comparison of the performance of these algorithms is also discussed. Manikandan (2017) illustrated the use of a binary classifier in the forecasting of a heart attack. The Gaussian NB classifier has been tried for the significant classification of instances.

Mohan, Thirumalai, and Srivastava (2019) illustrated the use of hybrid ML techniques for efficient heart disease prediction. The technique used in their work is the hybrid random forest method with the linear model (HRFLM). The data are preprocessed, and then feature selection based on DT entropy is used before classification. Gayathri, Sumathi, and Santhanam (2013, p. 111) proposed the use of ML algorithms for the diagnosis of breast cancer. According to the authors, not all hospitals are equipped with mammogram machines to diagnose breast cancer. Thus, they propose the use of ML techniques for the early detection of breast cancer and to improve the accuracy in predicting cancer. Muhammad Usman, Khalid, and Aslam (2020) explored the use of DL techniques in the prediction of epileptic seizures. The prediction of a seizure before it actually occurs is a crucial way to defeat this disease because of the difficulty in curing it.

Plis, Bunescu, Marling, Shubrook, and Schwartz (2014) explored a method of diabetes management with the help of the automatic prediction of blood glucose levels. The blood glucose levels of a patient with diabetes keep changing, and the level of insulin to be administered also keeps changing to make sure that the glucose levels are normal. Clustering based on differential evolution by Vankara and Lavanya Devi (2020) for clustering the identified label for the genome and the use of the bidirectional variance for feature selection was performed using the cuckoo search. Ayatollahi, Gholamhosseini, and Salehi (2019) conducted experiments over the cardiovascular dataset with various data mining methods such as SVM and artificial neural network (ANN). They determined that SVM had more accuracy, sensitivity, and positive predictive values.

Alarsan and Younes (2019) applied the heart disease classification using electrocardiogram (ECG) features. The dataset that was experimented with random forest, gradient boosting, and multiclassification resulted in higher accuracy. Nourmohammadi-Khiarak et al. (2020) focused on reducing the feature set and taking the dominant feature into consideration with optimization methods, yielding higher accuracy levels. Alkhasawneh (2019) applied a hybrid cascade forward neural network and achieved higher accuracy for that compared to the Elman neural network and the cascade forward neural network. The algorithms were tested across six dataset test benches. The Salleh and Samat (2017) approach was based on implying fuzzy C-means and particle swarm optimization (PSO) in the early stage of preprocessing to handle the missing values. They achieved higher accuracy with DT with their model.

A comprehensive literature survey divulges the fact that ML techniques are used extensively in healthcare for various applications such as disease prediction, diagnosis, etc. Thus, it is relevant and imperative to understand how ML methods can be applied to data related to healthcare. Also, it is essential to know the various steps involved right from collecting data for evaluating the models. Hence, the next sections of this chapter illustrate the use of ML algorithms on a heart disease prediction dataset.

11.3 **ML workflow**

As explained in the previous section, the various ML algorithms are extensively used for predictive analytics in healthcare systems. In this section, a heart disease prediction dataset is used as a case study, and a few ML algorithms are applied. Comprehensive coverage of the steps involved in the ML workflow, the tools used, and the results obtained are also provided. The underlying agenda for this exercise is to provide a reference point for readers who want to use ML for their work. Fig. 11.1 gives the steps used in ML.

11.3.1 **Data collection**

With data being the key component of ML, the first step in the ML workflow is data collection. The type or source of data depends on the task at hand. The data could be collected in real time or datasets available in various repositories. The Cleveland heart dataset used in our experiment is taken from the UCI ML repository (UCI ML Repository: Heart Disease Dataset, 1988). The dataset is multivariate in nature

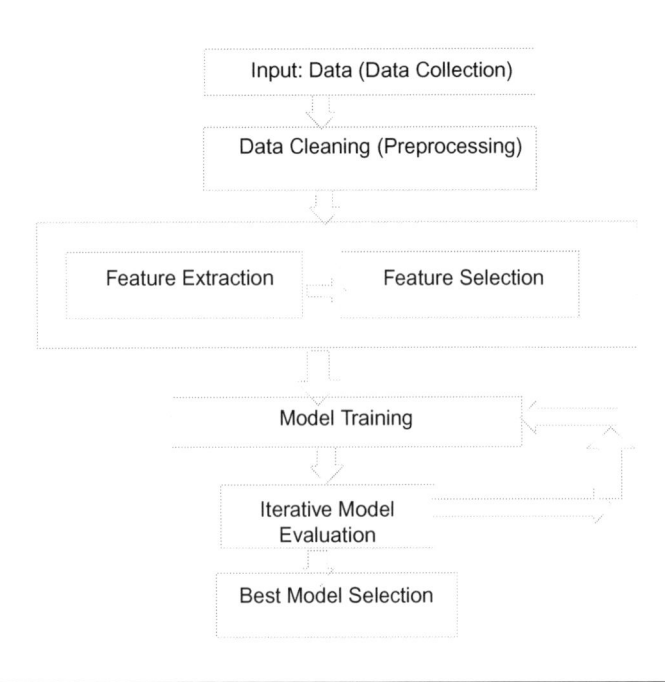

FIG. 11.1

Steps in ML.

with 75 attributes and 305 instances. It is labeled, which means that each instance is provided with the output indicating "yes" or "no." The label "yes" indicates a positive case of disease, and "no" is meant for negative cases. The sample dataset used for training is shown below in Table 11.1.

Table 11.2 provides the descriptions of a few of the attributes of the dataset.

11.3.2 Cleaning and preprocessing

The data collected from the source has to be preprocessed. Data cleaning takes care of any missing irrelevant or noisy data. Data preprocessing involves various aspects such as normalization, transformation, and instance selection (Singh, Singh, & Pant, 2017, p. 1). In our work, the dataset under consideration is preprocessed by eliminating the null values and also by filling in the missing values with the aggregated values of the other instances. The dataset is initially analyzed based on the various age groups affected with the disease and then analyzed against the gender. From the dataset, it can be observed that the major age group affected was between 41 and 60 years. Fig. 11.2 shows the plot of the number of patients suffering from the disease for various age groups.

11.3.3 Feature selection

After data cleaning and preprocessing, the feature selection methods are cast off to determine the various prominent subsets of attributes/features that have a large impact on the output. This is also known as picking up a subset of appropriate features for use in model building (Brownlee, 2020). This helps in reducing the dimensions by omitting irrelevant or redundant attributes that do not contribute much to the target variable.

The three general classes of feature selection methods are filter, wrapper, and embedded. In the case of filter methods, features are selected based on their ranks, which are assigned depending on the scores by applying statistical techniques. Filter methods select features without any prior knowledge or learning. Wrapper methods use learning techniques to evaluate the usefulness of a feature. The selection of various features is considered the searching problem. Embedded methods select features while the model is getting created (Hira & Gillies, 2015).

The dataset under consideration uses a subset of 15 attributes/features that has been selected out of 75 attributes based on the contribution score in predicting the presence or absence of the disease. The features selected are age, cp, restecg, oldpeak, trestbps, sex, fbs, ca, restef, chol, exang, thalach, xhypo, slope, and thal.

Table 11.1 Sample instances of dataset.

age	cp	restecg	exang	oldpeak	sex	fbs	trestbps	chol	xhypo	restef	slope	ca	thal	Disease Yes/no
63	3	0	0	2.3	1	1	145	233	0	3	0	0	1	1
37	2	1	0	3.5	1	0	130	250	0	2	0	0	2	1
41	1	0	0	1.4	0	0	130	204	1	2	2	0	2	1
56	1	1	0	0.8	1	0	120	236	1	2	2	0	2	1
57	0	1	1	0.6	0	0	120	354	1	2	2	0	2	1
57	0	1	0	0.4	1	0	140	192	1	1	1	0	1	1
56	1	0	0	1.3	0	0	140	294	0	0	1	0	2	1
44	1	1	0	0	1	0	120	263	0	1	2	0	3	1
52	2	1	0	0.5	1	1	172	199	1	0	2	0	3	1
57	2	1	0	1.6	1	0	150	168	0	0	2	0	2	1
49	2	1	0	2	1	0	120	188	0	0	1	3	3	0
59	0	1	1	0	1	0	140	177	1	1	2	1	3	0

The descriptions of a few of the attributes of the dataset are provided in Table 11.2.

Table 11.2 Attribute descriptions.

Name of the attribute	Value (type)	Description
Age	Continuous	Age in years
Slope	1 = upsloping 2 = flat 3 = downsloping	Slope of peak exercise
Cp	Continuous	Chest Pain
Exang	Continuous	Exercise induced angina
Fbs	1 = true, 0 = false	Fasting blood sugar level >120 mg/dL
Chol	Continuous	Serum cholesterol measured in mg/dL
Restecg	0 = measured as normal 1 = having ST-T 2 = value for left ventricular hypertrophy	Electrocardiographic results
Thaldur	3 = normal 6 = fixed defect 7 = reversible defect	Duration of Exercise
Sex	1: Male 0: Female	Gender
Oldpeak	1: yes 0: No	Depression induced by stress
Trestbps	Continuous	Blood Pressure
Thalach	Continuous	Max. Heart rate

The exact method used for selecting these features is not known. However, the gradual feature reduction (GFR) method was applied to verify that these 15 attributes/features used in the subset have high dominance and ranking compared to the remaining attributes. Fig. 11.3 shows the correlations among these 15 sets of features, which in turn can help us understand the relationships among these attributes and get different insights. This is further supported with the help of the Paris plot, as shown in Fig. 11.4.

11.3.4 Model selection

Model selection can have different meanings depending on the level of abstractions, such as selecting the hyper model parameters or learning algorithm. The primary concern of model selection is to choose one of the models among the set of candidate ML models for training based on the type of data (text, image,

C⁻ /usr/local/lib/python3.6/dist-packages/IPython/core/pylabtools.py:125: MatplotlibDeprecationWarning: Support for uppercase
    fig.canvas.print_figure(bytes_io, **kw)

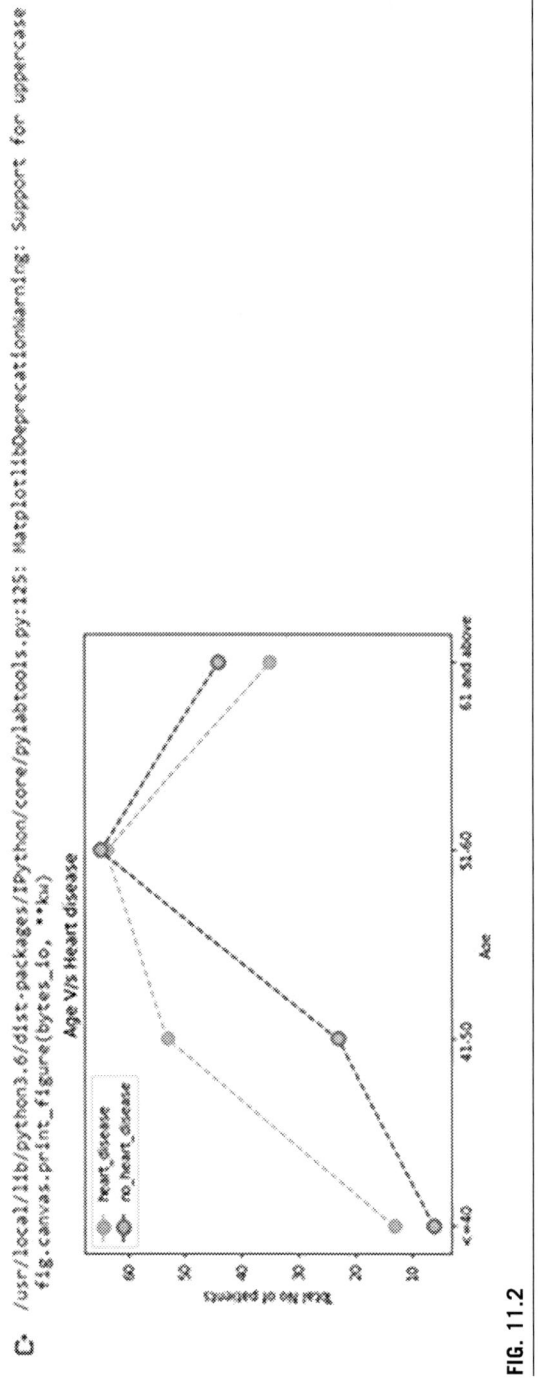

FIG. 11.2

Age group vs a positive case of the heart disease.

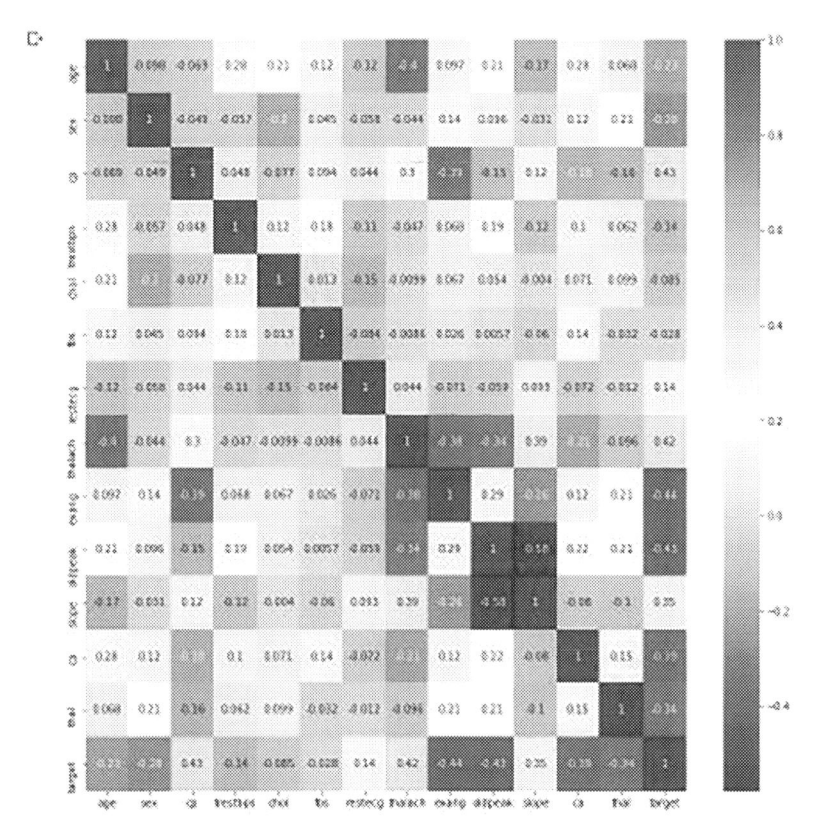

FIG. 11.3

Correlations among the set of features.

or sequence). Depending on the type of dataset available, an appropriate learning algorithm (supervised, unsupervised, or semisupervised) is selected. As the heart disease dataset under consideration is labeled one, supervised learning algorithms are deemed to be suitable.

11.3.5 Training and evaluation

The step next to model selection is training, followed by evaluation. The data that have been collected and prepared are used to incrementally improve the model's capability to predict. Eventually, it is used on unseen or test data.

At this stage, the dataset available is split into training and testing datasets. As a good rule of thumb, the training-evaluation split is in the ratio of 80:20 or 70:30, depending on the size of the dataset.

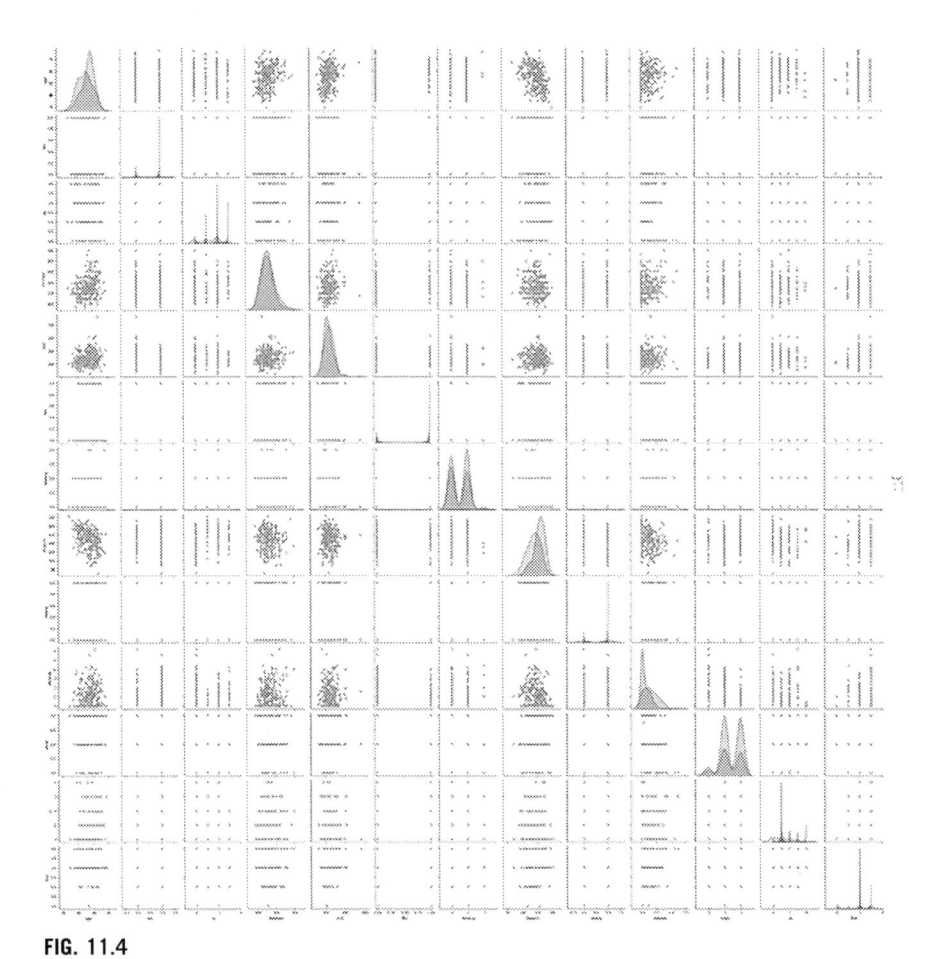

FIG. 11.4

Paris plot to study the distribution of each variable and relationship among variables.

No Permission Required.

11.4 Experimental setup

The experimentation is performed on a 16 GB RAM, Ubuntu 18, i5 fifth-generation machine. The Anaconda environment for Python has been used with packages Scikit-learn, Pandas, Seaborn, NumPy, and Matplotlib. Scikit learns the library to make available various classifiers such as SVM, NB, LR, and DT. The library's Pandas and NumPy are implemented for complex computations involved in the feature set determination and dimensionality reduction. Packages such as Seaborn and Matplotlib are used to visualize the data as well as the results at different levels in the models.

Table 11.3 Confusion matrix.

Actual class		Predicated class	
		Class = Yes	Class = No
	Class = Yes	True Positive (TP)	False Negative (FN)
	Class = No	False Positive (FP)	True Negative (TN)

As the heart disease dataset used in the experimentation is labeled, supervised machine algorithms, namely LR, DT, SVM, and NB are used. The performance of these algorithms is compared using the confusion matrix, which allows visualization of the performance. The various performance parameters, such as the recall, precision, support, and F1-score as in computed and confusion matrix are derived, are also used for comparison. These parameters are defined in Table 11.3.

Precision: It is defined as correctly predicted labels to the total correctly predicted labels.

$$Precision = \frac{Tp}{Tp + FP} \tag{11.1}$$

Recall: It is defined as the ratio of correctly predicted labels to the total number of instances of all labels.

$$Recall = \frac{Tp}{Tp + FN} \tag{11.2}$$

F1-score: It is computed as the linear average weight of the precision and the recall.

$$F1 - score = 2 * \frac{Precesion * Recall}{Precesion + Recall} \tag{11.3}$$

Support: Support is defined as the actual number of occurrences of a class in the dataset.

11.5 Supervised ML algorithms

In this section, supervised learning algorithms applied to the dataset are explained in brief, along with performance parameters.

11.5.1 Support vector machine

The base learner SVM is used over the dataset to maximize the distance among the features selected and lead to the generation of hyperplanes, which can classify the instances. The training test split of the instances is 70:30. In our SVM model, we take the output of linear functions. If the values are greater than 1, we assume the instance to belong to one class, and if the values are below -1, then the instances are classified to another class. The loss function, which accelerates to exhaust the margin, is referred to as the hinge loss. The hinge loss is calculated as:

$$c(x, y, f(x)) = \begin{cases} 0 & \text{if } y^*f(x) \geq 1 \\ 1 & -y^*f(x), \text{ else} \end{cases} \qquad c(x, y, f(x)) = (1 - y^*f(x)) \qquad (11.4)$$

The cost is computed to zero if predicted, and assigned values match. If the values don't match, then the loss is calculated. To balance between the margin boosting and loss, we sum up the parameters for regularization to cost function as below:

$$min_w \gamma \|w\|x^2 + \sum_{i=1}^{n} (1 - yi < xi, w >) \qquad (11.5)$$

The gradients for the functions are computed as below

$$\frac{\partial}{\partial wk} \gamma \|w2\| = 2\gamma wk \qquad (11.6)$$

$$\frac{\partial}{\partial wk}(1 - yi < \{xi, w\}) = \begin{cases} 0, & \text{if } yi < xi, w \geq 1 \\ -yixik, & \text{else} \end{cases} \qquad (11.7)$$

When the model classifies correctly, the update in gradient with the regularization parameter is

$$w = w - \alpha \cdot (2\gamma w) \qquad (11.8)$$

On the other hand, when there is misclassification, the loss has to be updated with the regularization parameter in gradient as

$$w = w + \alpha \cdot (yi \cdot xi - 2\gamma w) \qquad (11.9)$$

One of the major hyperparameters that can be tuned to advance the performance factor for SVM is the use of different kernels such as linear, polynomial, sigmoid, and RBF. Experimentations using all four types of kernels and the corresponding score are shown in Figs. 11.5–11.8.

After experimenting with various tuning factors of SVM, the RBF kernel was chosen because of the highest score achieved for the current dataset.

Fig. 11.9 shows the confusion matrix for the SVM classifier model.

In the current set-up, the RBF kernel with a gamma value is set to 10 and C is 0.1. The above SVM model was deployed for the heart disease dataset and the accuracy of the model was 83%. The performance parameters are shown in Table 11.4.

11.5.2 Logistic regression

LR is the supervised learning model used for classification, which categorizes the observations into discrete classes based on the selected feature sets. The binary LR classifies the observations into two discrete classes by using a more complex sigmoid function. The sigmoid function with the hypothesis

$$h_\theta(x) = \beta_0 + \beta_1 x$$

```
[45] modelSVMRaw2 = SVC(C=0.1, kernel='poly')
     modelSVMRaw2 = modelSVMRaw2.fit(X_new,Y)
     cnt1 = 0
     for i in modelSVMRaw2.predict(X_new):
       if i == Y[1]:
         cnt1 = cnt1 + 1
     print("Polynomial SVM score ")
     print(float(cnt1)/298)

     Polynomial SVM score
     0.19463087248322147
```

FIG. 11.5

Linear SVM score.

No Permission Required.

```
[46] modelSVMRaw2 = SVC(C=0.1, kernel='rbf')
     modelSVMRaw2 = modelSVMRaw2.fit(X_new,Y)
     cnt1 = 0
     for i in modelSVMRaw2.predict(X_new):
       if i == Y[1]:
         cnt1 = cnt1 + 1
     print("RBF SVM score ")
     print(float(cnt1)/298)

     RBF SVM score
     0.3959731543624161
```

FIG. 11.6

Sigmoid SVM score.

No Permission Required.

```
[34] modelSVMRaw2 = SVC(C=0.1, kernel='linear')
     modelSVMRaw2 = modelSVMRaw2.fit(X_new,Y)
     cnt1 = 0
     for i in modelSVMRaw2.predict(X_new):
         if i == Y[1]:
             cnt1 = cnt1 + 1
     print("Linear SVM score ")
     print(float(cnt1)/298)


 ⤷   Linear SVM score
     0.3926174496644295
```

FIG. 11.7

Polynomial SVM score.

No Permission Required.

The cost function defining the probability for observation into discrete classes can be computed as

$$Cost(h_\theta(x), y) = \begin{cases} -\log(h_\theta(x)) & \text{if } y = 1 \\ -\log(1 - h_\theta(x)) & \text{if } y = 0 \end{cases}$$

To reduce or minimize the cost value, we compute the gradient descents and run it on each parameter.

$$\theta_j = \theta_j - \alpha \frac{\partial}{\partial \theta_j} J\theta$$

LR was applied repetitively on the dataset; the test model accuracy is 80%. The performance parameters computed are as below in Table 11.5. Fig. 11.10 gives the confusion matrix of the model.

```
[44] modelSVMRaw2 = SVC(C=0.1, kernel='sigmoid')
     modelSVMRaw2 = modelSVMRaw2.fit(X_new,Y)
     cnt1 = 0
     for i in modelSVMRaw2.predict(X_new):
         if i == Y[1]:
             cnt1 = cnt1 + 1
     print("Sigmoid SVM score ")
     print(float(cnt1)/298)


 ⤷   Sigmoid SVM score
     0.3724832214765101
```

FIG. 11.8

RBF SVM score.

No Permission Required.

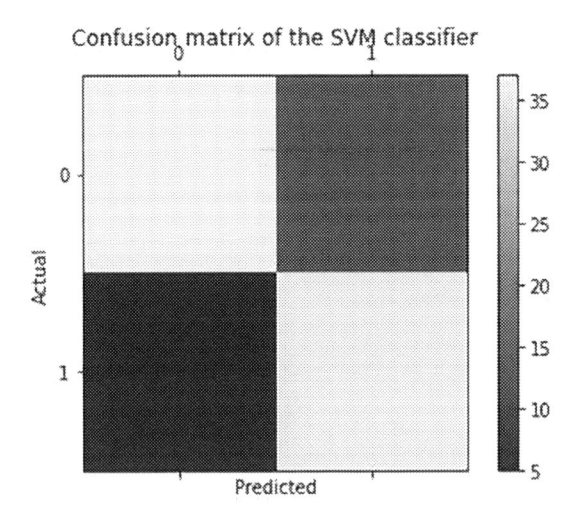

Confusion matrix of the SVM classifier

FIG. 11.9

Confusion matrix for the support vector machine.

11.5.3 Decision tree

The DT is a popularly used supervised learning method wherein a tree is constructed through an algorithmic approach that supports the split of a dataset based on different conditions. A node in a tree represents the split condition based on a specific attribute, and a terminating leaf represents a discrete class. Information gain is computed at every split with every attribute, and a split with the highest information is chosen iteratively, covering all the attributes. Because our dataset consists of instances belonging to two classes, we treat data as impure. For the impure dataset, "gini" impurity is determined, that is, the probability of misclassification of the new instance of data to a different class. The gini index for our classification problem is represented below:

$$C_T = \sum (P_k * (1 - P_k))$$

where C_T represents the cost, and P_k represents the proportion rate of training instances of particular class k in one particular node. If an error at a node is zero, then it represents the correct classification of the instance to its class. In our model to deal with the overfitting problem in the construction of the DT, pruning is performed, that is, the branches with less significance are eliminated. The accuracy was 78%. The performance parameters are shown in Table 11.6.

Table 11.4 Performance parameters for SVM.

Label	Precision	Recall	F1-score	Support
0	0.92	0.76	0.82	50
1	0.75	0.90	0.82	41

Table 11.5 Performance parameters for LR.

Label	Precision	Recall	F1-score	Support
0	0.92	0.74	0.82	50
1	0.75	0.92	0.82	41

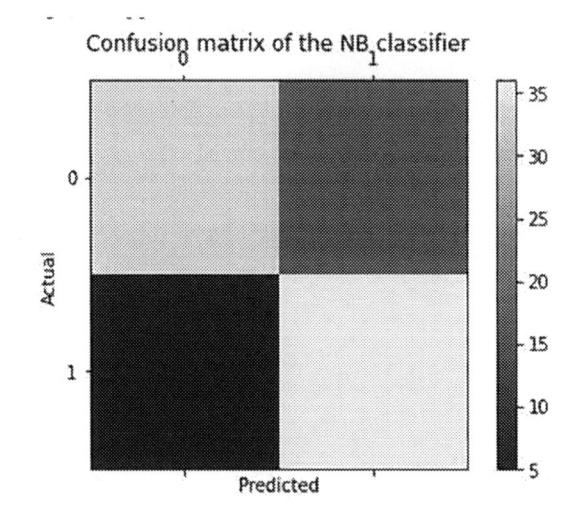

FIG. 11.10

Confusion matrix for LR classifier.

No Permission Required.

11.5.4 Naive Bayes classifier

The NB classifier is termed a probabilistic learning model. It is constructed on the Bayes theorem and is used for classification problems. The strong underlying assumption of this algorithm is the independence of the features or attributes involved in prediction. This algorithm makes predictions about any instance to a particular class by computing the class prior probability, the likelihood to a particular class, the posterior probability, and the predictor prior probability, as below:

$$P(D|X) = P(x|d)P(d)/p(x)$$

where,

$P(D|X)$ is the posterior probability for class D, predictor given as $(x,$ attributes$)$
$P(x|d)$ is the likelihood for a particular label class
$P(d)$ is the prior probability for a label class
$p(x)$ is the prior probability for the predictor

Table 11.6 Performance parameters for DT.

Label	Precision	Recall	F1-score	Support
0	0.85	0.72	0.78	50
1	0.71	0.85	0.78	41

Table 11.7 Performance parameters for NB.

Label	Precision	Recall	F1-score	Support
0	0.91	0.70	0.78	50
1	0.71	0.89	0.79	41

The likelihood for a set of attributes over any response variable is computed as below

$$P(X1, X2, ...Xn|Y) = \prod_{i=1}^{n} P(Xi|Y)$$

The accuracy obtained by the NB classifier is 80%. The performance factors are in Table 11.7.

After completing the training and evaluation, the performance for the learning model is analyzed, and different approaches can be used to accelerate the performance. When there is less data available, various approaches to improve the performance include hyperparameter tuning, feature engineering, and the use of ensemble techniques. The following section discusses the use of ensemble ML methods.

11.6 Ensemble ML models

Ensemble models combine the results from multiple learning algorithms to improve predictive performance. The result obtained using ensemble models is more accurate compared to the result from a single model. Factors such as noise, variance, and bias are responsible for the difference between the actual value and the value predicted by any ML model. Ensemble algorithms combine several ML models to form a new predictive model that reduces the bias and variance while improving the predictions. There are two methods of ensemble ML: homogenous ensemble learning where learners of the same style are aggregated and heterogeneous ensemble learning where learners of different styles are aggregated. In this work, heterogeneous ensemble learning models are implemented.

Another way to categorize ensemble learning models is sequential and parallel ensemble models. In the case of sequential ensemble methods, the base learners are generated sequentially. It exploits the reliance on the various base learning algorithms. The aggregated performance can be accelerated by weighing iteratively mislabeled examples with the highest weight. In parallel ensemble models, base learners are produced in parallel. This harnesses independence between the base learners because the error can be drastically reduced by averaging. The first step defines the formulation of different classification/regression models using the same predetermined training dataset. Every base model is trained with a dissimilar subset of training data for each same algorithm or by using the identical split of training data with different base learning models. In this work, ensemble ML models for the same split of training data with different algorithms are used. The following ensemble algorithms are implemented in this work: MV, weighted-average voting (WAV), bagging, and gradient boosting.

11.6.1 Majority voting

In the MV ensemble model, each base classifiers' prediction for each instance of data is fed to the model; the prediction with the majority of votes is considered the final output of the model. Here, the model is free from any bias that might have occurred in any base learner model due to some of the attributes or feature subsets. Fig. 11.11 demonstrates the working of our MV ensemble model.

The prediction by every ML algorithm for each instance is passed to the MV model; the model selects the prediction that has received the majority of votes. For the dataset under consideration, there are two labels, yes and no corresponding to 1 or 0. Here, the classifier $C_t|x$ is either 0 or 1 depending on whether our classifier t chooses x. The ensemble model chooses class X that has the largest vote in total.

$$\sum_{t=1}^{T} dtX(j) = max_{x=1,...,Y} \sum_{t=1}^{T} C_t x$$

Here, assuming our classifier output is completely independent, the MV performance leads to better predictions when there are more classifiers and they are independent. Assuming there are X classifiers for two-class problems, the ensemble learning model will have minimum $[X/2+1]$ classifiers predicting the correct label.

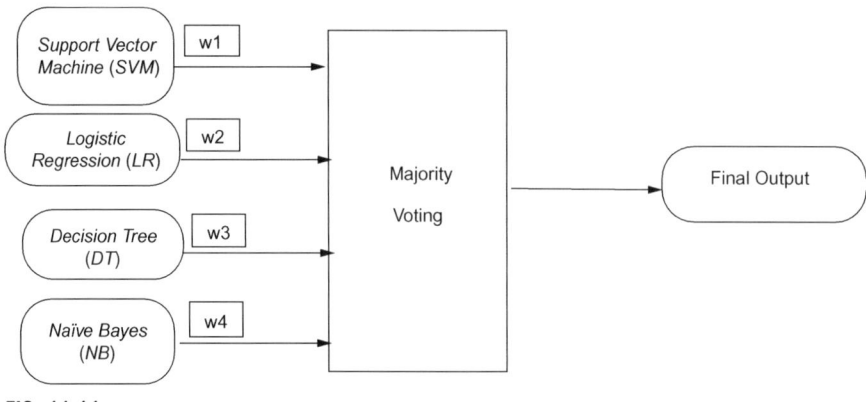

FIG. 11.11

Workflow for majority voting model.

Table 11.8 Performance parameters for MV.

Label	Precision	Recall	F1-score	Support
0	0.89	0.71	0.81	50
1	0.75	0.91	0.81	41

Let P_x be the probability of each classifier predicting the class correctly, then the total probability of the ensemble learning model of voting with classifiers $t > X/2 + 1$ will be

$$P_m = \sum_{x=\frac{x}{2}+1}^{X} \binom{T}{X} p^x (1-p)^{T-x}$$

P_m is 1 if $> T- > \infty$ and $p > 0.5$ and P_m is 0 if $T- > \infty$ and $p < 0.5$

Assuming the above parameters, the ensemble model is trained and fed with the output of each base learner such as SVM, LR, DT, and NB of each test instance. The model predicted the instances receiving the majority of the vote with an accuracy of 81%. The other performance factors are shown in Table 11.8.

11.6.2 **Weighted average voting**

In this type of ensemble model, all base learning models-classifiers are assigned with varying weights probing the importance of each classifier in the model. Here, the contributions of each model to the final predicted model are based on the weights assigned to their classifiers in the model. The working model is shown in Fig. 11.12.

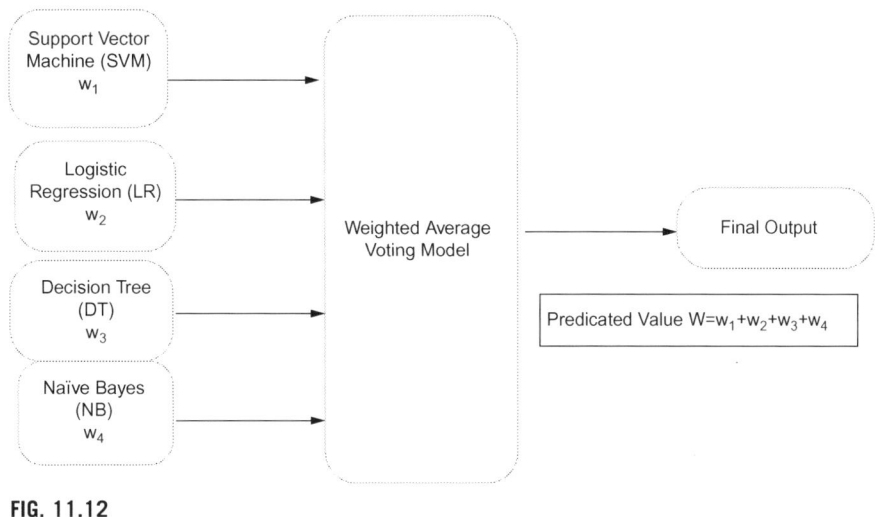

FIG. 11.12

Workflow of weighted average voting.

Table 11.9 Performance parameters for WAV.

Label	Precision	Recall	F1-score	Support
0	0.86	0.85	0.84	50
1	0.77	0.95	0.85	41

The weights w_1, w_2, w_3, and w_4 are allocated based on the contributions of classifiers in each model. Each model contributes to the final prediction based on the importance of each classifier in its base model. In our experimentation, we assigned various weights to all the base learners, and the final model accuracy was up to 83%. The performance factors are shown in Table 11.9.

11.6.3 Bagging

Bagging is a parallel ensemble learning model that stands for bootstrap aggregation. This method is used to reduce the variance in the ensemble compared to the high variance among its components. The majority of learning models are very sensitive to the data from which they have been trained; if the training data are scrambled, then the resulting predictions show a high level of variance. In this type of ensemble model, given the data, multiple bootstrapped subsets of training data are pulled for each learning model. This means that every model is trained with a different subset of training data, and the final prediction model aggregates the results from each model. Fig. 11.13 demonstrates our experimentation model.

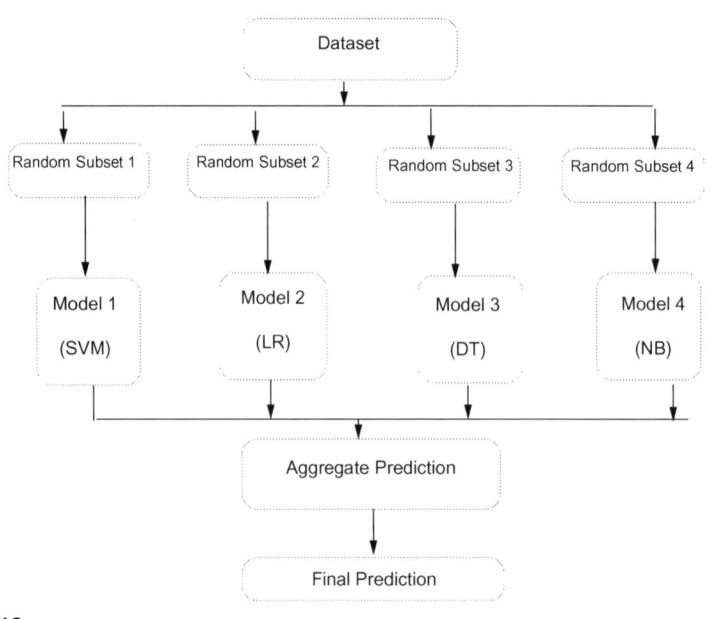

FIG. 11.13

Workflow of the bagging model.

Table 11.10 Performance parameters for bagging.

Label	Precision	Recall	F1-score	Support
0	0.91	0.75	0.82	50
1	0.73	0.91	0.82	41

Table 11.11 Performance parameters for GB.

Label	Precision	Recall	F1-score	Support
0	0.88	0.79	0.83	50
1	0.77	0.86	0.81	41

The random subsets of data are taken for training our models; then, the final aggregations are performed for the final prediction of the model. The accuracy obtained is 81%. The performance factors are shown in Table 11.10.

11.6.4 Gradient boosting

In this type of ensemble model, every weak base learner is trained iteratively to make it stronger by identifying the loss gradients in every model. The loss function determines how best our model coefficients are fitting to the underlying dataset. The loss of gradient is measured as:

$$C = ax + b + e, \quad \text{where } e \text{ is error}$$

The main aim of this method is to optimize the loss using the gradient optimization procedure while adding the models iteratively. Only those models are added that substantially reduce the loss in the prediction of labels. The gradients for minimizing the loss are computed as below:

$$y_i^p = y_i^p + \alpha * \delta \sum (y_i - y_i^p)^2 \Big|_{\gamma y_i^p}$$

where α is defined as the learning rate and $\sum (y_i - y_i^p)^2$ defines the sum of the residuals. In our experimentation, gradient boosting is performed with all four base learners and the iterative loss is minimized with the above gradients. The accuracy is 80% and the performance factors are shown in Table 11.11.

11.7 Results and discussion

This segment covers the outcomes and discusses performances for base ML algorithms and ensemble ML models.

11.7.1 Visualization of performance metrics of base learners

The four supervised learning algorithms were applied to the dataset, and the results are as listed in Tables 11.4–11.6. The performance metrics used for prediction were

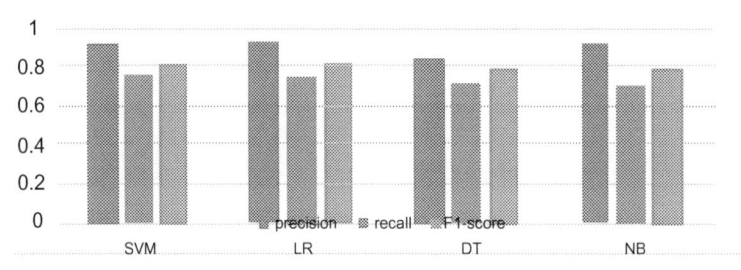

FIG. 11.14

Comparison of performance parameters for label 0.

recall, precision, support, and F1-score for two classes of output. The heart disease dataset has two labels, as mentioned in the description of the dataset, and the performance metrics for these two labels have been listed in tables. These parameters have been plotted, and a comparison of four algorithms is shown in Figs. 11.14 and 11.15 for label 0 corresponding to no disease and label 1 corresponding to the positive case of disease.

From Fig. 11.16, it is seen that the accuracy of SVM is the highest among all the base models. LR and NB have slightly lower values, and DT performance is poor. The performance of the base model may not be satisfactory due to overfitting or poor generalization. These issues, to a certain extent, can be alleviated using ensemble learning models, as explained earlier.

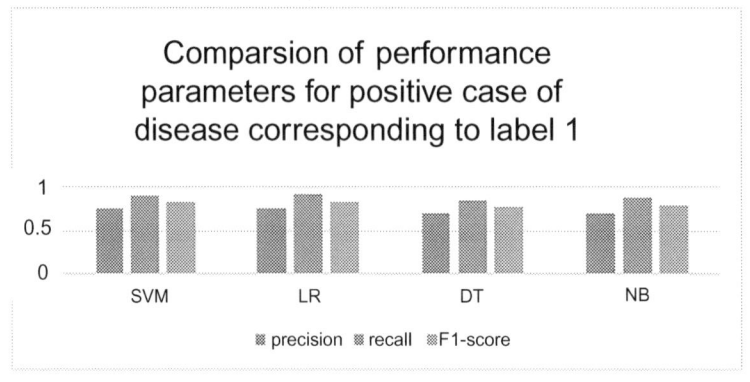

FIG. 11.15

Comparison of performance parameters for label 1.

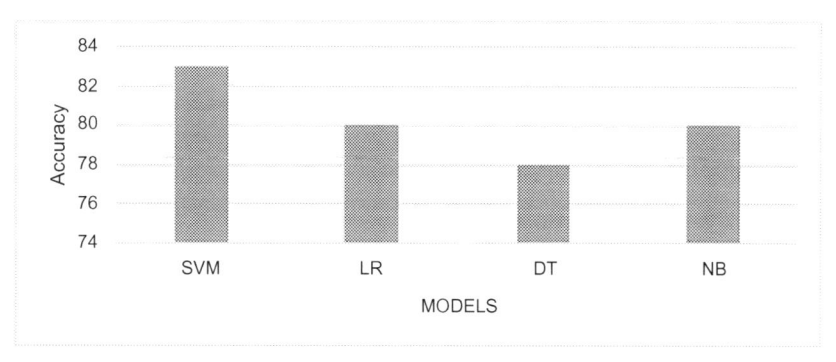

FIG. 11.16

Comparison of performance of base classifier models with respect to accuracy.

No Permission Required.

11.7.2 Visualization of performance metrics of ensemble learners

To ensure the model predictions independent of features, a set of ensemble ML algorithms such as MV, WAV, bagging, and gradient boosting has been applied on the heart disease dataset. Figs. 11.17 and 11.18 show the performance parameters and accuracy of five ensemble models: MV, WAV, bagging, and gradient boosting.

From the Fig. 11.19 results, the accuracy of WAV is the highest, and MV and bagging exhibit similar performances. The performance of the weighted average is treated as optimal as the weak classifiers are iteratively made strong by assigning varying votes as per the classifier performance.

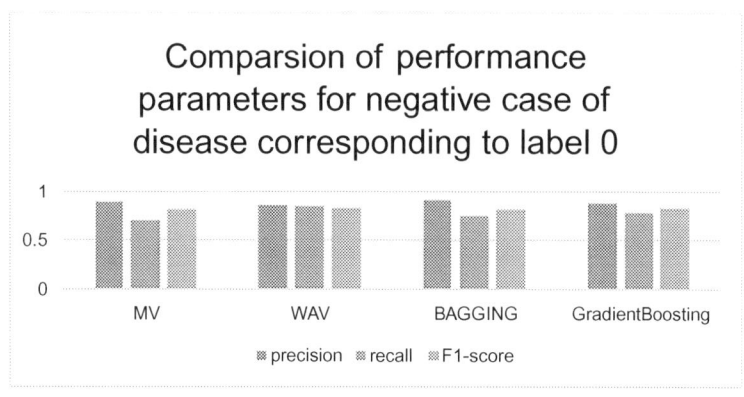

FIG. 11.17

Comparison of performance parameters for label 0.

No Permission Required.

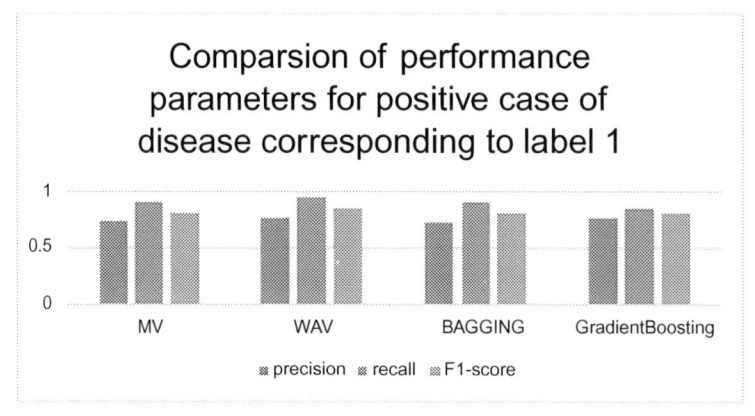

FIG. 11.18

Comparison of performance parameters for label 1.

No Permission Required.

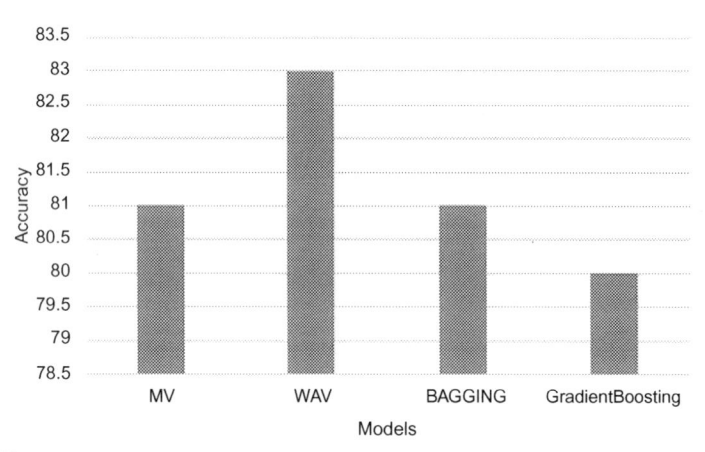

FIG. 11.19

Comparison of ensemble models with respect to accuracy.

No Permission Required.

11.8 Summary

In the broad sweep of artificial intelligence, the use of ML techniques for healthcare systems has been on the rise. ML applications in healthcare are many and they range from disease prediction to medical imaging diagnosis, outbreak prediction to personalized medicine recommendation, and so on. The limitations of human ability to analyze large-dimension complex data in healthcare can very well be replaced by machines. In order to provide guidelines as to how ML algorithms can be used on the healthcare-related dataset, in this chapter, four base learning algorithms—SVM, LR, NB, and DT—have been implemented on the heart disease dataset. The entire workflow of ML has been covered comprehensively, explaining all the steps involved. The results obtained with the experimental set-up have also been discussed, including some performance parameters that have been summarized and libraries used for visualizing/plotting these results. Apart from base learners, ensemble learning models and their significance as well as implementation results and analysis of their performance have also been included. This chapter and the efforts put into it will possibly serve as a guideline for readers interested in using ML techniques for any dataset of their interest.

Acknowledgments

We thank the Department of Computer Science and Engineering, Visvesvaraya Technological University, Belagavi, for providing the required infrastructure for carrying out the experiments. We would like to acknowledge the resources and infrastructure provided by the TEQIP supported research hub on "Data Science Techniques," VTU, Belagavi.

References

Alarsan, F. I., & Younes, M. (2019). Analysis and classification of heart diseases using heartbeat features and machine learning algorithms. *Journal of Big Data, 6*(1), 1. https://doi.org/10.1186/s40537-019-0244-x.

Alkhasawneh, M. S. (2019). Hybrid cascade forward neural network with Elman neural network for disease prediction. *Arabian Journal for Science and Engineering, 44*(11), 9209–9220. https://doi.org/10.1007/s13369-019-03829-3.

Auto, C. Q. (2019, January 15). *AI & machine learning news.* 14, January 2019. Retrieved from https://info.cloudquant.com/tag/machine-learning/page/8/.

Ayatollahi, H., Gholamhosseini, L., & Salehi, M. (2019). Predicting coronary artery disease: A comparison between two data mining algorithms. *BMC Public Health, 19*(1), 1. https://doi.org/10.1186/s12889-019-6721-5.

Brownlee, J. (2019, August 12). *Supervised and unsupervised machine learning algorithms.* Retrieved from https://machinelearningmastery.com/supervised-and-unsupervised-machine-learning-algorithms.

Brownlee, J. (2020, April 27). *An introduction to feature selection.* Retrieved from https://machinelearningmastery.com/an-introduction-to-feature-selection.

Gandhi, M., & Singh, S. N. (2015). Predictions in heart disease using techniques of data mining. In *2015 1st international conference on futuristic trends in computational analysis and knowledge management, ABLAZE 2015* (pp. 520–525). Institute of Electrical and Electronics Engineers Inc. https://doi.org/10.1109/ABLAZE.2015.7154917.

Gayathri, B. M., Sumathi, C. P., & Santhanam, T. (2013). Breast cancer diagnosis using machine learning algorithms—A survey. *International Journal of Distributed and Parallel Systems, 4*(3), 105–112. https://doi.org/10.5121/ijdps.2013.4309.

Hira, Z. M., & Gillies, D. F. (2015). A review of feature selection and feature extraction methods applied on microarray data. *Advances in Bioinformatics, 2015*, 1–13. https://doi.org/10.1155/2015/198363.

Latha, C. B. C., & Jeeva, S. C. (2019). Improving the accuracy of prediction of heart disease risk based on ensemble classification techniques. *Informatics in Medicine Unlocked, 16*, 100203. https://doi.org/10.1016/j.imu.2019.100203.

Manikandan, S. (2017). Heart attack prediction system. In *2017 international conference on energy, communication, data analytics and soft computing (ICECDS)*. https://doi.org/10.1109/icecds.2017.8389552.

Michalik, P., Štofa, J., & Zolotová, I. (2014a). Concept definition for Big Data architecture in the education system. In *2014 IEEE 12th international symposium on applied machine intelligence and informatics (SAMI)* (pp. 1–5). https://doi.org/10.1109/sami.2014.6822433.

Michalik, P., Štofa, J., & Zolotová, I. (2014b). Concept definition for Big Data architecture in the education system. In *SAMI 2014—IEEE 12th international symposium on applied machine intelligence and informatics, proceedings* (pp. 331–334). IEEE Computer Society. https://doi.org/10.1109/SAMI.2014.6822433.

Mohan, S., Thirumalai, C., & Srivastava, G. (2019). Effective heart disease prediction using hybrid machine learning techniques. *IEEE Access, 7*, 81542–81554. https://doi.org/10.1109/access.2019.2923707.

Muhammad Usman, S., Khalid, S., & Aslam, M. H. (2020). Epileptic seizures prediction using deep learning techniques. *IEEE Access, 8*, 39998–40007. https://doi.org/10.1109/access.2020.2976866.

Nourmohammadi-Khiarak, J., Feizi-Derakhshi, M.-R., Behrouzi, K., Mazaheri, S., Zamani-Harghalani, Y., & Tayebi, R. M. (2019). New hybrid method for heart disease diagnosis utilizing optimization algorithm in feature selection. *Health and Technology, 10*(3), 667–678. https://doi.org/10.1007/s12553-019-00396-3.

Plis, K., Bunescu, R., Marling, C., Shubrook, J., & Schwartz, F. (2014). *A machine learning approach to predicting blood glucose levels for diabetes management.* Association for the Advancement of Artificial Intelligence.

Purushottam, Saxena, K., & Sharma, R. (2016). Efficient heart disease prediction system. *Procedia Computer Science, 85*, 962–969. Elsevier B.V. https://doi.org/10.1016/j.procs.2016.05.288.

Salleh, M. N. M., & Samat, N. A. (2017). FCMPSO: An imputation for missing data features in heart disease classification. *IOP Conference Series: Materials Science and Engineering, 226*(1), 012102. Institute of Physics Publishing https://doi.org/10.1088/1757-899x/226/1/012102.

Santhana Krishnan, J., & Geetha, S. (2019). Prediction of heart disease using machine learning algorithms. In *2019 1st international conference on innovations in information and communication technology (ICIICT)*. https://doi.org/10.1109/iciict1.2019.8741465.

Singh, N., Singh, D. P., & Pant, B. (2017). A comprehensive study of big data machine learning approaches and challenges. In *International conference on next generation computing and information systems (ICNGCIS), Jammu, 2017* (pp. 80–85). doi. https://doi.org/10.1109/ICNGCIS.2017.14.

Vankara, J., & Lavanya Devi, G. (2020). PAELC: Predictive Analysis by Ensemble Learning and Classification heart disease detection using beat sound. *International Journal of Speech Technology, 23*(1), 31–43. https://doi.org/10.1007/s10772-020-09670-6.

Yadav, S. S., Jadhav, S. M., Nagrale, S., & Patil, N. (2020a). Application of machine learning for the detection of heart disease. In *2020 2nd international conference on innovative mechanisms for industry applications (ICIMIA)* (pp. 1–3). https://doi.org/10.1109/icimia48430.2020.9074954.

Yadav, S. S., Jadhav, S. M., Nagrale, S., & Patil, N. (2020b). Application of machine learning for the detection of heart disease. In *2nd international conference on innovative mechanisms for industry applications, ICIMIA 2020—Conference proceedings* (pp. 165–172). Institute of Electrical and Electronics Engineers Inc. https://doi.org/10.1109/ICIMIA48430.2020.9074954.

Convolutional Siamese networks for one-shot malaria parasite recognition in microscopic images

12

G. Madhu[a], B. Lalith Bharadwaj[a], B. Rohit[b], K. Sai Vardhan[a], Sandeep Kautish[c], and Pradeep N[d]

[a]*Department of Information Technology, VNR Vignana Jyothi Institute of Engineering and Technology, Hyderabad, Telangana, India,* [b]*Department of Computer Science, VNR Vignana Jyothi Institute of Engineering and Technology, Hyderabad, Telangana, India,* [c]*LBEF Campus, Kathmandu, Nepal (In Academic Collaboration with APUTI Malaysia),* [d]*Computer Science and Engineering, Bapuji Institute of Engineering and Technology, Davangere, Karnataka, India*

12.1 Introduction

Malaria is a life-threatening, mosquito-borne disease caused by a *Plasmodium* parasite that is commonly spread to humans from the bite of Anopheles mosquitoes (WHO, 2020). These *Plasmodium* parasites are of numerous types, but only some infect humans. Mainly, *Plasmodium* types such as *falciparum, vivax, ovale, malariae,* and *knowlesi* cause infections in humans (WHO, 2019). *Plasmodium vivax* is the routine type of malaria parasite that poses a huge threat to humans. It is widespread across the world, particularly in tropical and subtropical areas of Africa, South America, and Asia (WHO, 2019). As per the World Health Organization (WHO) survey reports on malaria, the mortality rate is high (0.4 million deaths) in 2019. The WHO was able to reduce the proximity of the infection globally. The research in malaria is still ongoing with funds from various countries (WHO, 2019) and the menace of malaria can be mitigated with collective efforts. This epidemic is health-endangering for mothers, infants, and children. Therefore, prevention and treatment methods are reliable to date, but with the increase in automation, we can provide fast and robust solutions with evolving technology. Commonly, there are several procedures to examine malaria through microscopic thin or thick blood smears that are popular and commonly used methods (Cuomo, Noel, & White, 2009; Devi, Roy, Singha, Sheikh, & Laskar, 2018). Thus, these techniques are extremely time-consuming and they are purely based on a clinical expert that has limited reliability (Cuomo et al., 2009; Dhiman, Baruah, &

Singh, 2010). To handle these issues, automatic diagnosis models will ensure an accurate diagnosis process for malaria detection and early diagnosis to prevent deaths. Artificial intelligence (AI) techniques have made huge contributions to achieve greater and accurate detection through various medical diagnoses techniques such as computerized tomography scan, magnetic resonance imaging, microscopy, and ultrasound analysis (Cuomo et al., 2009; Eun, Kim, Park, & Whangbo, 2015; Fan, Wei, & Cao, 2016; Sertel, Dogdas, Chiu, & Gurcan, 2011). This era of deep learning (DL) proven its significance by performing assorted tasks in computer vision such as image classification (Simonyan & Zisserman, 2015), localization, object detection (Ren, He, Girshick, & Sun, 2016), segmentation (Ronneberger, Fischer, & Brox, 2015), reconstruction, etc. With the increase in data, the model complexity increases. These sophisticated models tend to attain higher performance with a greater computational cost. In standard classification, the input image is filled with a series of layers, and finally, we generate a probability distribution over all the classes. One-shot classification requires only one training example for each class. In object detection, researchers investigated deeper and created novel visionary models for concepts such as few-shot, zero-shot, and one-shot detection.

Because of the aforementioned challenges, this research develops convolutional Siamese networks (CSN) for the one-shot diagnostic model for the parasite detection of malaria. In one-shot detection, Siamese neural networks have proven their competency in machine vision after a revival. Siamese neural networks tend to recognize objects by separating similar objects and contrary objects. CSN is implemented into two-phase criteria to detect similarity or dissimilarity. The first phase consists of extracting detailed features using a fully connected convolutional block. The second phase consists of a similarity check where inputs are discriminated against with similarity measures. CSN models are evaluated on generic metrics and performance is captured by attaining test accuracy scores of 87.10% and 87.38%. Finally, it shows decisive and consistent learning by tuning the hyperparameters in the CSN model carefully without the requirement of additional generalizations such as dropout (Srivastava, Hinton, Krizhevsky, Sutskever, & Salakhutdinov, 2014).

The chapter is organized as follows: A brief survey of the state-of-the-art Siamese neural network techniques and applications are presented in Section 12.2. Section 12.3 presents the materials and methods, Section 12.4 presents the proposed methodology for one-shot malaria parasite detection from microscopic images and a detailed description of Siamese neural architecture. The results and discussions are presented in Section 12.5. The conclusions are presented in Section 12.6.

12.2 Related works
12.2.1 State-of-the-art methods for one-shot learning

Recent works in the literature on the image recognition problem have shown more attention from the research community for generating successful models, which include works of convolutional neural networks (CNNs), AlexNet (Krizhevsky,

Sutskever, & Hinton, 2012), VGG (Simonyan & Zisserman, 2015), ResNet (He, Zhang, Ren, & Sun, 2016), and others. For the importance of the Siamese convolutional neural network and its ingenious potential to capture detailed variants for one-shot learning in object detection. Bromley, Guyon, LeCun, Säckinger, and Shah (1994) first invented the Siamese network to determine signature verification for image matching problems. This network contains twin networks used for verifying whether a signature is fraudulent. The data samples they considered were relatively small and the designed neural network was constructed. Fe-Fei (2003) presented a Bayesian framework for unsupervised one-shot learning in the object classification task. The authors proposed a hierarchical Bayesian program to solve one-shot learning for handwritten recognition. Chopra, Hadsell, and LeCun (2005) applied a selective technique for learning complex similarity measures. This was used to study a function that maps input patterns into target spaces; it was applied for face verification and recognition. Chen and Salman (2011) discussed a regularized Siamese deep network for the extraction of speaker-specific information from mel-frequency cepstral coefficients (MFCCs). This technique performs better than state-of-the-art techniques for speaker-specific information extraction. Liu (2013) presented a probabilistic-based Siamese network for learning representations on MNIST and COIL-100 images that express learning as maximizing the plausibility of binary similarity labels for a couple of input images. Cano and Cruz-Roa (2020) presented a review of one-shot recognition by the Siamese network for the classification of breast cancer in histopathological images. However, one-shot learning is used to classify the set of data features from various modules, in which there are few annotated examples. That permits us to combine new data from new classes without retraining.

12.2.2 Siamese network for face recognition and verification

Azadmanesh (2014) proposed Siamese networks for biometric hashing that is used to determine a similarity, preserving mapping for hashing. The angular distance metric used to train the verification rates achieved better verification rates. Khalil-Hani and Sung (2014) discussed a Siamese network of two CNNs for face verification on face images. This network was trained based on a stochastic gradient descent method with an annealed global learning rate. Taigman, Yang, Ranzato, and Wolf (2014) presented a Siamese network on face recognition. Once the networks learn face recognition, the network is replicated twice. These features are used for prediction, and the parameters of this network are used for cross-entropy loss and backpropagation. Lin, Cui, Belongie, and Hays (2015) presented Siamese networks on matching street-level and aerial view images to ground image matching and evaluated for the geo-localization. Berlemont, Lefebvre, Duffner, and Garcia (2015) proposed Siamese network-based nonlinear similarity measures for inertial gesture analysis and rejection. Ruiz, Linares, Sanchez, and Velez (2020) proposed a Siamese network that can do off-line handwritten signature verification based on random forgeries from a writer-independent context. It is useful for recent signers after any extra training phase required. Chakraborty, Dan, Chakraborty, and Neogy (2020) discussed a Siamese network for face recognition from the AT&T-ORL face databases. This work discussed the performance of the Siamese network with the help of various regularization and normalization methods.

12.2.3 Siamese network for scene detection and object tracking

Baraldi, Grana, and Cucchiara (2015) proposed a deep Siamese neural network for scene detection from broadcast videos. This model broadcast videos into comprehensible scenes by utilizing a distance metric between shots. Koch, Zemel, and Salakhutdinov (2015) implemented Siamese on Omnilog data with varying distribution of samples and depicted its performance on large data with maneuver optimization. Succeeding them, many researchers captured its applicability and started contributing excessively. Bertinetto, Valmadre, Henriques, Vedaldi, and Torr (2016) explicitly constructed a Siamese network comprised of embedding, which resulted in object tracking in an offline state. While considering problems such as human reidentification, pedestrian tracking, and instant search tracking, a lot of methods were proposed for human intended surveillance using Siamese networks. Varior, Haloi, and Wang (2016) proposed a gating function that captures local patterns through a resurgent mid-layered feature to strengthen the discriminative competence of features for human reidentification. Tao, Gavves, and Smeulders (2016) proposed a Siamese-based instance search tracker (SINT) for object reidentification with a matching mechanism by overcoming invariances in frames with exemplar invariances. Goncharov, Uzhinskiy, Ososkov, Nechaevskiy, and Zudikhina (2020) proposed deep Siamese networks with a k-NN classifier that are applied to plant disease detection; it helps the farming community.

12.2.4 Siamese network for two-stage learning and recognition

Leal-Taixe, Canton-Ferrer, and Schindler (2016) described two-stage learning on the lone problem of pedestrian tracking. In the first stage, they used Siamese networks for capturing spatial-temporal features. In the next stage, they used synchronized local patterns with a gradient-boosting classifier for a fused prediction. Zhang, Liu, Ma, and Fu (2016) developed a Siamese neural network-based gait recognition framework for the gait feature extraction of human recognition. Chung, Tahboub, and Delp (2017) recommended a two-stream-based CNN. Each stream represents a Siamese, and it learns spatial and temporal information distinctly. Sun, He, Gritsenko, Lendasse, and Baek (2017) presented a Siamese neural network for learning the point-wise correspondence metric, and this approach is used to improve the similarity of spectral descriptors for matching problems. Du, Fang, and Shen (2017) developed a Siamese network for the verification of the authorship of handwriting from the same author.

12.2.5 Siamese network for medical applications

Juan et al. (2017) presented a Siamese network for automated spinal metastases detection using human MRI images, and computed similarity between all the neighbor MRI slices by adopting a weighted averaging procedure to aggregate the results. Shen et al. (2017) discussed a Siamese neural network and multilevel-based similarity perception for person reidentification of multiple camera views. Jindal, Gupta, Yadav, Sharma, and Vig (2017) used a Siamese neural network for automatic karyotyping, that is, classifying human chromosomes by collecting data from local hospitals for building a decisive model. This network aims to closely collate the embeddings of samples coming from a similar label pair. Rafael-Palou et al. (2021) proposed a three-dimensional (3D) Siamese neural network for reidentification and growth detection of pulmonary nodules after image registration from CT scan images. Sampathkumar et al. (2020) proposed an efficient hybrid approach for the detection of the cancer-causing gene by using a modified bio-inspired algorithm, also known as a cuckoo search, with crossover for microarray data. Mahajan et al. (2020) proposed a Siamese network for the classification of radiomic features that are extracted from T2 MRI images. The Siamese network achieved an AUC of 0.853 and 0.894 and compared with discriminant analysis, it is achieved an AUC of 0.823 and 0.836, respectively. Hradel, Hudec, and Benesova (2020) developed a Siamese neural network-based diagnostic model for breast cancer detection from histological images. It is used to classify microscopy tissue images into normal, benign, in situ, and invasive cancer classes.

12.2.6 Siamese network for visual tracking and object tracking

Li, Yan, Wu, Zhu, and Hu (2018) suggested the Siamese network for visual tracking on large-scale image pairs. The Siamese subnetwork is used for the feature extraction and region proposal subnetwork that consists of classification and regression. He, Luo, Tian, and Zeng (2018) proposed a twofold-based Siamese network for real-time object tracking. It is built on a semantic branch and appearance branch. Each branch consists of a similarity learning Siamese network and automatically enhanced tracking performance. Zhong, Yang, and Du (2018) presented the Siamese network for palmprint recognition and employed two-parameter sharing by using VGG-16 networks that extract the two input image features. Higher layers have precisely attained the similarity of two input images by utilizing the convolutional features. Manocha et al. (2018) presented Siamese neural networks to retrieve semantically similar audio clips. This approach encodes the audio clips into a vector representation. It classifies the files belonging to the same audio class. Liu et al. (2019) proposed Siamese networks for the scene classification of remote sensing. This is associate with the identification and verification of CNN models. Zeng, Chen, Luo, and Ye (2019) presented CNNs with a Siamese network that is used for classifying color retinal fundus with Inception-V3 as a feature extractor comprised of a shared weight mechanism with the assistance of transfer learning. Caffo et al. (2019) proposed a Siamese network trained on a paired lateral interhemispheric neighborhood that is used to harness the discriminative power of total brain volumetric asymmetry.

12.2.7 **Siamese network for natural language processing**

Lei, Castaneda, Tirkkonen, Goldstein, and Studer (2019) proposed Siamese networks for supervised user equipment (UE) positioning and unsupervised channel charting in wireless systems. Ranasinghe, Orăsan, and Mitkov (2019) discuss the Siamese recurrent networks that are used to measure the semantic textual similarity between pairs of texts in natural language processing. Yuanyuan, Yuewei, Fan, Wenhui, and Jie (2019) proposed a Siamese network-based user identity linkage approach that studies and compares with higher-level features and uses input web-browsing activities. Hayale, Negi, and Mahoor (2019) presented deep Siamese networks for facial expression recognition that learn facial expression recognition by dynamically modulating verification signals. Vinayakumar and Soman (2020) presented a Siamese neural network by using recurrent networks for homoglyph attack detection. It learns the best features by using an input in the form of raw strings. Ostertag and Beurton-Aimar (2020) proposed a Siamese network for matching ostraca fragments; it is used to forecast the presence or absence of a match.

By observing the aforementioned studies, Siamese network architectures outperformed with conventional DL approaches. These promising results illustrate that there is a lot of impact on Siamese neural networks in real-world problems. This motivation helped lead to the construction of deeper models for an epidemic of infectious diseases such as malaria and to provide detailed research with systematic performance. However, to our knowledge, there is no available literature on Siamese neural network methods for malaria parasite detection.

12.3 **Materials and methods**

Malaria datasets were collected from the National Institutes of Health (NIH) (Rajaraman et al., 2018; Sivaramakrishnan, Stefan, & Antani, 2019), which are publicly accessible. This study utilized *Plasmodium falciparum*-infected and uninfected patient blood cell images that are equal to divide into infected and uninfected cell images. This research used segmented red blood cells. This consists of 27,598 cell images that comprise two categories, infected or parasitized (C0) and uninfected (C1), shown in Fig. 12.1 and annotated by medical experts. Each class consists of 13,799 cell image samples with varying blood-smeared imagery. The Siamese network required two separate inputs and a label set. The malaria image data are shown in Table 12.1.

From Fig. 12.1, it is observed that the main difference between infected or parasitized and uninfected images is the presence or absence of the parasite cells. However, the range of color present in these cell images should be noted.

Table 12.1 shows the Siamese features that are partitioned equally, comprising both parasitized and uninfected ones. The data are equally divided into train and test images to build a versatile model which improves the performance.

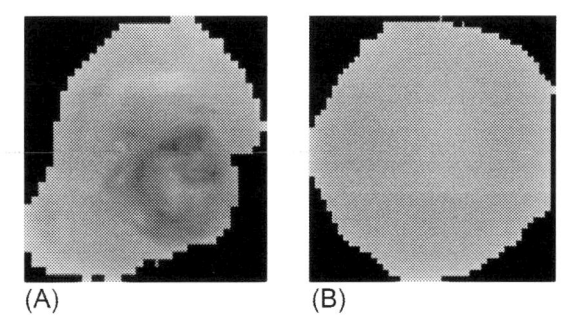

(A)   (B)

FIG. 12.1

Malaria images contain two different classes: (A) Parasitized cell, (B) uninfected cell.

No permission required.

Table 12.1 A complete description of parasitized and uninfected malaria images.

Dataset	Siamese feature set		Siamese label set
Training set [50%]	Input-1	6899	Labels: 6899
	Input-2	6899	
Testing set [50%]	Input-1	6899	Labels: 6899
	Input-2	6899	

No permission required.

The training and testing malaria cell images divided from malaria database which is picked randomly and each input for Siamese network has variability in samples and its distribution is plotted in bar graphs as shown in Figs. 12.2 and 12.3. Labels that are adaptable by utilizing the Siamese neural network are created with the help of the following algorithm, shown in Algorithm 12.1.

Algorithm 12.1: Siamese label creation

1. $\mathcal{D} \in \{X, Y\}$;   # # X: *Complete Feature Set & Y: Complete Label Set.*
2. $\mathcal{D}' \in \{(X_{S_1}, X_{S_2}), (Y_{S_1}, Y_{S_2})\}$ # # *Transition State.*
3. $Y_s' \leftarrow []$ # # *An Empty Array.*
4. *for* (y_{s_1}, y_{s_2}) *in* (Y_{s_1}, Y_{s_2}) *do*:
5.   *if* $y_{S_1} == y_{S_2}$ *do*:
6. $Y_s'. add(0)$;   # # *0 $\leftarrow$ Similar.*
7.   *else do*:
8. $Y_s'. add(1)$;   # # *1 $\leftarrow$ Dissimilar.*
9. $\mathcal{D}'' \in \{(X_{S_1}, X_{S_2}), Y_s'\}$ # # *Data Format to fit into a Siamese Neural Network.*
$X \leftarrow$ *Consists of full feature set.*
$Y \leftarrow$ *Consists of full label set.*
$X_{S_1} \leftarrow$ *Consists of half the samples from X.*
$X_{S_2} \leftarrow$ *Consists of the next half of samples from $X(X - X_{S_1})$) Similarly, for Y_{S_1} & Y.*
$Y_s' \leftarrow$ *Consists of labels that are partitioned according to the algorithm.*

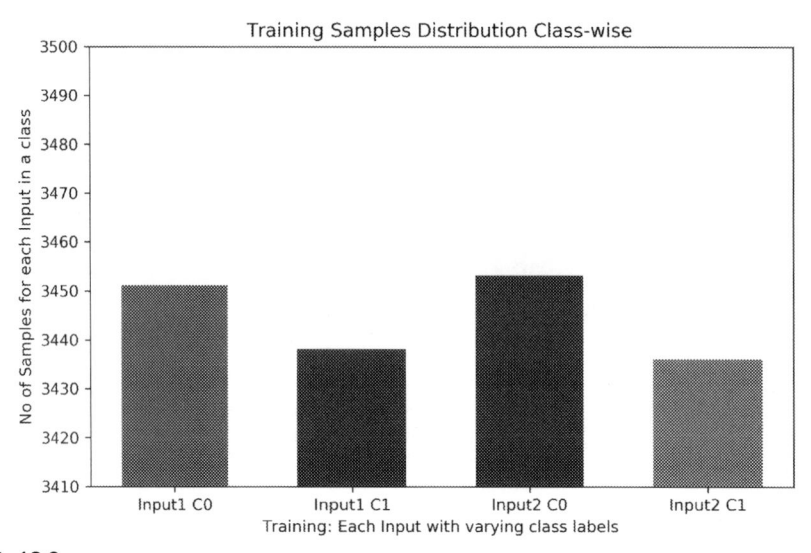

FIG. 12.2

Class-wise distribution of training dataset with varying class labels.

No permission required.

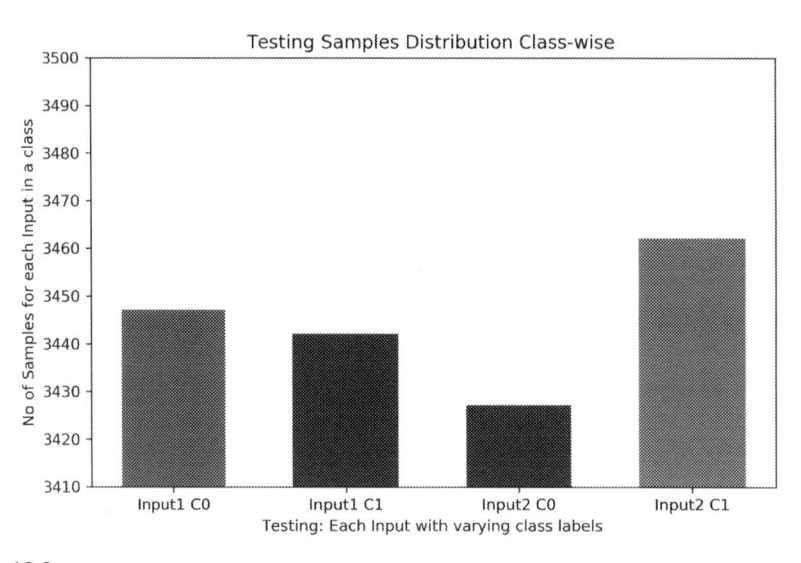

FIG. 12.3

Class-wise distribution of testing dataset with varying class labels.

No permission required.

12.4 **Proposed methodology**

Motivated by the aforementioned issues and challenges, this research proposes one-shot malaria parasite recognition from blood cell images by using the Siamese neural network (Koch et al., 2015).

12.4.1 **Siamese neural architecture**

Bromley et al. (1994) first presented Siamese neural networks for signature verification on the image matching problem. Our Siamese neural architecture contains a tandem input that constitutes segmented blood cells of $100 \times 100 \times 3$ pixels, and this is then followed by a feature extraction block. Input fed into this CNN block generates two outputs. With the help of a similarity measure (L_1 or L_2), we discriminate the outputs. This results in a single output, followed by an appropriate activation saying whether the sent inputs are similar or dissimilar; this is shown in Fig. 12.4.

Let us discuss every construction in the model in detail. The network is determined to differentiate between two inputs, that is, parasitized and uninfected malaria cells. It learns the similarity between them. This network contains two identical networks, each taking one of the two input images (parasitized and uninfected images) that share the same parameters; this is shown in Fig. 12.4. We've constructed our model by tweaking hyperparameters carefully, as convolution layers have a high capacity to convene lower- or higher-level features with alterations in receptive fields. In the literature, several authors have proven the significance of CNNs and their correspondence to build generic machine vision models (Ren et al., 2016; Ronneberger et al., 2015; Simonyan & Zisserman, 2015).

To reduce the scale of the network, CNNs are used for feature extraction. The feature extraction block is depicted in Fig. 12.5. These CNNs consist of an input layer, convolutional layer-1, an activation layer (ReLU), convolutional layer-2 with batch normalization, an activation layer (ReLU), convolutional layer-3 with batch normalization, identity, convolutional layer-4 with batch normalization, an activation layer (ReLU), convolutional layer-5 with batch normalization, sigmoid functions, and dense layers. A schematic diagram of these CNNs is shown in Fig. 12.5.

In CNNs, when input is fed into the first layer (we suppose the first layer to be the input layer) of the feature extraction block and transient through a sequence of convolutional layers and Batch Normalization layers with variants in feature maps, stride and size of the receptive field (Kernel size). The feed is forwarded with varying activations where each of the activation functions does have individual significance. The final layer in a feature extraction block contains a fully connected dense layer that linearizes the final convolutional layer feed with a sigmoid as activation. After completion of feature extraction, we visualize various feature maps that help us to know the learning at each convolution layer and to determine the consistency of the network. The cell image layer-wise representations are shown in Figs. 12.6–12.10.

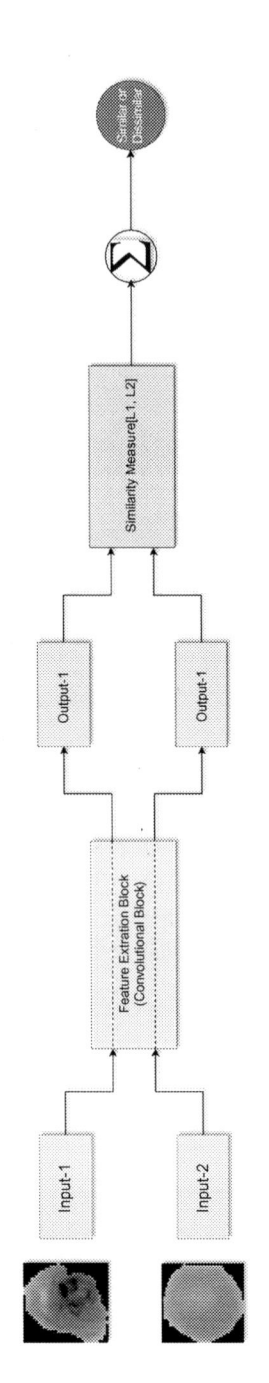

FIG. 12.4

Proposed flow diagram for the convolutional Siamese neural network model.

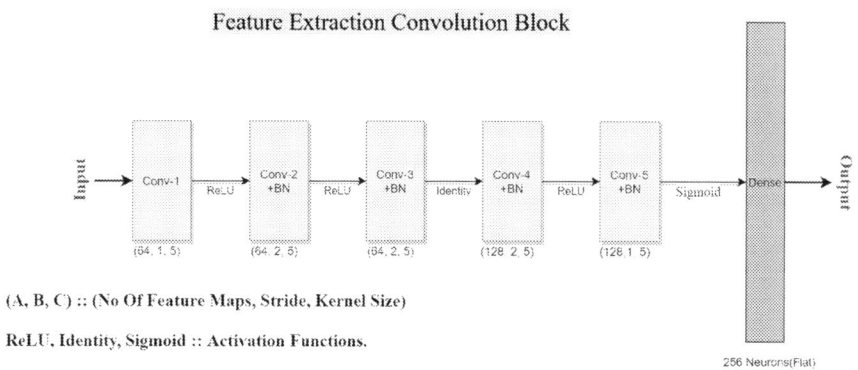

FIG. 12.5

Illustration of the fully connected convolutional neural network for feature extraction.

No permission required.

Figs. 12.6–12.10 show several illustrations of feature maps of the activation layers of the CNN model for a given input image to understand the model learning strategy. A parasitized cell image is fed into the network for visualizing the activations at different layers of the feature extraction stage.

$$\mathcal{L}^0 = (X_1^{train}, X_2^{train}) \tag{12.1}$$

$$\mathcal{L}^{i+1} = \mathcal{A}(\mathcal{L}^i . \mathcal{W}^i + \mathcal{B}^i) \tag{12.2}$$

$$ReLU(S) - \max_{x \in s}(0, x) \tag{12.3}$$

$$\sigma(x) = \frac{1}{1 + e^{-x}} \tag{12.4}$$

$$LeakyReLU(S) = \max_{x \in s}(\alpha x, x) \tag{12.5}$$

$$Identity(x) = x \tag{12.6}$$

where

$\mathcal{A}$ determines the activation function that is formulated.
$\mathcal{L}^0$ is the initial layer where data is fed as two inputs.
$\mathcal{L}^i$ is the *i*th layer having $\mathcal{W}^i$ as the weight matrix with the bias as $\mathcal{B}^i$.
ReLU, σ, *LeakyReLU*, *Identity* ∈ A.

The visualizations are constructed from a single input image mentioned (parasitized cell in Fig. 12.1). We did not visualize batch normalization layers as they do not show a large variation in cross-correlated feature maps and have negligible visual distinction.

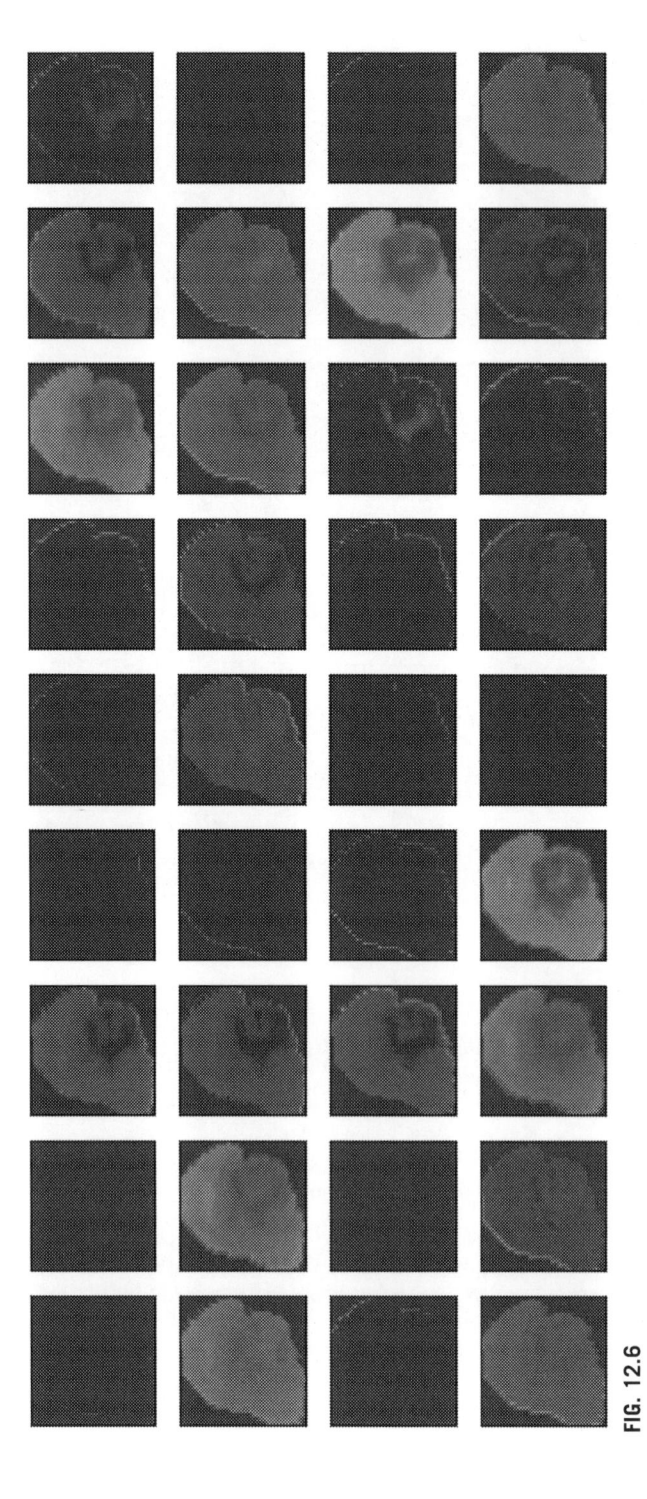

FIG. 12.6

Representation of first-layer feature map of convolution layer.

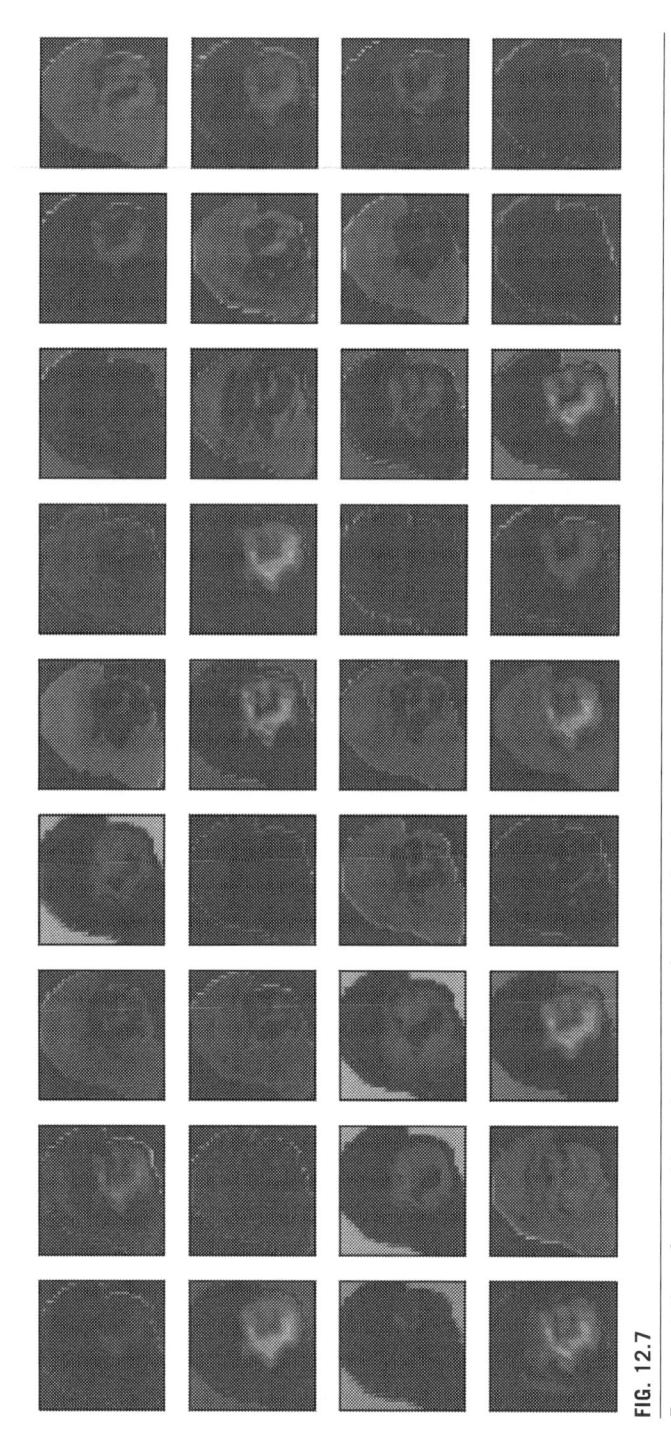

FIG. 12.7

Representation of second-layer feature map of convolution layer.

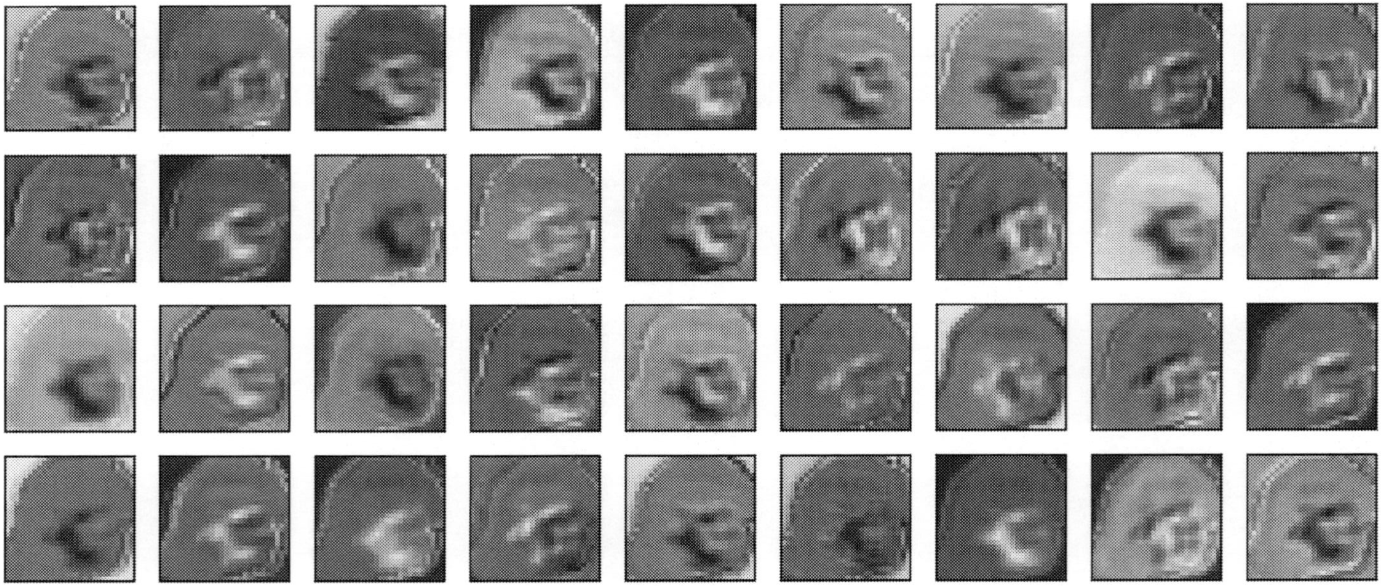

FIG. 12.8

Representation of third-layer feature map of convolution layer.

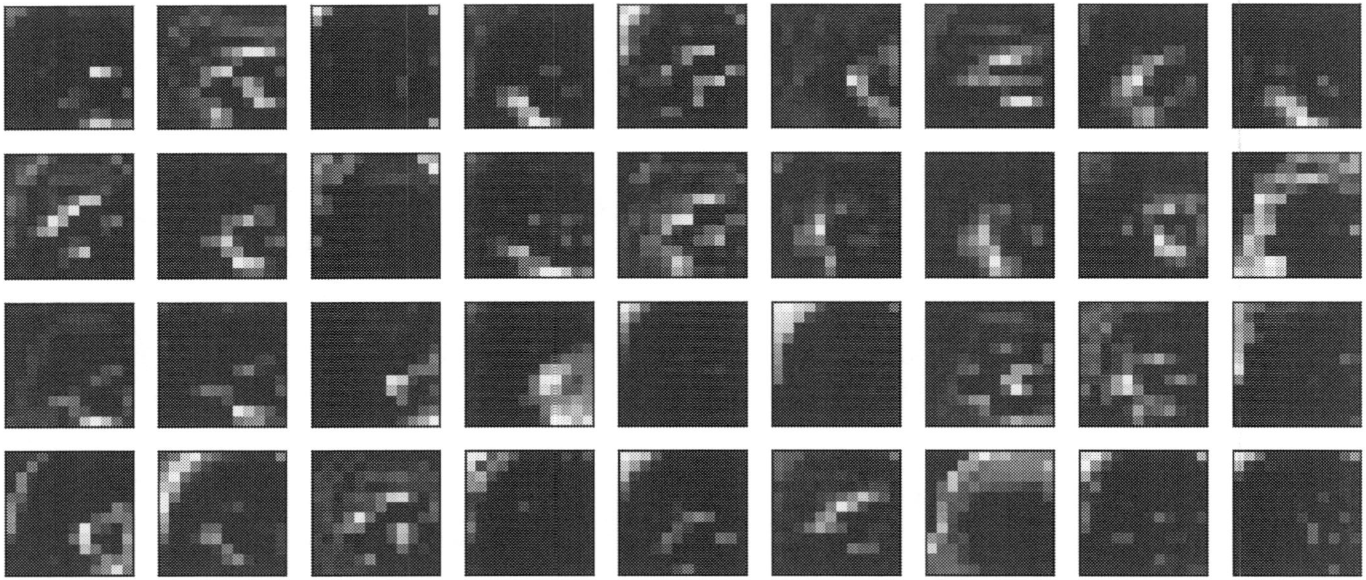

FIG. 12.9

Representation of fourth-layer feature map of convolution layer.

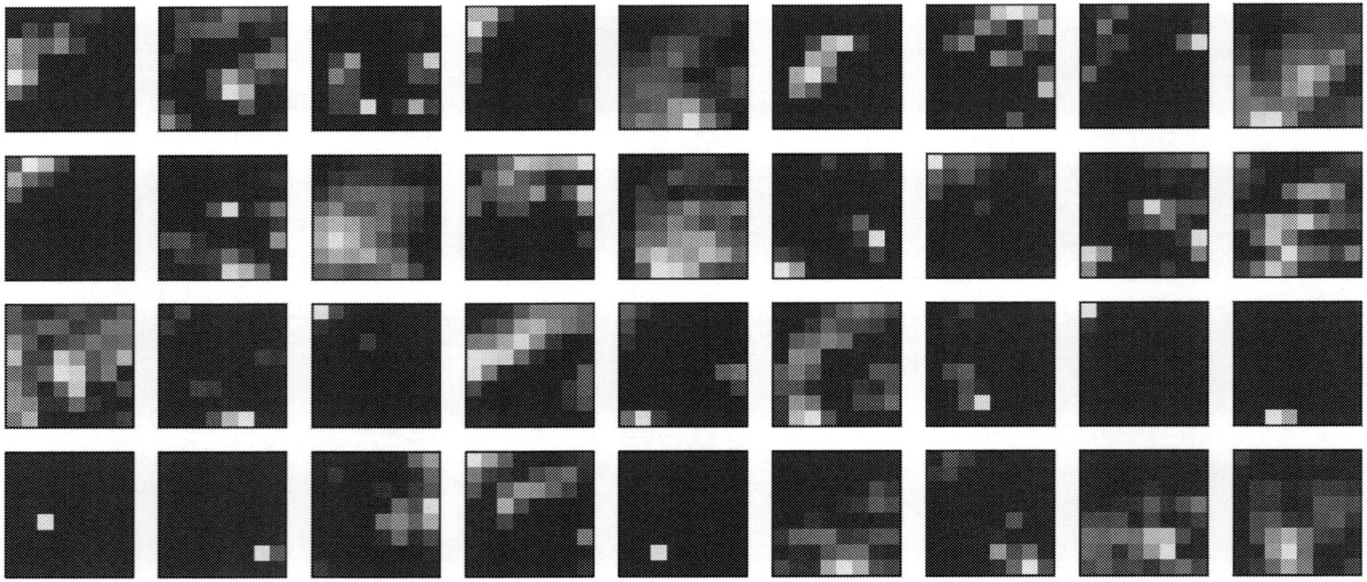

FIG. 12.10

Representation of fifth-layer feature map of convolution layer.

No permission required.

$$\mathcal{L}_1(\mathcal{F}_1, \mathcal{F}_2) = \|\mathcal{F}_1 - \mathcal{F}_2\| \tag{12.7}$$

$$\mathcal{L}_2(\mathcal{F}_1, \mathcal{F}_2) = |\mathcal{F}_1 - \mathcal{F}_2|^2 \tag{12.8}$$

where $\mathcal{F}_i$ is a linear feedforward from the feature extraction block for each input i ($i = 1, 2$). Subsequently, the feed is forwarded as linear units comprising extracted features from convolutional blocks. We additionally have not used pooling layers as they pool out the information of entities. Two outputs are generated serially one after the other. For these outputs, we find a similarity measure ($\mathcal{L}_1, \mathcal{L}_2$) to discriminate features. This discriminated feed is forwarded to the final layer with a sigmoid as activation, which depicts whether we have similarity or dissimilarity against the inputs considered. Now we compare the consistency of the constructed Siamese neural architecture and evaluate with classical metrics.

12.4.2 Training, parameter tuning, and evaluation

Before starting the training process, we must determine that we have not preprocessed data. They do not essentially require that step because data is considerably clean. To feed data into the network appropriately, we have isotopically reshaped data into $100 \times 100 \times 3$ and the sizes are not uniform. This study considered three channels for our input. The training process is initiated with a substantial amount of data, that is, 50% of the data samples were fed into the network. The remaining 50% of the samples are tested for thorough generalization. Modeling of the complete Siamese architecture comprises 3.29 million parameters. The training was carried out on 100 epochs with Adam (Kingma & Ba, 2014) as an optimizer with a learning rate of 0.0001. It is observed that decaying learning rate makes the convergence slow which has a greater impact on covering local mininma. We initialized learning with the Xavier uniform at each convolution block for faster convergence (Glorot & Bengio, 2010). We trained the network by sending six samples at a batch. The proper selection of batch size and learning rate initiates learning efficiently. We considered squared hinge as our loss function, which is defined as follows.

$$\mathcal{L} = \max_{i \in S}\left(0, \frac{1}{2} - y_{(i)} \times \hat{y}_{(i)}\right)^2 \tag{12.9}$$

where $y_{(i)}$ are true labels, $\hat{y}_{(i)}$ are predicted labels, and S are samples fed into the network for training.

Squared hinge loss evaluates the model by attaining convincing accuracy scores and improving the generalization of the model substantially. The selection of appropriate cost function would help in fast convergence with a reasonable generalization. We evaluated our model with base metrics such as accuracy scores, mean square error (MSE), and AUC-ROC. Here, MSE helps us to know the precise generalization gap for classical training. The AUC-ROC metric represents the learning rate significance at each instance of the model, and help us visualize how well our model is performing.

12.5 **Results and discussions**

In the proposed research, we compared our convolutional Siamese network with deep neural networks that incorporated with two-layered and three-layered networks. Further, we evaluated with a baseline k-NN ($k=1$) model with all the networks on both similarity measures, that is, L_1 and L_2, as shown in Figs. 12.11–12.16. Table 12.2 presents evaluated metrics such as accuracy scores, loss, and MSE, both in the training and testing phases. Each metric is evaluated for 100 epochs with a train and test pattern. CSN had the highest accuracy scores and the least loss and MSE scores.

We observed that k-NN ($k=1$) models were not able to differentiate well. When k-NN was performed with both distance measures, the accuracy scores attained were identical. But when the L_2 model was applied, it turned out to have the least loss comparatively (both training and testing). Eventually, k-NN had a poor performance. We used both kinds in neural networks, that is, two-layered and three-layered. The two-layered neural networks were constructed with an input layer and succeeded by one hidden layer and an output layer. We characterized the representation of the neural network as 1-64-128, meaning the first layer is the input layer, followed by a hidden layer with 64 neurons and a final layer that consists of 128 neurons.

Fig. 12.11 showed the CSN accuracy curve during the training and testing period with L_1 as a similarity measure. From this curve, the convergence for CSN+L_1 was faster that is, it started after the 10th epoch. Saturation is observed for the L_1 model from 30 epochs.

Fig. 12.12 showed the CSN accuracy curve during training and testing with L_2 as a similarity measure. The convergence for CSN+L_2 was slower compared

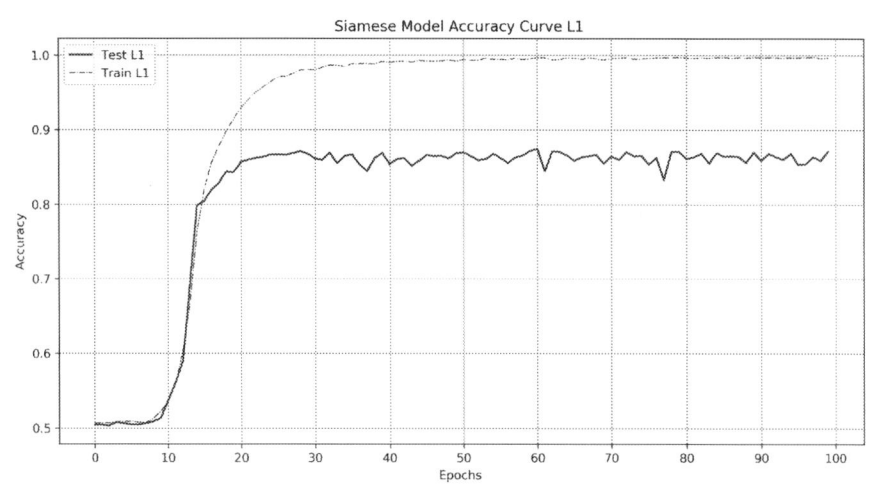

FIG. 12.11

Illustration of the CSN accuracy during training and testing with L_1 as a similarity measure.

No permission required.

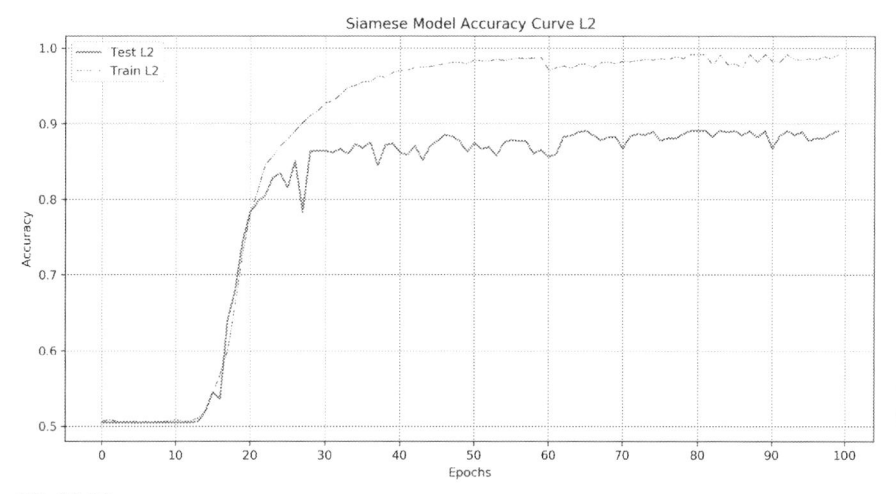

FIG. 12.12

Representation of the CSN accuracy during training and testing with L_2 as a similarity measure.

No permission required.

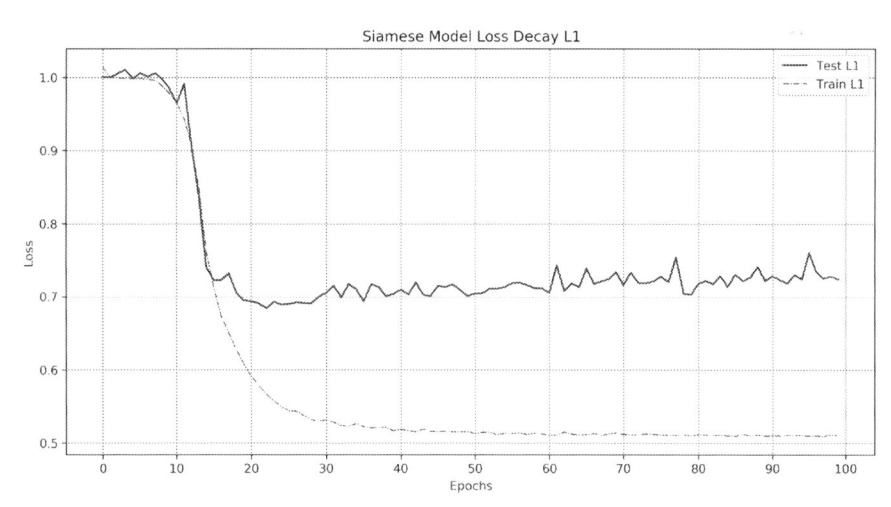

FIG. 12.13

Illustration of the CSN loss decay during training and testing with L_1 as a similarity measure.
No permission required.

to CSN + L_1, that is, convergence started around the 15th epoch. Saturation was observed after 40 epochs in CSN + L_2, but the model tried to learn better convergence and low performance to CSN + L_1.

Fig. 12.13 showed the CSN loss decay during training and testing with L_1 as a similarity measure. The loss started its decay just after the 7th epoch in CSN + L_1.

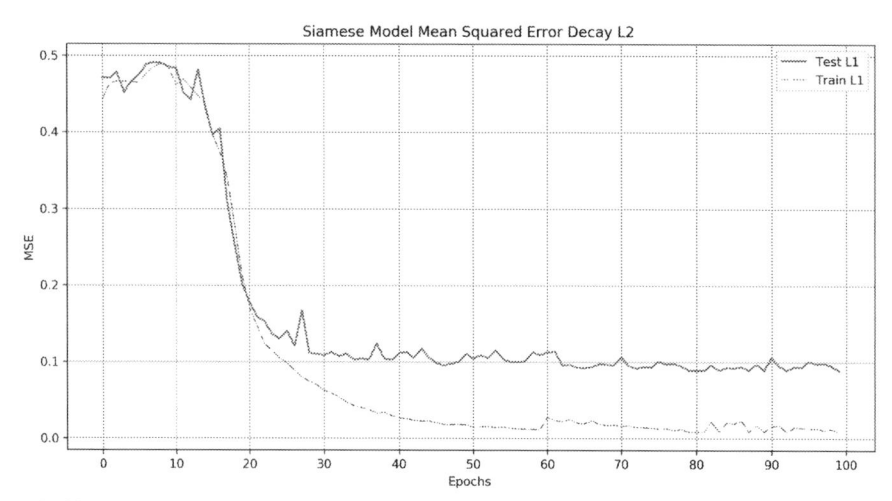

FIG. 12.14

CSN loss decay during training and testing with L_2 as the similarity measure.

No permission required.

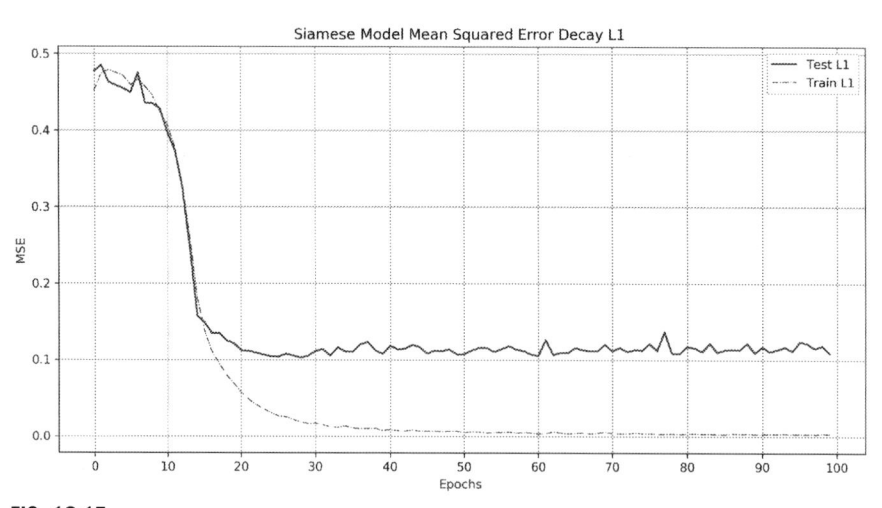

FIG. 12.15

Illustration of the CSN mean square error decay during training and testing with L_1 as the similarity measure.

No permission required.

The loss decay did not remain saturated at the 50th epoch (both training and testing) but there was little uplift after the 50th epoch.

Fig. 12.14 depicts CSN loss decay during training and testing with L_1 as a similarity measure. The loss started its decay around the 12th epoch in CSN + L_2. The loss decay saturated in the 50th epoch but has little training variants near the 60th

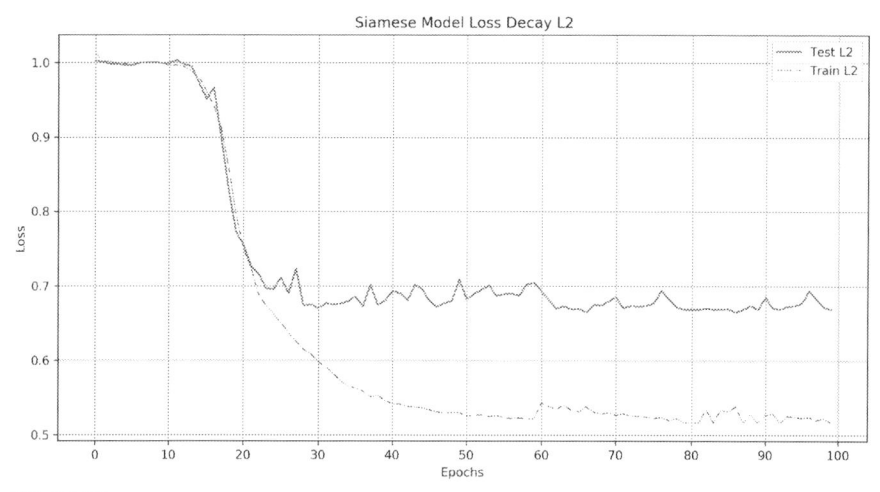

FIG. 12.16

Illustration of the CSN mean square error decay during training and testing with L_2 as the similarity measure.

Table 12.2 The performance of the models is assessed with different metrics including accuracy, loss values, and MSE.

Methods	Accuracy		Loss values		MSE	
	Train	**Test**	**Train**	**Test**	**Train**	**Test**
k-NN($k=1$)+L_1	50.65	50.47	2.0258	2.0189	0.4935	0.4953
k-NN($k=1$)+L_2	50.65	50.47	1.0772	1.0772	0.1466	0.4953
2 Layered neural network+L_1	80.97	52.75	0.7141	1.0772	0.1466	0.374
2 Layered neural network+L_2	77.96	52.13	0.7417	1.1284	0.1712	0.3903
3 Layered neural network+L_1	73.06	52.63	0.7886	1.1547	0.2159	0.3914
3 Layered neural network+L_2	68.88	53.77	0.8313	1.0926	0.2572	0.3864
CSN+L_1	99.67	87.10	0.5040	0.7240	0.0030	0.1093
CSN+L_2	98.42	87.38	0.5250	0.6953	0.0151	0.1053

epoch. In the testing curve, the model was able to still decay to a little extent even after the 50th epoch, which is a good sign for model generalization.

Fig. 12.15 illustrates CSN mean square error decay during training and testing with L_1 as a similarity measure. The MSE decay on model CSN+L_1 for training and testing was initiated around the 4th epoch and reached a saturated state after the 30th epoch. The final gap attained between training and testing is called the generalization gap, and the error is around 0.1 for the CSN $+L_1$ model.

Fig. 12.16 showed the CSN mean square error decay during training and testing with L_2 as a similarity measure. The MSE decay on model CSN $+L_2$ for training and testing was initiated after the 10th epoch and reached a saturated state after the 40th epoch. The final gap attained between training and testing is called the generalization gap, and the error is less than 0.1 for the CSN $+L_2$ model, which is a surprisingly good error rate. Finally, convergence was attained after 40 epochs but took a few more epochs to reach consistency (from 50 to 70). Also, we can see that consistency for CSN $+L_1$ is considerably stable after epoch 20. The test curve tends to slowly increase exponentially, which can lead to overfitting of the model if we do not stop training at the appropriate iteration. The CSN $+L_2$ model has good consistency in loss, which means that there is a proper fit for this model. Further, it shows that the generalization gap for both models is almost the same, that is, around 0.2 variances in the loss.

This study used ReLU (Agarap, 2018) as an activation function to process feed to successive layers. We also tried using LeakyReLU (Xu, Wang, Chen, & Li, 2015) by setting the leaking parameter (α) to 0.2. We observed that there was no significant variation with alterations. So, we assigned ReLU as our activation function between layers for both networks, but the final layer feed is activated with a sigmoid as activation. Similarly, we constructed three-layered networks with 1-64-128-256 as the pattern, which did not have any elevation in performance. When compared to two-layered networks, there is no slight generalization in the case of three-layered, but these improvements can be redundant without appropriate performance. Increasing the size of the neural network, tuning parameters, changing activations, and adjusting batch size did not have any improvement.

Finally, we carried out our experimentation even on large batches for both the networks. As they are deficient in assimilating spatial features and discriminating channel-wise correlations, proper feature extraction was aborted. Dissimilation of features from multichannels (RGB channels) would be subtle to understand for a regular neural network. To vanquish such difficulties, we propose convolutional Siamese networks as they verge to capture the spatial features. The evaluated metrics describe the stability of CSN (both L_1, L_2) by standing against others exceptionally. To realize the behavior of the CSN model, we cautiously calculated AUC-ROC curves and the confusion matrix for CSN models. The confusion matrices shown in Fig. 12.17 and Fig. 12.18 represent the class-wise performance of the model. The test accuracy of the CSN model achieved scores of 87.10% and 87.38% (L_1, L_2). The AUC-ROC shown in Fig. 12.19 gains attention by describing model performance with prudent generalizations. From Fig. 12.18, in the confusion matrix we see the number of false negatives to be 544. This means that out of the total number of parasitized classes, the CSN with the L_1 model has classified 544 of them as uninfected and 2868 of them as parasitized. Fig. 12.18 shows the confusion matrix, it can be observed that the number of false-negative classes is 635. This means that out of the total number of parasitized classes. Thus, The CSN $+L_2$ model was classified as 635 uninfected malaria images., The CSN model outperformed for the one-shot

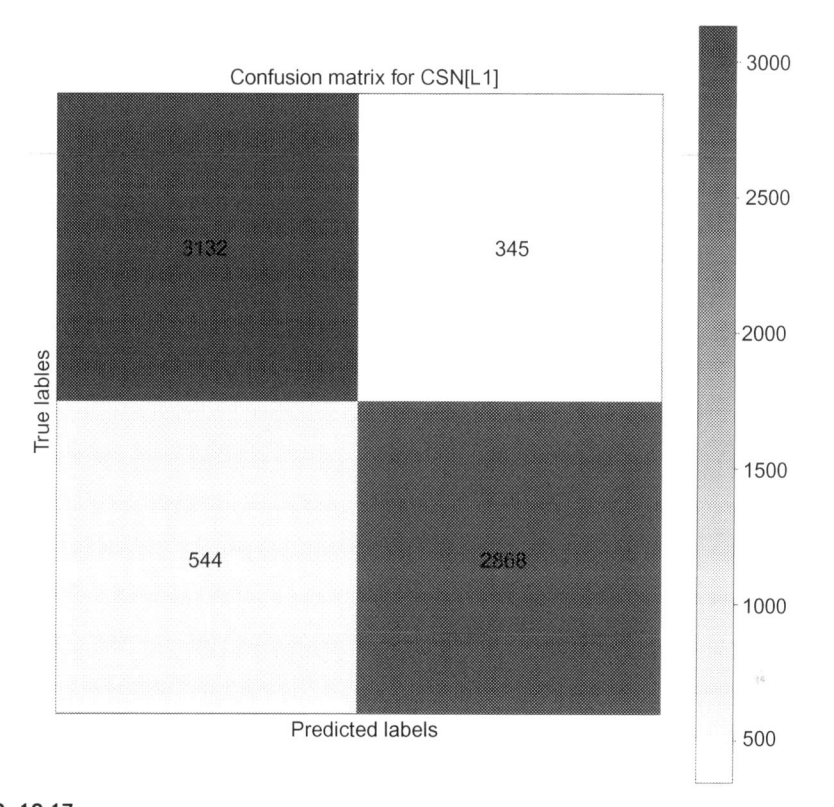

FIG. 12.17

The confusion matrices obtained for CSN with L_1.

recognition of parasites and classifying the malaria-infected and uninfected blood cells to assist in the diagnosis of malaria.

12.6 Conclusions

This research proposes one-shot malaria parasite recognition from thin blood cell images by using the convolutional Siamese network (CSN), which tends to detect and classify thin blood smear images of humans consisting of malaria. CSN is implemented in a two-phase criteria to detect similarity or dissimilarity. The starting phase consists of extracting detailed features by using fully connected convolutional blocks. The second phase consists of a similarity check where inputs are discriminated with similarity measures (L_1, L_2). The discriminated feature is feed forwarded to the final layer with a sigmoid function that depicts whether it is similar or dissimilar to the given input. Considering 50% for the test set, we are obliged to build a

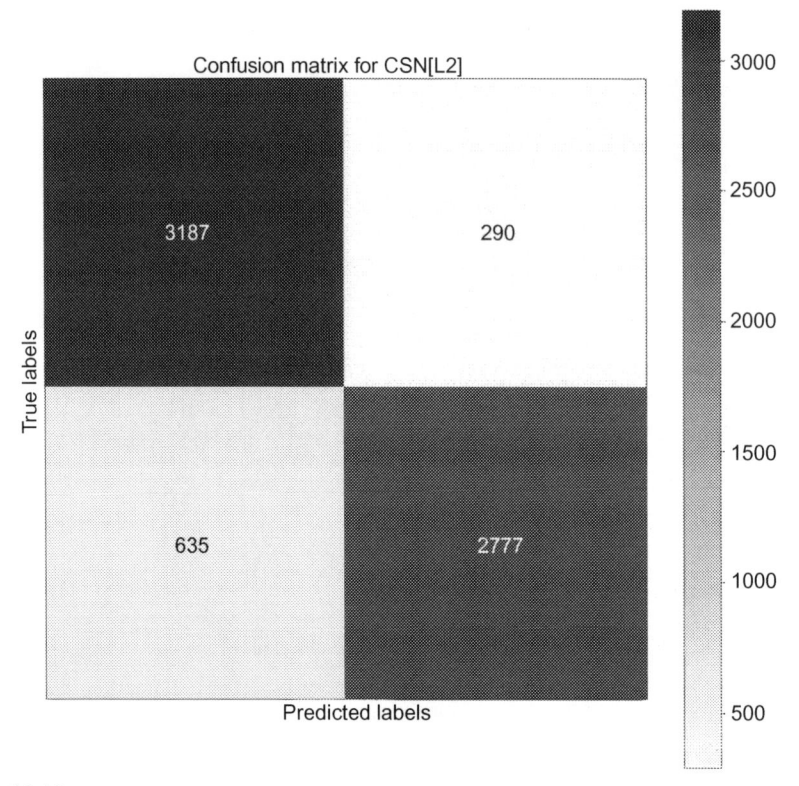

FIG. 12.18

The confusion matrices obtained for CSN with L_2.

No permission required.

streamlined model for malaria one-shot identification on two similarity metrics and depict the CSN performing with varying distance measures. CSN models are evaluated on generic metrics and performance is captured by attaining test accuracy scores of 87.10% and 87.38% (L_1, L_2). Further, we depicted confusion matrices for comparing model performance class-wise and to evaluate in which scenario CSN is making an inaccurate decision. We depicted decisive and consistent learning by tuning hyperparameters in the CSN model carefully without the requirement of additional generalizations such as dropout. To our knowledge, the SCN for the identification of malaria in thin blood smear imagery is the first attempt in automated medical diagnosis using deep learning. Finally, it is observed that the CSN model outperformed for the one-shot recognition of parasites as well as classifying the malaria-infected and uninfected blood cells to aid in the diagnosis of malaria. It will assist in diagnosing various infectious diseases rapidly. In the future, there is a chance of enhancing the model by the effective use of similarity metrics with high discriminative capacity and indulging deeper neural networks that would simulate outperforming models.

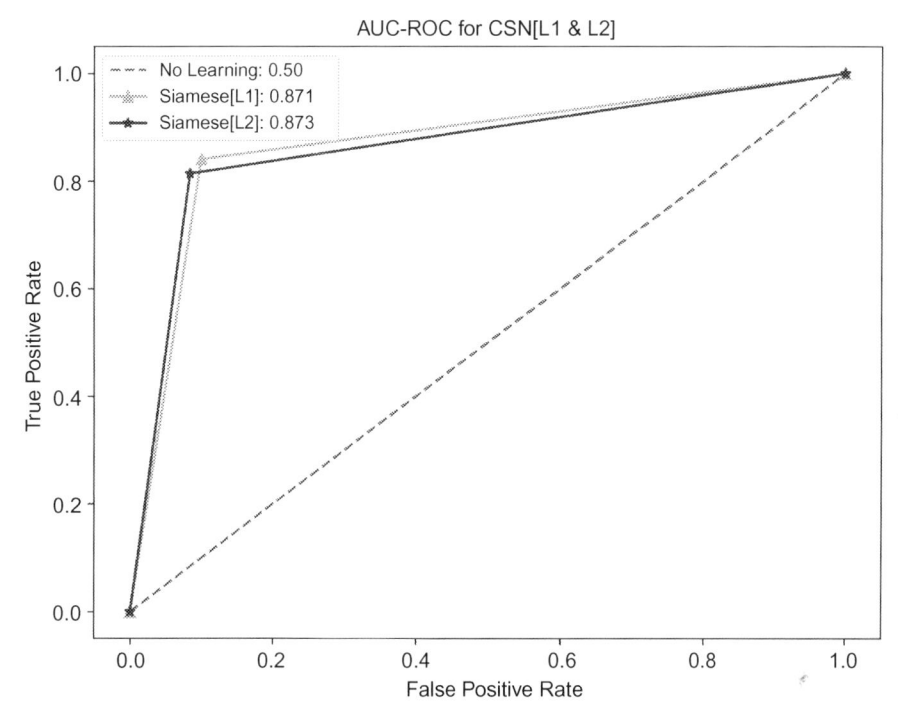

FIG. 12.19

ROC curves of the Siamese network curves for both L_1 and L_2.

References

Agarap, A. F. (2018). *Deep learning using rectified linear units (ReLU)*. ArXiv Preprint ArXiv:1803.08375. https://arxiv.org/pdf/1803.08375.pdf.

Azadmanesh, M. (2014). *Siamese neural networks for biometric hashing*. SABANCI University. http://research.sabanciuniv.edu/32328/1/MatinAzadmanesh_10050897.pdf.

Baraldi, L., Grana, C., & Cucchiara, R. (2015). A deep Siamese network for scene detection in broadcast videos. In *MM 2015—Proceedings of the 2015 ACM multimedia conference* (pp. 1199–1202). Association for Computing Machinery, Inc. https://doi.org/10.1145/2733373.2806316.

Berlemont, S., Lefebvre, G., Duffner, S., & Garcia, C. (2015). Siamese neural network-based similarity metric for inertial gesture classification and rejection. In *11th IEEE international conference and workshops on automatic face and gesture recognition (FG) (Vol. 1)* (pp. 1–6).

Bertinetto, L., Valmadre, J., Henriques, J. F., Vedaldi, A., & Torr, P. H. S. (2016). Fully-convolutional Siamese networks for object tracking. In *Lecture notes in computer science (including subseries Lecture notes in artificial intelligence and Lecture notes in bioinformatics): Vol. 9914* (pp. 850–865). Springer Verlag. https://doi.org/10.1007/978-3-319-48881-3_56.

Bromley, J., Guyon, I., LeCun, Y., Säckinger, E., & Shah, R. (1993). Signature verification using a Siamese time-delay neural network. In *NIPS'93: Proceedings of the 6th International Conference on Neural Information Processing Systems, November* (pp. 737–744).

Caffo, B., Albert, M., Miller, M. I., Bernal, A., Shi, L., Vaillant, M., ... Faria, A. V. (2019). Using deep Siamese neural networks for detection of brain asymmetries associated with Alzheimer's disease and mild cognitive impairment. *Magnetic Resonance Imaging, 64*, 190–199. https://doi.org/10.1016/j.mri.2019.07.003.

Cano, F., & Cruz-Roa, A. (2020). An exploratory study of one-shot learning using Siamese convolutional neural network for histopathology image classification in breast cancer from few data examples. In *Proceedings of SPIE—The International Society for Optical Engineering (Vol. 11330)* SPIE. https://doi.org/10.1117/12.2546488.

Chakraborty, N., Dan, A., Chakraborty, A., & Neogy, S. (2020). Effect of dropout and batch normalization in Siamese network for face recognition. In *Advances in intelligent systems and computing: Vol. 1059* (pp. 21–37). Springer. https://doi.org/10.1007/978-981-15-0324-5_3.

Chen, K., & Salman, A. (2011). Extracting speaker-specific information with a regularized Siamese deep network. In *Advances in neural information processing systems 24: 25th annual conference on neural information processing systems 2011, NIPS 2011.*

Chopra, S., Hadsell, R., & LeCun, Y. (2005). Learning a similarity metric discriminatively, with application to face verification. In *Proceedings—2005 IEEE Computer Society conference on computer vision and pattern recognition, CVPR 2005 (Vol. 1)* (pp. 539–546). IEEE Computer Society. https://doi.org/10.1109/CVPR.2005.202.

Chung, D., Tahboub, K., & Delp, E. J. (2017). A two stream Siamese convolutional neural network for person re-identification. In *Proceedings of the IEEE international conference on computer vision* (pp. 1983–1991). http://openaccess.thecvf.com/content_ICCV_2017/papers/Chung_A_Two_Stream_ICCV_2017_paper.pdf.

Cuomo, M. J., Noel, L. B., & White, D. B. (2009). *Diagnosing medical parasites: A public health officers guide to assisting laboratory and medical officers.* Air Education and Training Command Randolph AFB TX. http://www.phsource.us/PH/PARA/DiagnosingMedicalParasites.

Devi, S. S., Roy, A., Singha, J., Sheikh, S. A., & Laskar, R. H. (2018). Malaria infected erythrocyte classification based on a hybrid classifier using microscopic images of thin blood smear. *Multimedia Tools and Applications, 77*, 631–660. https://doi.org/10.1007/s11042-016-4264-7.

Dhiman, S., Baruah, I., & Singh, L. (2010). Military malaria in northeast region of India. *Defence Science Journal, 60*(2), 213–218. https://doi.org/10.14429/dsj.60.342.

Du, W., Fang, M., & Shen, M. (2017). Siamese convolutional neural networks for authorship verification. *Tech. Rep..* http://cs231n.stanford.edu/reports/2017/pdfs/801.pdf.

Eun, S. J., Kim, H., Park, J. W., & Whangbo, T. K. (2015). Effective object segmentation based on physical theory in an MR image. *Multimedia Tools and Applications, 74*(16), 6273–6286. https://doi.org/10.1007/s11042-014-2089-9.

Fan, D., Wei, L., & Cao, M. (2016). Extraction of target region in lung immunohistochemical image based on artificial neural network. *Multimedia Tools and Applications, 75*(19), 12227–12244. https://doi.org/10.1007/s11042-016-3459-2.

Fe-Fei, L. (2003). A Bayesian approach to unsupervised one-shot learning of object categories. In *Proceedings ninth IEEE international conference on computer vision* (pp. 1134–1141).

Glorot, X., & Bengio, Y. (2010). Understanding the difficulty of training deep feedforward neural networks. *J. Mach. Learn. Res., 9*, 249–256.

Goncharov, P., Uzhinskiy, A., Ososkov, G., Nechaevskiy, A., & Zudikhina, J. (2020). Deep Siamese networks for plant disease detection. In *EPJ web of conferences, mathematical modeling and computational physics 2019 (MMCP 2019) (Vol. 226)*. https://doi.org/10.1051/epjconf/202022603010.

Hayale, W., Negi, P., & Mahoor, M. (2019). Facial expression recognition using deep siamese neural networks with a supervised loss function. In *Proceedings—14th IEEE international conference on automatic face and gesture recognition, FG 2019*Institute of Electrical and Electronics Engineers Inc. https://doi.org/10.1109/FG.2019.8756571.

He, A., Luo, C., Tian, X., & Zeng, W. (2018). A twofold Siamese network for real-time object tracking. In *Proceedings of the IEEE conference on computer vision and pattern recognition* (pp. 4834–4843). http://openaccess.thecvf.com/content_cvpr_2018/papers/He_A_Twofold_Siamese_CVPR_2018_paper.pdf.

He, K., Zhang, X., Ren, S., & Sun, J. (2016). Deep residual learning for image recognition. In *2016 IEEE Conference on Computer Vision and Pattern Recognition (CVPR), Las Vegas, NV, USA* (pp. 770–778). https://doi.org/10.1109/CVPR.2016.90.

Hradel, D., Hudec, L., & Benesova, W. (2020). Interpretable diagnosis of breast cancer from histological images using Siamese neural networks. In *Twelfth international conference on machine vision, 2019, Amsterdam, Netherlands (Vol. 11433(21))*. https://doi.org/10.1117/12.2557802.

Jindal, S., Gupta, G., Yadav, M., Sharma, M., & Vig, L. (2017). Siamese networks for chromosome classification. In *Proceedings of the IEEE international conference on computer vision workshops* (pp. 71–81). http://openaccess.thecvf.com/content_ICCV_2017_workshops/papers/w1/Jindal_Siamese_Networks_for_ICCV_2017_paper.pdf.

Juan, W., Zhiyuan, F., Ning, L., Huishu, Y., Min-Ying, S., & Pierre, B. (2017). A multiresolution approach for spinal metastasis detection using deep Siamese neural networks. *Computers in Biology and Medicine*, 137–146. https://doi.org/10.1016/j.compbiomed.2017.03.024.

Khalil-Hani, M., & Sung, L. S. (2014). A convolutional neural network approach for face verification. In *Proceedings of the 2014 international conference on high performance computing and simulation, HPCS 2014* (pp. 707–714). Institute of Electrical and Electronics Engineers Inc. https://doi.org/10.1109/HPCSim.2014.6903759.

Kingma, D. P., & Ba, J. (2014). *Adam: A method for stochastic optimization*. arXiv preprint arXiv:1412.6980 https://arxiv.org/pdf/1803.08375.pdf.

Koch, G., Zemel, R., & Salakhutdinov, R. (2015). Siamese neural networks for one-shot image recognition. In *ICML deep learning workshop (Vol. 2)*. http://www.cs.toronto.edu/~gkoch/files/msc-thesis.pdf.

Krizhevsky, A., Sutskever, I., & Hinton, G. E. (2012). ImageNet classification with deep convolutional neural networks. In *Advances in neural information processing systems: Vol. 2* (pp. 1097–1105). Curran Associates, Inc.

Leal-Taixe, L., Canton-Ferrer, C., & Schindler, K. (2016). Learning by tracking: Siamese CNN for Robust Target Association. In *IEEE Computer Society conference on computer vision and pattern recognition workshops* (pp. 418–425). IEEE Computer Society. https://doi.org/10.1109/CVPRW.2016.59.

Lei, E., Castaneda, O., Tirkkonen, O., Goldstein, T., & Studer, C. (2019). Siamese neural networks for wireless positioning and channel charting. In *2019 57th annual Allerton conference on communication, control, and computing, Allerton 2019* (pp. 200–207). Institute of Electrical and Electronics Engineers Inc. https://doi.org/10.1109/ALLERTON.2019.8919897.

Li, B., Yan, J., Wu, W., Zhu, Z., & Hu, X. (2018). High performance visual tracking with Siamese region proposal network. In *Proceedings of the IEEE Computer Society conference on computer vision and pattern recognition* (pp. 8971–8980). IEEE Computer Society. https://doi.org/10.1109/CVPR.2018.00935.

Lin, T., Cui, Y., Belongie, S., & Hays, J. (2015). Learning deep representations for ground-to-aerial geolocalization. In *2015 IEEE Conference on Computer Vision and Pattern Recognition (CVPR), Boston, MA, USA* (pp. 5007–5015). https://doi.org/10.1109/CVPR.2015.7299135.

Liu, C. (2013). *Probabilistic Siamese networks for learning representations.* (Doctoral dissertation). Retrieved from https://tspace.library.utoronto.ca/handle/1807/43097.

Liu, X., Zhou, Y., Zhao, J., Yao, R., Liu, B., & Zheng, Y. (2019). Siamese convolutional neural networks for remote sensing scene classification. *IEEE Geoscience and Remote Sensing Letters, 16*(8), 1200–1204. https://doi.org/10.1109/LGRS.2019.2894399.

Mahajan, A., Dormer, J., Li, Q., Chen, D., Zhang, Z., & Fei, B. (2020). Siamese neural networks for the classification of high-dimensional radiomic features. In *Medical imaging 2020: Computer-aided diagnosis (Vol. 11314).*

Manocha, P., Badlani, R., Kumar, A., Shah, A., Elizalde, B., & Raj, B. (2018). Content-based representations of audio using Siamese neural networks. In *2018 IEEE International Conference on Acoustics, Speech and Signal Processing (ICASSP)* (pp. 3136–3140). Institute of Electrical and Electronics Engineers Inc. https://doi.org/10.1109/ICASSP.2018.8461524.

Ostertag, C., & Beurton-Aimar, M. (2020). Matching ostraca fragments using a Siamese neural network. *Pattern Recognition Letters, 131*, 336–340. https://doi.org/10.1016/j.patrec.2020.01.012.

Rafael-Palou, X., Aubanell, A., Bonavita, I., Ceresa, M., Piella, G., Ribas, V., & Ballester, M. A. G. (2021). Re-identification and growth detection of pulmonary nodules without image registration using 3D Siamese neural networks. In *Medical Image Analysis: Vol. 67.* Springer Verlag. https://doi.org/10.1016/j.media.2020.101823.

Rajaraman, S., Antani, S. K., Poostchi, M., Silamut, K., Hossain, M. A., Maude, R. J., ... Thoma, G. R. (2018). Pre-trained convolutional neural networks as feature extractors toward improved malaria parasite detection in thin blood smear images. *PeerJ, 2018* (4). https://doi.org/10.7717/peerj.4568.

Ranasinghe, T., Orăsan, C., & Mitkov, R. (2019). Semantic textual similarity with Siamese neural networks. In *RANLP, 2019.* https://wlv.openrepository.com/bitstream/handle/2436/622709/PDFsam1_proceedings-ranlp-2019.pdf?sequence=2.

Ren, S., He, K., Girshick, R., & Sun, J. (2016). Faster R-CNN: Towards real-time object detection with region proposal networks. *IEEE Trans. Pattern Anal. Mach. Intell., 39* (6), 1137–1149.

Ronneberger, O., Fischer, P., & Brox, T. (2015). U-net: Convolutional networks for biomedical image segmentation. In *Lecture notes in computer science (including subseries Lecture notes in artificial intelligence and Lecture notes in bioinformatics): Vol. 9351* (pp. 234–241). Springer Verlag. https://doi.org/10.1007/978-3-319-24574-4_28.

Ruiz, V., Linares, I., Sanchez, A., & Velez, J. F. (2020). Off-line handwritten signature verification using compositional synthetic generation of signatures and Siamese neural networks. *Neurocomputing, 374*, 30–41. https://doi.org/10.1016/j.neucom.2019.09.041.

Sampathkumar, A., Rastogi, R., Arukonda, S., Shankar, A., Kautish, S., & Sivaram, M. (2020). An efficient hybrid methodology for detection of cancer-causing gene using CSC for micro array data. *Journal of Ambient Intelligence and Humanized Computing*, 1–9. https://doi.org/10.1007/s12652-020-01731-7.

Sertel, O., Dogdas, B., Chiu, C. S., & Gurcan, M. N. (2011). Microscopic image analysis for quantitative characterization of muscle fiber type composition. *Computerized Medical Imaging and Graphics*, *35*(7–8), 616–628. https://doi.org/10.1016/j.compmedimag.2011.01.009.

Shen, C., Jin, Z., Zhao, Y., Fu, Z., Jiang, R., Chen, Y., & Hua, X. S. (2017). Deep siamese network with multi-level similarity perception for person re-identification. In *MM 2017—Proceedings of the 2017 ACM multimedia conference* (pp. 1942–1950). Association for Computing Machinery, Inc. https://doi.org/10.1145/3123266.3123452.

Simonyan, K., & Zisserman, A. (2015). Very deep convolutional networks for large-scale image recognition. In *3rd international conference on learning representations, {ICLR} 2015, San Diego, CA, USA, May 7–9, 2015, conference track proceedings* (pp. 1409–1556). http://arxiv.org/abs/1409.1556.

Sivaramakrishnan, R., Stefan, J., & Antani, S. K. (2019). Performance evaluation of deep neural ensembles toward malaria parasite detection in thin-blood smear images. *PeerJ*. https://doi.org/10.7717/peerj.6977, e6977.

Srivastava, N., Hinton, G., Krizhevsky, A., Sutskever, I., & Salakhutdinov, R. (2014). Dropout: A simple way to prevent neural networks from overfitting. *Journal of Machine Learning Research*, *15*, 1929–1958. http://jmlr.org/papers/volume15/srivastava14a/srivastava14a.pdf.

Sun, Z., He, Y., Gritsenko, A., Lendasse, A., & Baek, S. (2017). *Deep spectral descriptors: Learning the point-wise correspondence metric via Siamese deep neural networks*.

Taigman, Y., Yang, M., Ranzato, M., & Wolf, L. (2014). DeepFace: Closing the gap to human-level performance in face verification. In *Proceedings of the IEEE Computer Society conference on computer vision and pattern recognition* (pp. 1701–1708). IEEE Computer Society. https://doi.org/10.1109/CVPR.2014.220.

Tao, R., Gavves, E., & Smeulders, A. W. M. (2016). Siamese instance search for tracking. In *Proceedings of the IEEE Computer Society conference on computer vision and pattern recognition* (pp. 1420–1429). IEEE Computer Society. https://doi.org/10.1109/CVPR.2016.158.

Varior, R. R., Haloi, M., & Wang, G. (2016). Gated Siamese convolutional neural network architecture for human re-identification. In *Lecture notes in computer science (including subseries Lecture notes in artificial intelligence and Lecture notes in bioinformatics): Vol. 9912* (pp. 791–808). Springer Verlag. https://doi.org/10.1007/978-3-319-46484-8_48.

Vinayakumar, R., & Soman, K. P. (2020). Siamese neural network architecture for homoglyph attacks detection. *ICT Express*, *6*(1), 16–19. https://doi.org/10.1016/j.icte.2019.05.002.

WHO. (2019). *World malaria report*. https://www.who.int/publications-detail/world-malaria-report-2019.

WHO. (2020). *Fact sheet about malaria*. https://www.who.int/news-room/fact-sheets/detail/malaria (Accessed 14 January 2020).

Xu, B., Wang, N., Chen, T., & Li, M. (2015). *Empirical evaluation of rectified activations in convolutional network*. ArXiv Preprint ArXiv:1505.00853 https://arxiv.org/pdf/1505.00853.pdf.

Yuanyuan, Q., Yuewei, W., Fan, D., Wenhui, L., & Jie, Y. (2019). Siamese neural networks for user identity linkage through web browsing. *IEEE Transactions on Neural Networks and Learning Systems*, 1–11. https://doi.org/10.1109/tnnls.2019.2929575.

Zeng, X., Chen, H., Luo, Y., & Ye, W. (2019). Automated diabetic retinopathy detection based on binocular Siamese-like convolutional neural network. *IEEE Access, 7*, 30744–30753. https://doi.org/10.1109/ACCESS.2019.2903171.

Zhang, C., Liu, W., Ma, H., & Fu, H. (2016). Siamese neural network based gait recognition for human identification. In *2016 IEEE International Conference on Acoustics, Speech and Signal Processing (ICASSP)* (pp. 2832–2836). IEEE.

Zhong, D., Yang, Y., & Du, X. (2018). Palmprint recognition using Siamese network. In *Lecture notes in computer science (including subseries Lecture notes in artificial intelligence and Lecture notes in bioinformatics): Vol. 10996* (pp. 48–55). Springer Verlag. https://doi.org/10.1007/978-3-319-97909-0_6.

Kidney disease prediction using a machine learning approach: A comparative and comprehensive analysis

13

Siddhartha Kumar Arjaria[a], Abhishek Singh Rathore[b], and Jincy S. Cherian[c]

[a]Rajkiya Engineering College, Banda, Uttar Pradesh, India, [b]Computer Science & Engineering, Shri Vaishnav Vidyapeeth Vishwavidyalaya, Indore, Madhya Pradesh, India, [c]The Bhopal School of Social Sciences, Bhopal, Madhya Pradesh, India

13.1 Introduction

Staying healthy is everyone's desire and this can be achieved through proper diet and workouts. If someone becomes ill, then the focus is on the early detection of their disease and its proper care. Chronic kidney disease (CKD) is one of the foremost issues these days. This disease causes more deaths than some cancers. It affects millions of people and many of those remain undetected. It is the ninth-deadliest disease in the United States. Hence, the early diagnosis of this disease is a major concern.

The first question that arises is, "What is CKD?" In CKD, the kidneys are damaged so the patient's body is unable to filter out waste. This may damage other organs of the body. Proper treatment of this disease can be prescribed only if it is detected in the early stages. Treatments are constantly improving, but still, the problem is detection. The death rates can be decreased only if the symptoms can show the presence or absence of a kidney disorder. Hence there is a need for a data mining strategy that takes various symptoms (features) as input and predicts whether the patient is affected by a kidney disorder.

13.1.1 Causes of chronic kidney disease

Blood sugar and high blood pressure are the foremost reasons for kidney disorders (Chronic Diseases in America, 2019) in adults, as shown in Fig. 13.1. Some other factors are cardiac problems, obesity, a family history of kidney disease, and sometimes older age. The possible symptoms are listed in Table 13.1. The first objective

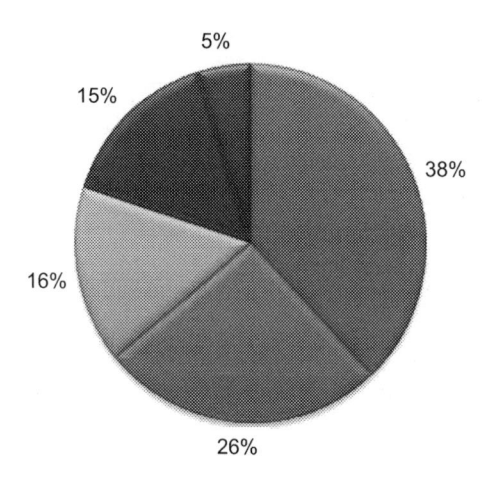

5%

15%

38%

16%

26%

■Diabetes ■High Blood Pressure ■Glomerulonephiritis ■Other Causes ■Unknown Cause

FIG. 13.1

Causes of end-stage kidney disease in the United States.Reported causes of end-stage kidney disease in the United States with $N = 726,331$.

From Chronic Diseases in America. *(c. 2019). https://www.cdc.gov/chronicdisease/resources/infographic/
chronic-diseases.htm.*

of the study is to find more relevant symptoms (features) of CKD. Different feature selection algorithms are applied for that are discussed in the section on quality measurement.

13.1.2 Detection of chronic kidney disease

To detect CKD, screenings are done on patients, especially those who are diabetic, have high blood pressure, or a history of kidney failure in their family. Early screening helps to identify the root cause of the disease. Various research is done on the detection of CKD, where most researchers have used the UCI dataset. According to this dataset, the features that can help detect CKD include age, bacteria, blood pressure, blood glucose random, specific gravity, pus cell, albumin, sugar, red blood cells, pus cell clumps, red blood cell count, coronary artery disease, hypertension, diabetes mellitus, blood urea, serum creatinine, potassium, sodium, hemoglobin, packed cell volume, white blood cell count, appetite, pedal edema, and anemia, as shown in Table 13.1. The parameters of these attributes can detect the presence of CKD.

CKD has five distinct stages, as illustrated in Table 13.2. The National Kidney Foundation (NKF) has given instructions to identify the seriousness of kidney disease. Each stage requires a different treatment. Any kind of kidney malfunction can be detected by calculating the glomerular filtration rate (GFR). It helps to identify how effectively the kidneys can clean the blood. GFR is measured based on blood tests. The level of serum creatinine or serotonin in the blood helps in

Table 13.1 Dataset attributes.

S.No.	Attribute
1	age
2	al
3	ane
4	appet
5	ba
6	bgr
7	bp
8	bu
9	cad
10	dm
11	hemo
12	htn
13	pc
14	pcc
15	pcv
16	pe
17	pot
18	rbc
19	rc
20	sc
21	sg
22	sod
23	su
24	wc
25	Ckd. Non ckd

Table 13.2 Different stages of CKD.

Stage	GFR (mL/min)
Stage 1 Normal CKD	>90
Stage 2 Mild CKD	=60–89
Stage 3A Moderate CKD	=45–59
Stage 3B Moderate CKD	=30–44
Stage 4 Severe CKD	=15–29
Stage 5 End CKD	<15

From Stages of Chronic Kidney Disease. *(c. 2020). https://www.davita.com/education/kidney-disease/stages.*

estimating GFR. Higher levels of creatinine denote a decline in kidney function. GFR is calculated on the basis of serum creatinine, age, race, and gender. Some other factors include serum albumin, weight, and blood urea nitrogen.

13.1.3 Treatments for chronic kidney disease

The main treatments involve medical management, dialysis, and kidney transplant. If the disease is detected at an early stage, then medical management will be sufficient to overcome this disease. Kidney dialysis help a patient's body excrete any waste and extra fluids from the blood. It is usually done with the help of a machine to clean the blood. A kidney transplant is a surgery to replace the damaged kidney with a healthy one from someone else's body. It can be taken from a live donor or a deceased donor. The number of patients with CKD is continuously increasing while the availability of transplant organs is not growing at the same pace. In last decade, the number of patients who are on a list for a kidney transplant approximately doubled in the United States (Kaballo et al., 2017). Hence every year, waiting times are getting longer, which is the main cause of an increase in death rates due to CKD or end-stage kidney disease (ESKD). This can be minimized only if the disease is detected in stage 1 or 2, followed by proper medical treatment.

13.2 Machine learning importance in disease prediction

The correct diagnosis of disease is the most important factor for treatment. The expertise of the doctor plays an important role in this. With the availability of various high-performance artificial intelligence/machine learning (AI/ML) models, researchers and practitioners are applying the same in every domain of daily human life. These models have significant contributions in healthcare for the early and accurate prediction of diseases. These AI/ML models exploit digital pathology and help medical practitioners. Every year, nearly 175,000 new cases of Stage V CKD (Mehta, 2019) appear. Hence, the early and easier diagnosis of CKD is crucial. This section includes various ML models applied by different researchers for a more accurate diagnosis of CKD.

Radiologists generally prefer ultrasound images for the examination of a kidney's features as they have no side effects. Zheng, Furth, Tasian, and Fan (2019) applied the transfer learning method to extract features from ultrasound scans of patient kidneys and had promising results for early prediction of the disease. Hao et al. (2019) developed a texture branch-based convolution neural network (CNN) to study the morphological and pathological features. Acharya et al. (2019) extracted higher-order attributes and lengthened the quandary patterns from ultrasound images for the early detection of CKD.

Vauchel et al. (2018) applied targeted ML to analyze the characteristics and sources of infection in kidney events. Tran et al. (2019) applied the KNN algorithm weighted by NGAL measurements to classify acute kidney injuries within 24 hours of patient admission. Senanayake et al. (2019) assessed the quality (reproducibility, robustness, generalizability, and clinical significance) of different ML models for the prediction of kidney transplants. Neves et al. (2015) used an artificial neural network (ANN) based on gender, age, family history, primary and secondary risk factors, and renal biomarkers for the early detection of CKD to avoid life-threatening conditions.

Liu, Tseng, Wang, Huang, and Randhawa (2019) used an ML-based model to perform RNA-seq on biopsies from the kidney. Hussain, Hamarneh, O'Connell, Mohammed, and Abugharbieh (2016) developed a dual regression forest model to estimate the patient's kidney size, shape, and orientation for the early prediction of CKD. Hore, Chatterjee, Shaw, Dey, and Virmani (2017) developed a neural network that was trained by the genetic algorithm to efficiently detect CKD at earlier stages. Almansour et al. (2019) gave optimal parameters of SVM and ANN for CKD prediction. Abdolkarimzadeh, FazelZarandi, and Castillo (2018) used the fuzzy rough quick reduct algorithm for feature selection and applied a type-II fuzzy system to classify features for CKD detection.

The world is facing a pandemic, and in the near future, telemedicine will be popular worldwide. The use of ML algorithms for disease diagnosis will be a revolutionary step in this direction.

13.3 **ML models used in the study**

In this study, KNN, SVM, random forest, naïve Bayes, ANN, logistic regression, and AdaBoost classifiers are used for the early detection of CKD. Their comparative performances are recorded.

13.3.1 **KNN classifier**

K-nearest neighbor is a supervised ML algorithm used in the classification of multiclass dependent variables. It estimates the class conditional densities in a supervised manner. The density is characterized by a set of population parameters (such as mean and variance), but in many cases, the formal structure of density is not prescribed. It is a nonparametric density estimation method.

Suppose $X = \{X_1, X_2, ...X_n\}$ are random samples, Y is the dependent variable defined as $Y = f(X)$, and the objective is to estimate the density of f. Because the parameters of the population are unknown, the density is divided into small intervals on X-axis (as shown in Fig. 13.2).

Let's say the region under the small cut, that is, interval (a,b), is R_n. So, the probability that the sample falls in that region is:

$$p(x) = \int_{R_n} f(x) \times dx \qquad (13.1)$$

For a small interval, $f(x)$ is constant:

$$p(x) = f(x) \times V \qquad (13.2)$$

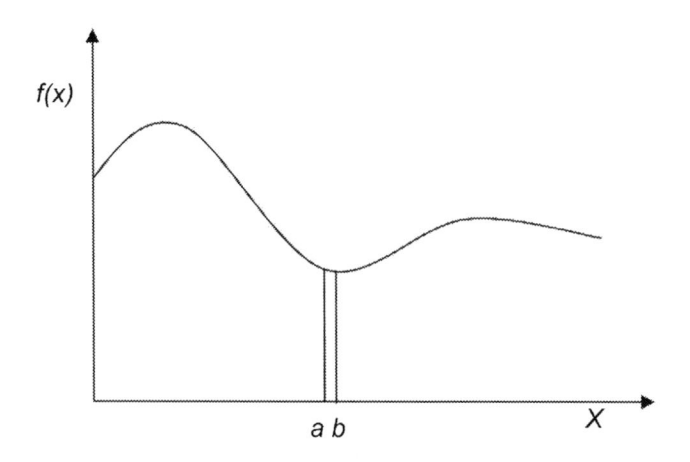

FIG. 13.2

Density estimation of $f(X)$. Density estimation of function with hyperplane.

where V is the volume of R_n:

$$f(x) = p(x)/V \tag{13.3}$$

The probability of observing k points in the interval is:

$$p(k) = \binom{n}{k} \times p(x)^k \times (1 - p(x))^{(n-k)} \tag{13.4}$$

The expectation is:

$$E[k] = n \times P(x) \sim k \tag{13.5}$$

Therefore,

$$f(x) = k/(n \times V) \tag{13.6}$$

For fixing the number of samples in the region, density can easily be calculated, as shown in Fig. 13.3.

13.3.2 Logistic regression

It is used when the response/dependent variable is qualitative in comparison to linear regression, where the response variable is quantitative. Logistic regression is a non-linear function and is defined as:

$$y = f(x) = \frac{1}{1 + e^{-\beta^T x}} \tag{13.7}$$

where β is a vector and the coefficient of the regressor variable is x. The shape of the logistic function is shown in Fig. 13.4.

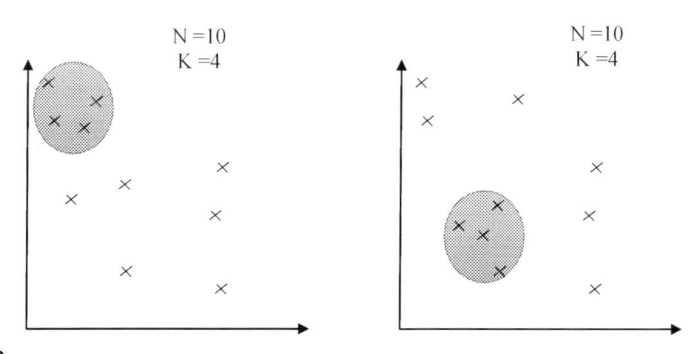

FIG. 13.3

KNN classifier.KNN classifier with $n = 10$ and $k = 3$ over different regions.

From Raschka, S. (c. 2014). Kernel density estimation via the Parzen-Rosenblatt window method. *https:// sebastianraschka.com/Articles/2014_kernel_density_est.html.*

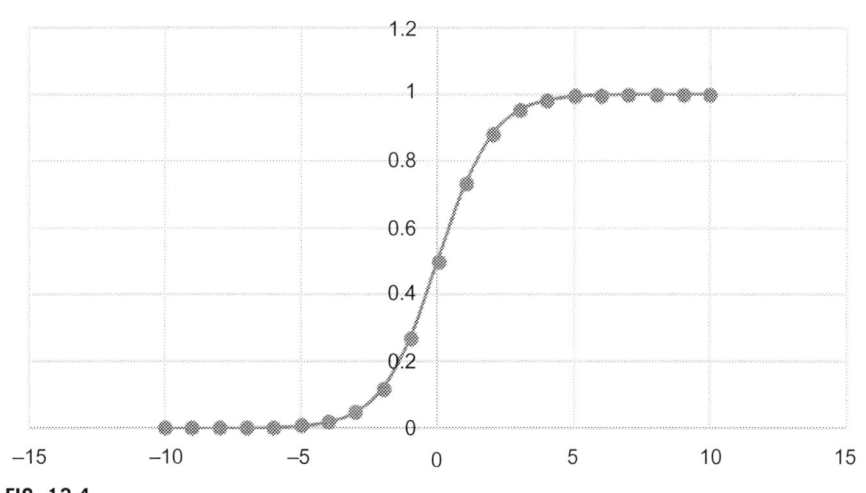

FIG. 13.4

Logistic function.The logistic function crosses the y axis on $x = 0$.

No permission required.

Assuming the response variable takes either 0 or 1, thus it follows Bernoulli distribution. Therefore, the loss function is defined as:

$$L(\beta) = \prod_i p(y_i | x_i, \beta) \tag{13.8}$$

$$L(\beta) = \prod_i f(x_i)^{y_i} \times (1 - f(x_i))^{(1-y_i)} \tag{13.9}$$

The log likelihood of Formula (13.9) is:

$$l(\beta) = \sum_i [y_i \times \log f(x_i) + (1 - y_i) \times \log(1 - f(x_i))] \qquad (13.10)$$

and the coefficients updated using gradient descent are:

$$\beta = \beta + a \times \nabla_\beta \times l(\beta) \qquad (13.11)$$

where α is the learning rate. The regression line obtained is used as a decision boundary classifying qualitative response variables.

13.3.3 Support vector machine

SVM is a supervised ML model to classify two class problems. It uses logistic regression as a classifier between two classes. The logistic regression predicts 1 when, suppose $f(x) \geq 0.5$; otherwise it predicts 0. As $f(x)$ moves far from the decision boundary, the confidence will be higher. To gain maximum confidence, SVM introduced support vectors such that the support vectors are the set of samples on the hyperplane that is closest to the decision boundary on both sides. This thus creates a margin with the decision boundary, as illustrated in Fig. 13.5. The objective is to maximize the margin width for a higher level of confidence.

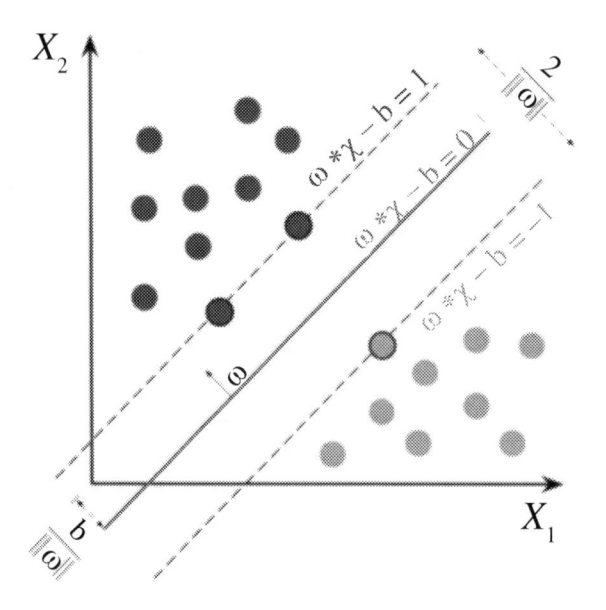

FIG. 13.5

Support vector machine. The classification margin of a support vector machine.

From Larham. (c. 2018). SVM margin. https://commons.wikimedia.org/wiki/File:SVM_margin.png.

13.3.4 **Random forest**

As the name suggests, the random forest tree is a collection of a relatively large number of interrelated decision trees, applying ensemble learning methods for the supervised learning tasks. All trees as a group/forest outperform any of the member trees. The constituent trees are small in size and are supposed to be uncorrelated with other trees of the forest. If one tree performs an error on any sample, the remaining trees are uncorrelated and have a very low chance to repeat that error.

Suppose there are nine decision trees (although the number should be very high) in the forest. Six of them predict 1 on any input, and the remaining three predict 0 on the same input. The verdict will be 1 as a higher number of trees predicts 1 and they are uncorrelated, thus having a low chance of wrong predictions.

13.3.5 **Naïve Bayes**

The naïve Bayes classifier is grounded on the Bayesian Rule:

$$P(Y|X) = (P(X|Y) \cdot P(Y))/(P(X)) \tag{13.12}$$

where Y is the response variable and $X = \{X_1, X_2, \ldots X_n\}$ are controlled variables. So, estimating the posterior probability of Y_j, and $\max(Y_j)$ is the predicted class of the response variable.

$$P(Y_j|X_1, X_2, \ldots, X_N)$$

$$= \frac{P(X_1|X_2, X_3, \ldots, X_N, Y_j) P(X_2|X_1, X_3, \ldots, X_N, Y_j) \ldots P(X_N|X_1, X_2, \ldots, X_{N-1}, Y_j) P(Y_j)}{\sum_{j=1}^{m} P(X_1|Y_j) \cdot P(X_2|Y_j) \ldots P(X_n|Y_j)}$$

$$\tag{13.13}$$

In the naïve Bayes classifier, the controlled variables are assumed to be independent of each other and contribute equally in the outcome, therefore:

$$P(Y_j|X_1, X_2, \ldots, X_N) = \frac{P(X_1|Y_j) P(X_2|Y_j) \ldots P(X_N|Y_j) P(Y_j)}{\sum_{j=1}^{m} P(X_1|Y_j) \cdot P(X_2|Y_j) \ldots P(X_n|Y_j)} \tag{13.14}$$

For the classification, the denominator/marginal evidence remains the same and thus acts as a constant. Therefore, the equation can be written as:

$$P(Y_j|X_1, X_2, \ldots, X_N) \propto \max_j P(X_1|Y_j) P(X_2|Y_j) \ldots P(X_N|Y_j) P(Y_j) \tag{13.15}$$

$$iP(Y_j|X_1, X_2, \ldots, X_N) \propto \max_j \left\{ P(Y_j) \prod_{i=1}^{n} P(X_i|Y_j) \right\} \tag{13.16}$$

13.3.6 **Artificial neural network**

This is a human brain-inspired supervised ML model to replicate human behavior for decision making. A single-layer neural network is shown in Fig. 13.6.

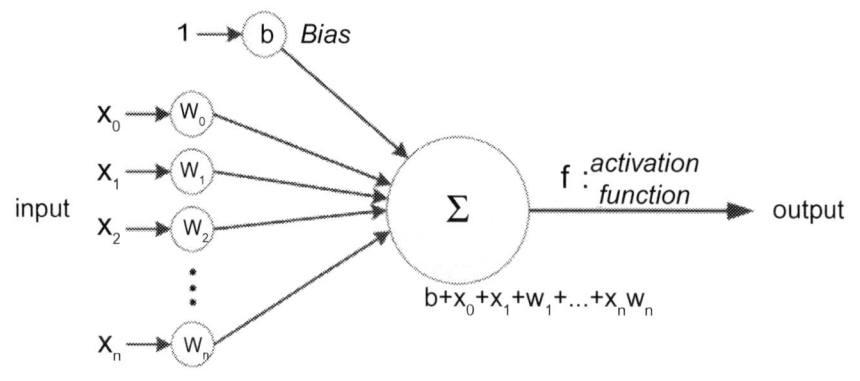

FIG. 13.6

Artificial neural network.A single-layer neural network.
From Chauhan, N. S. (c. 2019). Introduction to artificial neural networks (ANN). https://towardsdatascience. com/introduction-to-artificial-neural-networks-ann-1aea15775ef9.

It consists of a vector of controlled variable X that is weighted with synaptic weights W and a bias to adjust weights. All weighted inputs are summed up with bias and transferred to the activation function. The activation functions decide which neuron to activate concerning the labeled output. Because data are usually nonlinear, the activation function adds linearity in the models. Commonly used activation functions are:

$$\text{Sigmoidal function}: \quad g(z) = \frac{1}{1 + e^{-z}} \tag{13.17}$$

$$\text{Hyperbolic tangent function}: \quad g(z) = \frac{e^{z} - e^{-z}}{e^{z} + e^{-z}} \tag{13.18}$$

$$\text{Rectified linear unit (ReLU)}: \quad g(z) = \begin{cases} 1, & z > 0 \\ 0, & \text{otherwise} \end{cases} \tag{13.19}$$

13.3.7 AdaBoost

Adaptive boosting (AdaBoost) is an iterative ensemble supervised classifier that combines multiple poor classifiers to make an outperforming classifier. It randomly samples a subset of the dataset. It trains the model on the training set that provides better classification accuracy. Wrongly classified observations are prioritized by assigning higher weights. The model iterates until there are no more (acceptable) errors in the fitted model.

However, it is susceptible to overfitting and cannot handle outliers as well as other learning algorithms.

13.4 **Results and discussion**

This section includes the comparative results obtained by various ML algorithms explained in the previous section.

13.4.1 **Quality measurement**

Six types of feature selection techniques are used in this study: information gain, information gain ratio, Gini Index, X^2, relief, and FCBF (fast correlation-based feature for feature selection) are used in this study. The attributes are selected based on their scores on different statistical tests and this is independent of the type of ML algorithm used.

13.4.1.1 *Information gain*

To find which features are useful for the classification of the response variables, information gain is used. The change in entropy from the prior state to information gained in some states is the expected information gain. For a random variable X, with probability $P(x)$, the entropy is defined as:

$$H(x) = -\sum_i P(x_i) \cdot \log_2 P(x_i) \tag{13.20}$$

The conditional uncertainty/entropy $H(X|Y)$ is:

$$H(X|Y) = -\sum_j P(y_j) \sum_i P(x_i|y_j) \log_2 P(x_i|y_j) \tag{13.21}$$

$$\therefore \text{Information gain } IG(X, Y) = H(X) - H(X|Y) \tag{13.22}$$

13.4.1.2 *Gain ratio*

For a smaller number of features with a large number of values, information gain works fine. When the number of variables increases, information gain is divided by its intrinsic values to make it unbiased. The gain ratio is calculated as:

$$\text{Gain ratio} = \frac{\text{Information gain}(X, Y) \sum_j P(y_j)}{H(X|Y)} \tag{13.23}$$

13.4.1.3 *Gini Index*

This is used to measure the inequality between features of a dataset based on the frequency distribution. $0 \le \text{Gini Index } (X, Y) \le 1$ where 0 indicates equality while 1 suggests total inequality between two features. For n number of instances, Gini Index is calculated as:

$$\text{Gini Index}(X, Y) = \frac{\sum_{i=1}^n \sum_{j=1}^n |x_i - y_j|}{2n^2 \bar{x}} \tag{13.24}$$

13.4.1.4 Chi-squared distribution

X^2 distribution is used to measure the goodness of fit of the proposed model. The basic assumption is that the observation is standard normally distributed. It is the sum of squares of the aforementioned random variables. It is used in this study to test the independence between features.

$$\chi^2 = \sum_{ij} \frac{(O_{ij} - E_{ij})^2}{E_{ij}} \tag{13.25}$$

where O_{ij} is the observed count of observation i with observation j. E_{ij} is the expected count.

13.4.1.5 FCBF

FCBF (Yu & Liu, 2003) is a multivariate feature selection algorithm to find dependency and correlation between different features of the dataset. The dependencies between features are calculated by finding the symmetric uncertainty (SU) between features using a backward selection strategy. The SU is defined as:

$$SU(X, Y) = 2\left[\frac{H(X) - H(X|Y)}{H(X) + H(X|Y)}\right] \tag{13.26}$$

The value of SU lies between $(0,1)$, where 0 signifies no dependency between two features while the second feature can be predicted with the help of the first feature if its value is 1.

13.4.2 Evaluation techniques

The efficiency of the classification method can be compared using the confusion matrix and the ROC curve.

13.4.2.1 Confusion matrix

A confusion matrix can be used to perfectly analyze the potential of a classifier. All the diagonal elements denote correctly classified outcomes. The misclassified outcomes are represented on the off diagonals of the confusion matrix. Hence, the best classifier will have a confusion matrix with only diagonal elements and the rest of the elements set to zero. A confusion matrix generates actual values and predicted values after the classification process. The effectiveness of the system is determined according to the following values generated in the matrix. The classifiers for two classes have the following confusion matrix, as depicted in Table 13.3.

Table 13.3 Confusion matrix.

	Predicted class	
Actual class	**Positive**	**Negative**
Positive	True positive (TP)	False negative (FN)
Negative	False positive (FP)	True negative (TN)

The entries in the confusion matrix are defined as the following:

- True positive rate (TP) is the total number of correct results or predictions when the actual class was positive.
- False positive rate (FP) is the total number of wrong results or predictions when the actual class was positive.
- True negative rate (TN) is the total number of correct results or predictions when the actual class was negative.
- False negative rate (FN) is the total number of wrong results or predictions when the actual class was negative.

The accuracy calculation (AC) is used to compare the efficiency of the system. It takes into account the total number of correct predictions made by the classifier. It is calculated by the following equation:

$$\text{Accuracy} = \frac{TP + TN}{TP + TN + FP + FN} \qquad (13.27)$$

The recall is calculated by taking the proportion of correctly identified positive inputs. It is the TP rate and is measured by the given equation:

$$\text{Recall} = \frac{TN}{FP + TN} \qquad (13.28)$$

Precision is the correctly predicted positive cases by the classifier. It is measured by the given equation:

$$\text{Precision} = \frac{TN}{FN + TN} \qquad (13.29)$$

F1 score or F measure is also a measure of the test's accuracy. It is defined as a weighted mean of precision and recall. It has its maximum value at 1 and worst at 0.

$$\text{F1 score} = \frac{2 \times \text{Precision} \times \text{Recall}}{\text{Precision} + \text{Recall}} \qquad (13.30)$$

13.4.2.2 Receiver operating characteristic (ROC) curve

An ROC curve is a graphical plot that demonstrates the capability of a binary classifier system at different threshold values. The ROC curve is plotted according to the false positive rate (FPR) and true positive rate (TPR) of various classifiers used.

$$\text{TPR or sensitivity} = \frac{TP}{TP + FN} \qquad (13.31)$$

and,

$$\text{FPR} = 1 - \text{Specificity} = \frac{FP}{FP + TN} \qquad (13.32)$$

It is an efficient method to observe the performance of the classification technique used.

13.4.3 Dataset description

The Kidney_Disease dataset (Dua & Graff, 2017) is used in this research to assess the performance of classical ML algorithms for the detection of CKD. This is a multi-variate dataset, as shown in Table 13.1, that has 25 features, among which 11 were numeric data and 14 were nominal ones. There are 400 instances in the dataset, where 250 instances belong to CKD and 150 instances to non-CKD.

Table 13.4 shows a comparison of the scores of attributes when calculated using different feature selection methods discussed in the section on quality measurement.

As shown in Table 13.4, it can be observed from the above comparisons that a few attributes such as ps, hemo, RC, sg, al, htn, and dm have comparatively higher scores than the other attributes. These attributes hence can help to predict CKD. Now, these attributes along with some ML algorithms can be trained and tested to predict chronic

Table 13.4 Scores obtained by different attributes using various feature selection techniques.

	Info.gain	Gain ratio	Gini	χ^2	Relief	FCBF
id	0.704	0.352	0.344	217.778	0.296	0.912
Pcv	0.589	0.296	0.313	197.739	0.116	0.666
Hemo	0.574	0.289	0.310	198.313	0.130	0.000
Rc	0.565	0.113	0.303	238.482	0.100	0.000
Sq	0.532	0.268	0.288	163.342	0.259	0.568
Al	0.465	0.263	0.248	224.602	0.124	0.519
Sc	0.453	0.227	0.243	154.174	0.011	0.442
Htn	0.338	0.356	0.163	88.200	0.316	0.550
Sod	0.315	0.080	0.168	108.883	0.068	0.000
Dm	0.306	0.330	0.147	82.200	0.276	0.483
Pot	0.237	0.057	0.130	30.992	0.180	0.000
Bqr	0.225	0.113	0.108	68.295	0.019	0.180
Bu	0.217	0.108	0.102	60.718	0.019	0.000
Bp	0.191	0.097	0.092	25.189	0.020	0.150
Su	0.171	0.134	0.084	105.131	0.004	0.000
Appet	0.161	0.220	0.073	49.200	0.074	0.237
Pe	0.148	0.210	0.066	45.600	0.086	0.217
Pc	0.148	0.210	0.066	10.696	0.164	0.217
Wc	0.126	0.064	0.074	15.402	0.003	0.000
Ane	0.113	0.185	0.050	36.000	0.060	0.169
Rbc	0.086	0.165	0.037	3.755	0.028	0.000
Pcc	0.076	0.158	0.033	25.200	0.096	0.000
Cad	0.061	0.145	0.026	20.400	0.026	0.000
Aqe	0.058	0.029	0.038	25.088	0.030	0.000
ba	0.039	0.126	0.016	13.200	−0.004	0.000

diseases. These features are the main reasons for CKD. In comparing the results, different classifiers are used such as SVM, KNN, neural network, random forest, AdaBoost, logistic regression, and naïve Bayes. The results are generated using the different number of features taken for classification. The most relevant features are selected with the help of various feature selection algorithms. The cross-validation with 10-fold is carried out with all the classifiers. The potentials of various classic ML models are compared before and after applying feature selection. The accuracy of ML is obtained by the AC (accuracy classification score), precision, recall, and area under the curve (AUC).

13.4.4 Model configurations

This section includes the configurations of all the models used for classification tasks.

KNN classifier
1. Number of neighbors: 5
2. Metric: Euclidean
3. Weight: uniform

Random forest
1. Number of trees: 10
2. Maximal number of considered features: unlimited
3. Replicable training: no
4. Maximal tree depth: unlimited
5. Stop splitting nodes with maximum instances: 5

SVM
1. SVM type: SVM, $C=1.0$, $\varepsilon=0.1$
2. Kernel: RBF, $\exp(\text{-auto}|x\text{-}y|^2)$
3. Numerical tolerance: 0.001
4. Iteration limit: 100

Neural network
1. Hidden layers: 100
2. Activation: ReLu
3. Solver: SGD
4. Alpha: 0.0001
5. Max iterations: 300
6. Replicable training: true

Logistic regression
Regularization: Ridge (L2), $C=1$

AdaBoost
Base estimator: tree
Number of estimators: 50
Algorithm (classification): Samme.r
Loss (regression): linear.

13.4.5 Result analysis with information gain

Table 13.5 shows the list of scores obtained by different attributes. These scores are arranged in descending order according to their scores.

The attributes pcv, hemo, rc, sg, al, sc, and htn can be used along with different ML models and the results can be compared, as depicted in Tables 13.6–13.10.

From Tables 13.6–13.10, it is clear that the various ML models are capable of predicting severe kidney disease with the help of five best-ranked or seven

Table 13.5 Information gain on different attributes.

Features	Info.gain
id	0.704
Pcv	0.589
Hemo	0.574
Rc	0.565
Sq	0.532
Al	0.465
Sc	0.453
Htn	0.338
Sod	0.315
Dm	0.306
Pot	0.237
Bqr	0.225
Bu	0.217
Bp	0.191
Su	0.171
Appet	0.161
Pe	0.148
Pc	0.148
Wc	0.126
Ane	0.113
Rbc	0.086
Pcc	0.076
Cad	0.061
Aqe	0.058
ba	0.039

Table 13.6 Attribute scores using information gain using three best features.

Model	AUC	CA	F1	Precision	Recall
Neural network	0.995	0.963	0.963	0.963	0.963
SVM	1.000	0.993	0.993	0.993	0.993
Naïve Bayes	1.000	0.993	0.993	0.993	0.993
Random forest	**1.000**	**0.998**	**0.998**	**0.998**	**0.998**
KNN	0.993	0.993	0.993	0.993	0.993
Logistic regression	1.000	0.993	0.993	0.993	0.993
AdaBoost	0.997	0.998	0.997	0.998	0.998

The bold values indicate the best results.

Table 13.7 Attribute scores using information gain using four best features.

Model	AUC	CA	F1	Precision	Recall
KNN	**0.568**	**0.625**	**0.481**	**0.397**	**0.625**
Naïve Bayes	0.555	0.625	0.481	0.391	0.625
Random forest	0.568	0.625	0.481	0.391	0.625
SVM	0.55	0.53	0.537	0.553	0.530
Neural network	0.555	0.625	0.481	0.391	0.625
Logistic regression	0.555	0.665	0.481	0.391	0.625
AdaBoost	0.555	0.625	0.481	0.391	0.625

The bold values indicate the best results.

Table 13.8 Attribute scores using information gain using five best features.

Model	AUC	CA	F1	Precision	Recall
KNN	0.998	0.995	0.995	0.995	0.995
SVM	0.997	0.980	0.980	0.980	0.980
Random forest	**1.000**	**0.998**	**0.998**	**0.998**	**0.998**
Naïve Bayes	0.995	0.978	0.978	0.978	0.978
Neural network	0.998	0.985	0.985	0.958	0.985
Logistic regression	1.000	0.993	0.993	0.993	0.993
AdaBoost	0.997	0.998	0.997	0.998	0.998

The bold values indicate the best results.

best-ranked features. Hence, using only the best-ranked attributes gives the same results with different classifiers. With a smaller number of features given by information gain, ensemble learning models (random forest and AdaBoost) give better results. With a higher number of features, KNN outperforms, as it has a higher value of dimensions for the hyperplane.

Table 13.9 Attribute scores using information gain using six best features.

Model	AUC	CA	F1	Precision	Recall
KNN	**1.000**	**1.000**	**1.000**	**1.000**	**1.000**
SVM	0.998	0.998	0.988	0.988	0.988
Random forest	1.000	0.998	0.998	0.998	0.998
Naïve Bayes	0.997	0.980	0.980	0.980	0.980
Neural network	1.000	0.983	0.983	0.983	0.983
Logistic regression	0.995	0.995	0.995	0.995	0.995
AdaBoost	0.997	0.998	0.997	0.998	0.998

The bold values indicate the best results.

Table 13.10 Attribute scores using information gain using seven best features.

Model	AUC	CA	F1	Precision	Recall
KNN	**1.000**	**1.000**	**1.000**	**1.000**	**1.000**
SVM	0.998	0.985	0.985	0.985	0.985
Random forest	**1.000**	**1.000**	**1.000**	**1.000**	**1.000**
Naïve Bayes	0.997	0.978	0.978	0.978	0.978
Neural network	0.999	0.985	0.985	0.983	0.985
Logistic regression	1.000	0.995	0.995	0.995	0.995
AdaBoost	0.997	0.998	0.997	0.998	0.998

The bold values indicate the best results.

13.4.6 Result analysis with information gain ratio

The next type of feature selection method used is the information gain ratio. Table 13.11 lists the scores obtained by different attributes in descending order.

The attributes such as htn, dm, pcv, hemo, sg, al, and sc have good scores as compared to other attributes. These attributes are hence applied to ML algorithms and the results are compared.

It can be observed from Tables 13.12–13.16 that the classifiers give the same accuracy, precision, and recall when five, six, and seven best-ranked features are used for classification. As discussed in the section on quality measurement gain ratio, it is an unbiased estimator of the information gain and the classification results of all ML algorithms give approximately the same results.

13.4.7 Result analysis with Gini Index

The Gini Index can also be used to find the best attributes that can predict the output of a classifier correctly. Using the Gini Index, the following attributes are scored higher than the other attributes: id, pcv, hemo, rc, sg, al, and sc, as shown in Table 13.17.

Table 13.11 Information gain ratio on different attributes.

Features	Information gain ratio
Htn	0.356
Id	0.352
Dm	0.330
Pcv	0.296
Hemo	0.289
Sq	0.268
Al	0.263
Sc	0.227
Appet	0.220
Pe	0.210
Pc	0.210
Ane	0.185
Rbc	0.165
Pcc	0.158
Cad	0.145
Su	0.134
Rc	0.133
Ba	0.126
Bqr	0.113
Bu	0.108
Bp	0.097
Sod	0.080
Wc	0.064
Pot	0.057
Aqe	0.029
Htn	0.356

Table 13.12 Attribute scores using information gain ratio using three best features.

Model	AUC	CA	F1	Precision	Recall
KNN	1.000	0.998	0.998	0.998	0.998
SVM	1.000	0.985	0.985	0.986	0985
Random forest	**1.000**	**1.000**	**1.000**	**1.000**	**1.000**
Neural network	1.000	0.985	0.985	0.986	0.985
Naïve Bayes	0.991	0.968	0.968	0.970	0.968
Logistic regression	1.000	0.985	0.985	0.986	0.985
AdaBoost	0.997	0.998	0.997	0.998	0.998

The bold values indicate the best results.

Table 13.13 Attribute scores using information gain ratio using four best features.

Model	AUC	CA	F1	Precision	Recall
KNN	**1.000**	**1.000**	**1.000**	**1.000**	**1.000**
SVM	1.000	0.998	0.998	0.998	0.998
Random forest	1.000	0.998	0.997	0.998	0.998
Neural network	1.000	0.998	0.998	0.998	0.998
Naïve Bayes	0.999	0.985	0.985	0.985	0.985
Logistic regression	1.000	0.995	0.995	0.995	0.995
AdaBoost	0.997	0.998	0.997	0.998	0.998

The bold values indicate the best results.

Table 13.14 Attribute scores using information gain ratio using five best features.

Model	AUC	CA	F1	Precision	Recall
KNN	1.000	0.998	0.997	0.998	0.998
SVM	**1.000**	**1.000**	**1.000**	**1.000**	**1.000**
Random forest	1.000	0.998	0.997	0.998	0.998
Neural network	1.000	0.998	0.998	0.998	0.998
Naïve Bayes	0.999	0.978	0.978	0.978	0.978
Logistic regression	1.000	0.998	0.998	0.998	0.998
AdaBoost	0.997	0.998	0.997	0.998	0.998

The bold values indicate the best results.

Table 13.15 Attribute scores using information gain ratio using six best features.

Model	AUC	CA	F1	Precision	Recall
KNN	0.998	0.995	0.995	0.995	0.995
SVM	1.000	0.998	0.998	0.998	0.998
Random forest	**1.000**	**1.000**	**1.000**	**1.000**	**1.000**
Neural network	1.000	0.998	0.998	0.998	0.998
Naïve Bayes	1.000	0.990	0.990	0.990	0.990
Logistic regression	1.000	0.995	0.995	0.995	0.995
AdaBoost	0.997	0.998	0.997	0.998	0.998

The bold values indicate the best results.

From the above comparison as depicted in Tables 13.18–13.22, it is clear that with the smaller number of features given by the Gini Index, ensemble learning models (random forest and AdaBoost) give better results while with a higher number of features, KNN outperforms again.

Table 13.16 Attribute scores using information gain ratio using seven best features.

Model	AUC	CA	F1	Precision	Recall
KNN	0.998	0.998	0.998	0.998	0.998
SVM	**1.000**	**0.998**	**0.998**	**0.998**	**0.998**
Random forest	1.000	0.998	0.997	0.998	0.998
Neural network	**1.000**	**0.998**	**0.998**	**0.998**	**0.998**
Naïve Bayes	1.000	0.995	0.995	0.995	0.995
Logistic regression	**1.000**	**0.998**	**0.998**	**0.998**	**0.998**
AdaBoost	0.997	0.998	0.997	0.998	0.998

The bold values indicate the best results.

Table 13.17 Gini Index.

Features	Gini Index
Id	0.344
Pcv	0.313
Hemo	0.310
Rc	0.303
Sq	0.288
Al	0.248
Sc	0.243
Sod	0.168
Htm	0.163
Dm	0.147
Pot	0.130
Bqr	0.108
Bu	0.102
Bp	0.092
Su	0.084
Wc	0.074
Appet	0.073
Pe	0.066
Pc	0.066
Ane	0.050
Aqe	0.038
Rbc	0.037
Pcc	0.033
Cad	0.026
Ba	0.016

Table 13.18 Attribute scores using the Gini Index using three best features.

Model	AUC	CA	F1	Precision	Recall
KNN	0.993	0.993	0.993	0.993	0.993
SVM	**1.000**	**0.993**	**0.993**	**0.993**	**0.993**
Random forest	0.996	0.998	0.997	0.997	0.998
Neural network	**1.000**	**0.993**	**0.993**	**0.993**	**0.993**
Naïve Bayes	0.995	0.963	0.963	0.963	0.963
Logistic regression	**1.000**	**0.993**	**0.993**	**0.993**	**0.993**
AdaBoost	0.997	0.998	0.997	0.998	0.998

The bold values indicate the best results.

Table 13.19 Attribute scores using the Gini Index using four best features.

Model	AUC	CA	F1	Precision	Recall
KNN	0.994	0.988	0.988	0.988	0.988
SVM	0.996	0.960	0.960	0.960	0.960
Random forest	**1.000**	**0.998**	**0.998**	**0.998**	**0.998**
Neural network	0.992	0.975	0.975	0.976	0.975
Naïve Bayes	0.996	0.958	0.958	0.960	0.958
Logistic regression	1.000	0.993	0.993	0.993	0.993
AdaBoost	0.997	0.998	0.997	0.998	0.998

The bold values indicate the best results.

Table 13.20 Attribute scores using the Gini Index using five best features.

Model	AUC	CA	F1	Precision	Recall
KNN	0.998	0.995	0.995	0.995	0.995
SVM	0.997	0.980	0.980	0.980	0.980
Random forest	1.000	0.993	0.992	0.993	0.993
Neural network	0.995	0.978	0.978	0.978	0.978
Naïve Bayes	0.998	0.985	0.985	0.960	0.958
Logistic regression	**1.000**	**0.993**	**0.993**	**0.993**	**0.993**
AdaBoost	0.997	0.998	0.997	0.998	0.998

The bold values indicate the best results.

13.4.8 **Result analysis with chi-square**

The next type of feature selection method used is the chi-square. The best features selected using chi-square include rc, al, hemo, pcv, sod, and sg, as shown in Table 13.23. These features are hence tested with some classifiers and the results are compared.

Table 13.21 Attribute scores using the Gini Index using six best features.

Model	AUC	CA	F1	Precision	Recall
KNN	**1.000**	**1.000**	**1.000**	**1.000**	**1.000**
SVM	0.998	0.985	0.985	0.985	0.985
Random forest	1.000	0.998	0.997	0.998	0.998
Neural network	0.997	0.978	0.978	0.978	0.978
Naïve Bayes	0.999	0.985	0.985	0.960	0.958
Logistic regression	1.000	0.995	0.995	0.995	0.995
AdaBoost	0.997	0.998	0.997	0.998	0.998

The bold values indicate the best results.

Table 13.22 Attribute scores using the Gini Index using seven best features.

Model	AUC	CA	F1	Precision	Recall
KNN	**1.000**	**1.000**	**1.000**	**1.000**	**1.000**
SVM	0.998	0.998	0.998	0.998	0.998
Random forest	**1.000**	**1.000**	**1.000**	**1.000**	**1.000**
Neural network	0.997	0.980	0.980	0.980	0.980
Naïve Bayes	1.000	0.983	0.983	0.983	0.983
Logistic regression	1.000	0.995	0.995	0.995	0.995
AdaBoost	0.997	0.998	0.997	0.998	0.998

The bold values indicate the best results.

Because the classification is a binary one, chi-square is having monotonic power function, and thus it also unbiased in nature. The classification using chi-square as feature selection has produced the almost same result with the different number of best features that are with 3, 4, 5, 6, or 7 features taken for classification as shown in Tables 13.24–13.28.

13.5 **Conclusion**

The chapter presents a systematic study of various models used in ML for the accurate diagnosis of chronic kidney diseases to assist in a doctor's decision making about treatment. The presented work applies various feature selection algorithms and ML algorithms on the given UCI data set. The performance of different ML models on a different number of features has been analyzed in the work. In most cases, the classification accuracy of different ML algorithms is greater than 96%. The performance is satisfactory in terms of accurate disease diagnosis. The current system is also useful in telemedicine systems in remote rural areas.

Table 13.23 Chi-square distribution of attributes.

Features	x^2
rc	238.482
Al	224.602
id	217.778
Hemo	198.313
Pcv	197.739
Sod	180.883
Sq	163.342
Sc	154.174
Su	105.131
Htn	88.200
Dm	82.200
Bqr	68.295
Bu	60.718
Appet	49.200
Pe	45.600
Ane	36.000
Pot	30.992
Pcc	25.200
Bp	25.189
Aqe	25.088
Cad	20.400
Wc	15.402
Ba	13.200
Pc	10.696
Rbc	3.755

Table 13.24 Attribute scores using chi-square using tree best features.

Model	AUC	CA	F1	Precision	Recall
KNN	0.996	0.988	0.988	0.988	0.988
SVM	0.995	0.965	0.965	0.965	0.965
Random forest	**1.000**	**0.998**	**0.997**	**0.998**	**0.998**
Neural network	0.992	0.983	0.983	0.983	0.983
Naïve Bayes	0.996	0.960	0.960	0.962	0.960
Logistic regression	1.000	0.990	0.990	0.990	0.990
AdaBoost	0.997	0.998	0.997	0.998	0.998

The bold values indicate the best results.

Table 13.25 Attribute scores using chi-square using four best features.

Model	AUC	CA	F1	Precision	Recall
KNN	0.996	0.988	0.988	0.988	0.988
SVM	0.996	0.973	0.973	0.973	0.973
Random forest	**1.000**	**0.998**	**0.998**	**0.998**	**0.998**
Neural network	0.995	0.980	0.980	0.980	0.980
Naïve Bayes	0.999	0.978	0.978	0.978	0.978
Logistic regression	1.000	0.993	0.992	0.993	0.993
AdaBoost	0.997	0.998	0.997	0.998	0.998

The bold values indicate the best results.

Table 13.26 Attribute scores using chi-square using five best features.

Model	AUC	CA	F1	Precision	Recall
KNN	1.000	0.993	0.993	0.993	0.993
SVM	0.996	0.978	0.978	0.978	0.978
Random forest	**1.000**	**1.000**	**1.000**	**1.000**	**1.000**
Neural network	0.995	0.980	0.980	0.980	0.980
Naïve Bayes	0.998	0.970	0.970	0.971	0.970
Logistic regression	1.000	0.993	0.993	0.993	0.993
AdaBoost	0.997	0.998	0.997	0.998	0.998

The bold values indicate the best results.

Table 13.27 Attribute scores using chi-square using six best features.

Model	AUC	CA	F1	Precision	Recall
KNN	**1.000**	**0.988**	**0.987**	**0.987**	**0.988**
SVM	0.995	0.968	0.968	0.968	0.968
Random forest	1.000	0.995	0.995	0.995	0.995
Neural network	0.994	0.965	0.965	0.965	0.965
Naïve Bayes	0.999	0.980	0.980	0.980	0.980
Logistic regression	1.000	0.993	0.993	0.993	0.993
AdaBoost	0.997	0.998	0.997	0.998	0.998

The bold values indicate the best results.

Despite the significant results obtained, CKD prediction in the early stages is still open to further studies. In future work, the ultrasound images of the kidney can be analyzed based on inter and intrafeature relationships with deep learning models. The size of the available dataset is also a concern. Different case -based medical data can also be collected and presented on UCI for research purposes.

Table 13.28 Attribute scores using chi-square using seven best features.

Model	AUC	CA	F1	Precision	Recall
KNN	**1.000**	**0.995**	**0.995**	**0.995**	**0.995**
SVM	0.997	0.975	0.975	0.975	0.975
Random forest	**1.000**	**0.995**	**0.995**	**0.995**	**0.995**
Neural network	0.996	0.975	0.975	0.975	0.975
Naïve Bayes	0.999	0.993	0.993	0.993	0.993
Logistic regression	1.000	0.993	0.993	0.993	0.993
AdaBoost	0.997	0.998	0.997	0.998	0.998

The bold values indicate the best results.

References

Abdolkarimzadeh, M., FazelZarandi, M. H., & Castillo, O. (2018). Interval type II fuzzy rough set rule based expert system to diagnose chronic kidney disease. In G. Barreto, & R. Coelho (Eds.), *Communications in computer and information science: Vol. 831*. Cham: Springer. https://doi.org/10.1007/978-3-319-95312-0_49.

Acharya, U. R., Meiburger, K. M., Koh, J. E. W., Hagiwara, Y., Oh, S. L., Leong, S. S., … Ng, K. H. (2019). Automated detection of chronic kidney disease using higher-order features and elongated quinary patterns from B-mode ultrasound images. *Neural Computing and Applications*, *32*(15), 11163–11172. https://doi.org/10.1007/s00521-019-04025-y.

Almansour, N. A., Syed, H. F., Khayat, N. R., Altheeb, R. K., Juri, R. E., Alhiyafi, J., … Olatunji, S. O. (2019). Neural network and support vector machine for the prediction of chronic kidney disease: A comparative study. *Computers in Biology and Medicine*, *109*, 101–111. https://doi.org/10.1016/j.compbiomed.2019.04.017.

Chronic Diseases in America. (2019). https://www.cdc.gov/chronicdisease/resources/infographic/chronic-diseases.htm.

Dua, D., & Graff, C. (2017). *UCI machine learning repository*. https://archive.ics.uci.edu/ml/datasets/Chronic_Kidney_Disease.

Hao, P., Xu, Z., Tian, S., Wu, F., Chen, W., Wu, J., & Luo, X. (2019). Texture branch network for chronic kidney disease screening based on ultrasound images. *Frontiers of Information Technology & Electronic Engineering*, *21*(8), 1161–1170. https://doi.org/10.1631/fitee.1900210.

Hore, S., Chatterjee, S., Shaw, R. K., Dey, N., & Virmani, J. (2017). Detection of chronic kidney disease: A NN-GA-based approach. In *Nature inspired computing* (pp. 109–115). Singapore: Springer. https://doi.org/10.1007/978-981-10-6747-1_13.

Hussain, M. A., Hamarneh, G., O'Connell, T. W., Mohammed, M. F., & Abugharbieh, R. (2016). Segmentation-free estimation of kidney volumes in CT with dual regression forests. In *Machine learning in medical imaging* (pp. 156–163). Springer International Publishing. https://doi.org/10.1007/978-3-319-47157-0_19.

Kaballo, M. A., Canney, M., O'Kelly, P., Williams, Y., O'Seaghdha, C. M., & Conlon, P. J. (2017). A comparative analysis of survival of patients on dialysis and after kidney transplantation. *Clinical Kidney Journal*, *11*(3), 389–393. https://doi.org/10.1093/ckj/sfx117.

Liu, P., Tseng, G., Wang, Z., Huang, Y., & Randhawa, P. (2019). Diagnosis of T-cell–mediated kidney rejection in formalin-fixed, paraffin-embedded tissues using RNA-Seq–based machine learning algorithms. *Human Pathology*, *84*, 283–290. https://doi.org/10.1016/j.humpath.2018.09.013.

Mehta, A. R. (2019). *World Kidney Day 2019: Important aspects for chronic kidney disease in modern time*. https://www.narayanahealth.org/blog/world-kidney-day-2019-important-aspects-for-chronic-kidney-disease-in-modern-time/#:~:text=In%202019%2C%20World%20Kidney%20Day,Kidney%20Health%20for%20Everyone%20Everywhere%E2%80%9D.&text=Chronic%20kidney%20disease%20is.

Neves, J., Martins, M. R., Vicente, H., Neves, J., Abelha, A., & Machado, J. (2015). An assessment of chronic kidney diseases. In A. Rocha, A. Correia, & S. Costanzo (Eds.), *Advances in intelligent systems and computing: Vol. 353*. Cham: Springer. https://doi.org/10.1007/978-3-319-16486-1_18.

Senanayake, S., White, N., Graves, N., Healy, H., Baboolal, K., & Kularatna, S. (2019). Machine learning in predicting graft failure following kidney transplantation: A systematic review of published predictive models. *International Journal of Medical Informatics*, *130*. https://doi.org/10.1016/j.ijmedinf.2019.103957.

Tran, N. K., Sen, S., Palmieri, T. L., Lima, K., Falwell, S., Wajda, J., & Rashidi, H. H. (2019). Artificial intelligence and machine learning for predicting acute kidney injury in severely burned patients: A proof of concept. *Burns*, *45*(6), 1350–1358. https://doi.org/10.1016/j.burns.2019.03.021.

Vauchel, T., Pirracchio, P., Chaussard, M., Lafaurie, M., Rouveau, M., Rousseau, C., … Legrand, M. (2018). Impact of an Acinetobacter baumannii outbreak on kidney events in a burn unit: A targeted machine learning analysis. *American Journal of Infection Control*, *46*(4), 435–438. https://doi.org/10.1016/j.ajic.2018.09.010.

Yu, L., & Liu, H. (2003). Feature selection for high-dimensional data: A fast correlation-based filter solution. In *The twentieth international conference on machine learning* (pp. 856–863). https://www.aaai.org/Papers/ICML/2003/ICML03-111.pdf.

Zheng, Q., Furth, S. L., Tasian, G. E., & Fan, Y. (2019). Computer aided diagnosis of congenital abnormalities of the kidney and urinary tract in children based on ultrasound imaging data by integrating texture image features and deep transfer learning image features. *Journal of Pediatric Urology*, *15*(1), 75.e1–75.e7. https://doi.org/10.1016/j.jpurol.2018.10.020.

Index

Printed in the United States
by Baker & Taylor Publisher Services